Lecture Notes in Networks and Systems 1220

The series "Lecture Notes in Networks and Systems" publishes the latest developments in Networks and Systems—quickly, informally and with high quality. Original research reported in proceedings and post-proceedings represents the core of LNNS.

Volumes published in LNNS embrace all aspects and subfields of, as well as new challenges in, Networks and Systems.

The series contains proceedings and edited volumes in systems and networks, spanning the areas of Cyber-Physical Systems, Autonomous Systems, Sensor Networks, Control Systems, Energy Systems, Automotive Systems, Biological Systems, Vehicular Networking and Connected Vehicles, Aerospace Systems, Automation, Manufacturing, Smart Grids, Nonlinear Systems, Power Systems, Robotics, Social Systems, Economic Systems and other. Of particular value to both the contributors and the readership are the short publication timeframe and the world-wide distribution and exposure which enable both a wide and rapid dissemination of research output.

The series covers the theory, applications, and perspectives on the state of the art and future developments relevant to systems and networks, decision making, control, complex processes and related areas, as embedded in the fields of interdisciplinary and applied sciences, engineering, computer science, physics, economics, social, and life sciences, as well as the paradigms and methodologies behind them.

Indexed by SCOPUS, EI Compendex, INSPEC, WTI Frankfurt eG, zbMATH, SCImago.

All books published in the series are submitted for consideration in Web of Science.

For proposals from Asia please contact Aninda Bose (aninda.bose@springer.com).

Mostapha Zbakh · Mohammed Essaaidi ·
Claude Tadonki · Abdellah Touhafi ·
Dhabaleswar K. Panda
Editors

Artificial Intelligence and High Performance Computing in the Cloud

Research and Application Challenges

Springer

Editors
Mostapha Zbakh
ENSIAS College of Engineering
Mohammed V University in Rabat
Rabat, Morocco

Mohammed Essaaidi
ENSIAS College of Engineering
Mohammed V University in Rabat
Rabat, Morocco

Claude Tadonki
Mines ParisTech-PSL
Paris, France

Abdellah Touhafi
Vrije Universiteit Brussel
Brussels, Belgium

Dhabaleswar K. Panda
The Ohio State University
Columbus, OH, USA

ISSN 2367-3370 ISSN 2367-3389 (electronic)
Lecture Notes in Networks and Systems
ISBN 978-3-031-78697-6 ISBN 978-3-031-78698-3 (eBook)
https://doi.org/10.1007/978-3-031-78698-3

This Springer imprint is published by the registered company Springer Nature Switzerland AG
The registered company address is: Gewerbestrasse 11, 6330 Cham, Switzerland

Preface

Artificial intelligence (AI) has a noticeable impact in many areas including experimental science, medicine, finance, industry and many others. In scientific research, AI analyzes vast quantities of data and helps to seek discoveries in various fields like physics, genomics, biology and chemistry. It accelerates scientific progress and opens up new research perspectives. In medicine, AI is used to diagnose and predict diseases, enabling early detection and rapid intervention. It is also applied to pharmaceutical research to accelerate drug discovery and improve treatments. AI is also used to spot interesting sites in explorative activities like in geophysics or cosmology. One of the most impressive applications of AI is on natural language processing, where the use of so-called large language models (LLM) has led to impressive human-computer interaction. Implementing and running cutting-edge AI requires powerful HPC infrastructures including cloud systems.

Cloud computing has gained great attention from both academia and IT industry as a new infrastructure requiring little to no investment in hardware platform, staff training or software/tools licensing. Cloud computing services are offered on a pay-as-you-go basis, thus offering a flexible solution to run common workloads on a shared platform. Its consideration is also beneficial from the energy standpoint through concurrent executions of end-user's applications under throughput optimization and shared storage.

High performance computing (HPC) offers means and techniques to accelerate various methods and models in AI. Within the ecosystem of a cloud, this benefit is obtained in a seamless manner from the platform standpoint.

This book represents the extended versions of the best papers submitted and presented at the 6th International Conference on Cloud Computing and Artificial Intelligence: Technologies and Applications (IEEE CloudTech 2023) held from 21 to 23 November 2023 in Marrakesh, Morocco. Additional contributions have been considered from recognized experts.

The contributions published in this book address new challenges and discuss many topics related to artificial intelligence and high performance computing considering the cloud computing paradigm. In addition to surveys and position papers on HPC-AI with cloud computing technologies, it also contains significant fundamental and experimental results.

Topics

- Cloud computing
- Resource allocation
- HPC
- Energy Efficiency
- Artificial Intelligence
- Security and Privacy

Target Audience

Information Systems Directors, academicians, researchers, students, developers and policy makers will find in this book very useful, through its twenty-four chapters that cover several theoretical and experimental studies and researches in the fields of cloud computing, big data and security.

Organization of the Book

This book covers several concepts and features related with cloud computing and artificial intelligence theoretical background, technologies and applications. It also addresses some issues related to high performance computing, microservices and containers. It is organized as follows:

Chapter 1 describes a Data Analytics Platform called IMAGEN, which supports the development of deep learning models for animal phenotyping and addresses various data challenges through a DataOps approach.
Chapter 2 presents an AI-oriented and multi-layered framework for intelligent generation of standardized privacy policies in cloud systems.
Chapter 3 presents CAReNet (Convolutional Attention Residual Network), an innovative architecture designed to effectively address the challenges of computing resources consumption and the reliance on extensive annotated datasets in modern deep learning architectures.
Chapter 4 focuses on challenges and opportunities in designing HPC and artificial intelligence (AI) middleware on modern/heterogeneous computing systems with both scale-up and scale-out strategies.
Chapter 5 provides an accessible discussion on the strengths, weaknesses and security concerns associated with prevalent identity management strategies employed by cloud service providers.
Chapter 6 explores and illustrates the problem of power consumption in HPC-AI systems so as to make the issue clearer to the reader and highlight the main solutions and perspectives.
Chapter 7 proposes an innovative cloud architecture called logical IoT cloud (LIC), which provides integrated systems management for cloud and IoT resources.
Chapter 8 illustrates the importance of HPC in AI investigations and solutions through a number of selected applications. For each application or AI-oriented study, critical aspects of the execution time, considering training and inference, are described.
Chapter 9 highlights scientific and technical issues of neural networks for large-scale numerical simulation.
Chapter 10 presents a solution for leveraging high performance computing (HPC) infrastructures and investigates the integration of distributed deep learning (DDL) techniques to address Industry 4.0 challenges across three distinct applications: intrusion detection with multi-layer perceptron, defect identification with convolutional neural networks and predictive maintenance with recurrent neural network.

Chapter 11 studies the sentiment analysis with Moroccan Arabic dialect text collected from social media platforms with the aim of achieving close-to-optimal accuracies for a large dataset using different techniques for feature extraction together with machine learning approaches.
Chapter 12 explores research directions to deal with anti-patterns while migrating to microservices, thereby aiming to improve migration quality and facilitate the creation of efficient microservices architectures.
Chapter 13 describes data imputation process at the edge level considering the limitations in processing capabilities of edge devices and compares several popular lightweight machine learning and deep learning algorithms in terms of inference time and accuracy.
Chapter 14 introduces a sophisticated end-to-end solution to the problem of occlusion in object detection that utilizes point cloud density analysis for precise occlusion rate assessment and combines depth information with RGB data to effectively separate occluded objects.
Chapter 15 presents a study on sentiment analysis with Arabic Twitter comments written in the Moroccan dialectal Arabic and modern standard Arabic.

Acknowledgments

The editors would like to thank all authors who submitted their chapters to this book. We thank also all reviewers for their time and tangible work they have made to successfully complete the reviewing process. We also sincerely thank Dr. Thomas Ditzinger, Springer Executive Editor, Interdisciplinary and Applied Sciences & Engineering, and Mrs Varsha Prabakaran, Springer Project Coordinator in Books Production service for the opportunity of having this book, for their assistance during its preparation process and for giving the authors the opportunity to publish their works in Springer book in LNNS series. Many thanks also to the Editorial Board and Springer's staff for their support. Finally, we would like to thank the following editorial committee members for professional and timely reviews: Dan Grigoras (UK), Takayuki Kushida (Japon), Abdelmounaam Rezgui (USA), Helen Karatza (Greece), Ali Kartit (Morocco), Cristophe Cérin (France), Ma Kun (China), Youssef Jabri (Morocco), Sidi Ahmed Mahmoudi (Belgium), Said Tazi (France), Francois Pinet (France), Pierre Manneback(Belgium), Bouchaib Ferrahi (Morocco), Nouredine Melab (France), Mahmoud Nassar (Morocco), Sédrick Stassin (Belgium), Claude Tadonki (France) and Mostapha Zbakh (Morocco).

Contents

IMAGEN Data Analytics Platform for Animal Phenotype Detection

Manu Agarwal(✉), Enkhzol Dovdon(✉), Haftom Meles Hailu, Luis Roma Barge, Yanja Dajsuren, and Jakob de Vlieg

Mathematics and Computer Science Department, Eindhoven University of Technology, Eindhoven, The Netherlands
{m.agarwal,e.d.dovdon,h.m.hailu,l.roma.barge,y.dajsuren, j.d.vlieg}@tue.nl

Abstract. In Precision Livestock Farming (PLF), deep learning-based approaches are increasingly employed to study animal behavior on farms. These behavioral studies enable animal phenotyping, which can be used for genetic selection and social network analysis in large animal groups. While considerable attention is often given to models in the deep learning field, the models themselves do not function in isolation. Efficient deep learning workflows require systems that bridge research and prototyping with production operations.

In this chapter, we introduce the IMAGEN Data Analytics Platform. This platform supports the development of deep learning models for animal phenotyping and addresses various data challenges through a DataOps approach. Although initially designed for the animal phenotype detection domain, the platform is domain-neutral and can be applied to similar cases in other application domains.

Keywords: Data analytics platform · Deep learning · HPC · Pilot-job framework · DataOps · Animal phenotype detection · Precision livestock farming

1 Introduction

Smart farming or "Data-driven farming" is an approach to farming that involves the process of collecting digital data from the farms and applying data analytics and machine learning for combating efficiency-related problems, innovative farm interventions, and profitability of agriculture [16]. In particular, precision livestock farming (PLF) is a sub-field of smart farming that employs the technology-enabled, data-driven approach to deal with the various challenges of livestock agriculture [5].

In the context of animal breeding, genomic selection is the process of estimating the future performance of breeding individuals for the traits of interest, such as pecking, tail biting, and temperament, via a large number of markers distributed across the whole genome [21]. Genomics selection provides a great venue for genetically improving animal welfare [9].

M. Zbakh et al. (Eds.): CloudTech 2023, LNNS 1220, pp. 1–12, 2024.
https://doi.org/10.1007/978-3-031-78698-3_1

Various barriers impede the adoption of data-driven efforts in genomic selection programs [26]. Apart from developing new algorithms and techniques to mine and analyze the data, a data analytics platform is needed for efficient data storage, meta-data management, data analysis, and data serving. A data analytics platform is necessary not just for harnessing data for new insights. However, it is also needed to facilitate the execution of data mining algorithms time-efficiently [23]. In addition, how the analysis results are served to the decision-makers is very important for a successful selection program [15]. Perakis et al. indicate that the end-users in precision livestock farming find it difficult to apply or comprehend the analysis result without appropriate data visualizations [23].

By implementing data-driven animal welfare measures in the livestock housing facilities, the efficiency of the livestock-based protein production systems can be improved significantly [9]. To this end, data-driven genomic selection is increasingly used to improve livestock social behavior, resilience to disease, and other stress factors. In particular, the data-driven genomic selection aims to leverage a data-driven approach to produce nutritious food and fiber for humans using healthy, efficient livestock [18]. The base requirement for such data-driven genomic selection is access to novel behavioral and welfare traits [9]. Thus, a reliable collection of individual animal-based behavioral and welfare traits becomes essential.

Technological advancements over the years have made it easier to obtain data from livestock housing facilities. Different data sources that can be acquired from a housing facility can include, amongst others, online sound, video observations, feeding intake, drinking behavior data, data from sensors on the animals, and data from milking robots [14]. Therefore, by leveraging this data, it is now possible that genomic selection programs are becoming increasingly data-driven.

To further the research and innovation efforts in the field of precision livestock farming, the larger IMAGEN [22] and SmartTurkeys [8] programs were initiated under the research grants from the Netherlands Organization for Scientific Research (NWO). A multidisciplinary consortium of leading Dutch universities and industrial partners associated with animal breeding came together to jointly research and develop technology for analyzing animal behavior in enterprise settings. Both programs use multi-modal big data such as videos, sensors, and RFID readers to understand animal social behavior better and link it to individual genes. For this purpose, the IMAGEN Data Analytics Platform is being developed to detect and study animal behaviors. In our earlier work [2], we presented the high-level layered architecture of this platform. This chapter presents our platform's features from the standpoint of usability, DataOps, and automated pipelines.

The High-performance computing (HPC) cluster-based IMAGEN Data Analytics Platform is a versatile solution designed for data scientists, researchers, and engineers. It comprises three core components:

1. Intuitive User Interface: IMAGEN Data Analytics Platform's user-friendly interface simplifies access to platform features. Whether you're an expert or a novice, the UI abstracts complexities, allowing seamless task submission,

compute configuration, and progress monitoring. Role-based access control enhances security.
2. Integration with a Pilot Job Framework: Traditional job submission methods can lead to inefficiencies. IMAGEN Data Analytics Platform addresses this by encapsulating smaller components within a Pilot-Job. This decouples task submission from resource assignment, streamlining execution and simplifying job management.
3. DataOps Principles: Automation is key in modern data analytics. IMAGEN Data Analytics Platform implements automated data pipelines for data movement, quality assurance, and visualization. These pipelines enhance efficiency and empower researchers in their work.

This chapter's outline is as follows: Sect. 2 presents the background and related work, Sect. 3 details the main components of our IMAGEN data analytics platform, and in Sect. 4, we summarize the conclusion and future work.

2 Background and Related Work

Several HPC cluster-based frameworks and seamless software development approaches have been proposed in the literature for infrastructure and associated data analytics platform architecture to facilitate data-driven efforts. This section explores some of the most widely known Pilot-Job frameworks, relevant context, and DataOps. As early as 2013, Zhu et al. proposed how cloud computing can be applied to smart farming [31]. In the context of the agriculture sector, Oliver et al. provided a detailed review of various cloud and distributed architectures for data management in agriculture [11]. Also, over the last few years, there has been an increase in using both public cloud as well as national HPC environments for hosting the smart farming platforms [11,19]. The following subsections detail the related work from the Pilot-Job frameworks and the DataOps perspective.

2.1 Pilot-Job Frameworks

In recent years, several Pilot-Job frameworks have gained widespread adoption for efficiently managing computing tasks on HPC clusters. For instance, RADICAL-Pilot (RP) [20], a scalable runtime system for executing large workloads comprised of many heterogeneous tasks on leadership-class HPC platforms, is a flexible Pilot System written in Python. RP effectively supports the execution of homogeneous and heterogeneous workloads, showing promising scalability and performance improvements at extreme scales. It enables users to execute numerous concurrent simulations and manage their data requirements across clusters, grids, and clouds. Boyer et al. [7] explore the efficient allocation of computational resources within the Large Hadron Collider (LHC) computing grid. The LHC, a massive particle accelerator, generates enormous amounts of data that require distributed processing across a global network. To address this challenge, the authors propose a Pilot-Job provisioning mechanism using CREAM (Computing Resource Execution and Management) computing elements. The CREAM system acts as an intermediary between user jobs and the underlying grid resources. These pilot jobs are dynamically created and dispatched to available computing nodes, reducing overhead and improving job execution efficiency. The study evaluates the performance of the proposed approach, considering factors such as job submission time, resource availability, and overall system throughput. In a study [17], researchers address the immense data processing challenge posed by the Large Hadron Collider (LHC) at CERN. The ATLAS experiment, a massive collaboration, plays a pivotal role in LHC research. ATLAS relies on the PanDA (Production and Data Analysis) Workload Management System to handle the unprecedented multi-petabyte data. PanDA manages data processing across 140 distributed data centers worldwide. As LHC data runs demand more resources than Grid computing can offer, the project integrates PanDA with supercomputers in the United States, Europe, and Russia. Casajus et al. [10] present the DIRAC Pilot Framework and its integration with the DIRAC Workload Management System (WMS). The DIRAC system is designed to manage computing tasks for user communities, where tasks may vary. The Pilot Jobs, implemented as a specialized type of DIRAC Compute Element, play a crucial role. They provide an overlay layer to computing resources, ensuring uniform presentation to the central WMS server for payload matching. Falkon [24] is a novel framework designed for rapid task execution on compute clusters. The primary goal of Falkon is to address the limitations of existing batch schedulers by providing a streamlined and efficient approach. Falkon employs multi-level scheduling, a technique borrowed from operating systems research. It separates resource acquisition (such as requesting resources from batch schedulers) from task dispatch.

2.2 DataOps and Automated Data Pipelines

With immense diversity in terms of data formats, as well as a lack of standardized practices for data and system integration [4], a data analytics platform's architecture should cater to the specific data integration, processing, and data serving challenges of the use-case at hand. Mark Underwood [29] explores the significance of continuous metadata in the context of continuous integration, stream processing, and enterprise DataOps. Traditionally, metadata has been perceived as static "data about data". However, recent trends in big data, cloud-native architectures, and streaming platforms demand a more dynamic view of metadata. These trends include DevOps, CI/CD, DataOps, and the need for improved data access and governance. DataOps is distinct from DevOps [30]. While DevOps focuses on software development, DataOps extends control over how data evolves. As data forms the underlying "currency" for roles like Data Analysts, Data Scientists, and AI Engineers, governance and data quality are critical. DataOps ensures tangible, meaningful outcomes and insights by orchestrating changing data throughout the entire data pipeline. Data versioning involves keeping track of changes made to a dataset over time. Each dataset version is saved and can be accessed even after subsequent changes. It's like maintaining a historical record of the data's evolution, allowing you to understand why specific data points have certain values. This practice is essential for efficiently sharing, reusing, and mapping datasets to deployed models [3]. In addition, data pipelines [6] are stages that interact with data. These stages include acquisition, management, analysis, and reasoning. The data pipeline is essential for software systems' descriptive, predictive, and prescriptive analytics.

The IMAGEN Data Analytics Platform incorporates a pilot-job framework, which, along with a user-friendly web interface, addresses end user concerns about usability and task submission. Additionally, by implementing automated data pipelines and a training dataset versioning scheme, our platform adheres to DataOps principles. The following section provides a more detailed exploration of these features.

3 IMAGEN Data Analytics Platform: Main Components

This section highlights the three significant components of the IMAGEN Data Analytics Platform: 1) An intuitive User Interface to our platform, 2) Integration with a Pilot Job Framework, and 3) DataOps principles adopted on the platform.

3.1 An Intuitive User Interface to Our Platform

We developed an intuitive user interface to enhance our data platform's user experience and usability. This interface allows platform users to operate with minimal learning curve. Additionally, it abstracts the complex SLURM-based HPC system from the end users. For example, to submit a deep learning training task, researchers only need to provide their containerized training code, task-specific configuration data, the input training data location on HPC, the output storage location (i.e., the learned model), and the compute requirements via the user interface. The interface then submits the training task to the HPC using the Pilot Job framework. Moreover, the user interface enforces an additional security layer by implementing role-based access control. Figure 1 showcases the platform's UI layer.

Key features of the user interface include:

1. A task selection interface for various tasks such as deep learning training, inferencing, and data quality checks.
2. An interface that enables users to fill task-specific configuration data.
3. An interface that enables users to track task status or state.
4. A user management interface for system administrators to oversee and track user activities and set up role-based access.

3.2 Integration with a Pilot Job Framework

Executing a complex simulation or experiment often involves generating numerous jobs to gain meaningful insight. The conventional approach of submitting

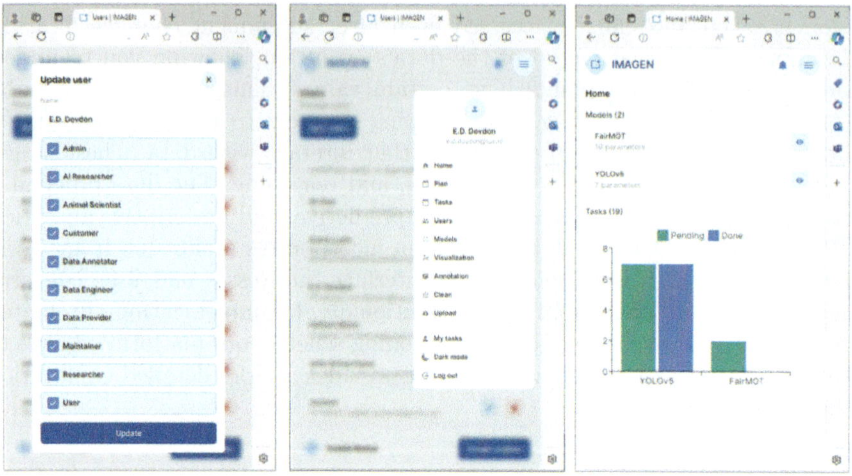

Fig. 1. Intuitive User Interface to our platform

individual jobs to the HPC cluster can lead to scheduling inefficiencies, with hundreds or thousands of jobs waiting in the batch queuing system [28]. Additionally, tracking the success or failure of each job becomes a challenging and tedious task. Pilot-Job frameworks offer an alternative strategy by introducing a flexible execution model that separates the task submission from resource assignment. A Pilot-Job serves as a container for an array of smaller jobs, whether independent or coupled and is treated as a single job by the computing resource's job scheduler [28]. Treating Pilot-Jobs as a single entity in the computing resource's job scheduler decouples the assignment of resources from the actual task submission, allowing job execution without queueing each task and addressing the inefficiencies associated with traditional job submission methods.

On our Data Platform, we adopted PiCaS (Pitch-catch System) as a pilot job framework. The PiCaS [27] is developed by SURF, and it is a task management tool designed to distribute and handle numerous jobs across diverse computing resources efficiently. In the context of PiCaS, tasks are represented as tokens, each encapsulating specific parameters. Operating as a Token Pool Server, PiCaS facilitates requesting the next available token, with token content customizable to suit the requirements of different applications. The PiCaS server is built on CouchDB. It maintains information about token availability, locks, and completion status. The adaptable and flexible data structure accommodates diverse task-related details, including work descriptions, output logs, and progress indicators [27]. A complete overview of the system, including the PiCaS client and PiCaS server, is illustrated in Fig. 2. PiCaS features a dedicated client library to fetch the tokens from the PiCaS server. This library enables users to upload, fetch, and modify tokens easily. Implemented as a Python Iterator, the PiCaS client library operates as a loop, continuously running as long as work is to be done.

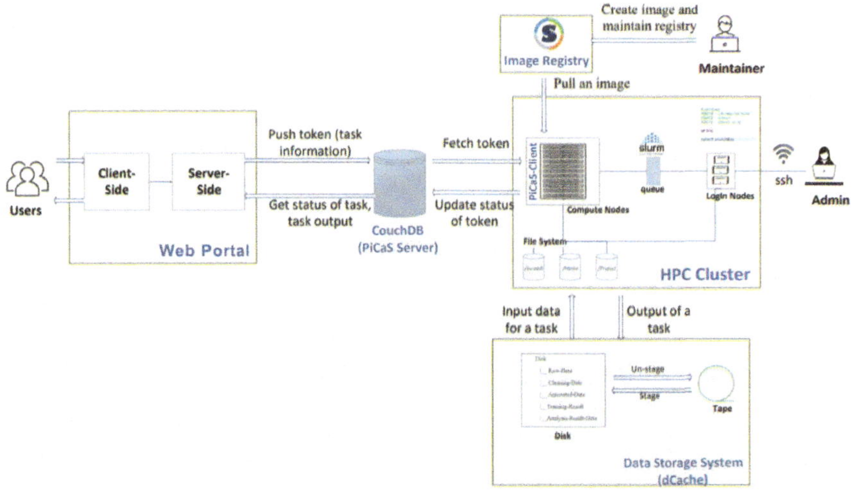

Fig. 2. System Overview: Module Interaction.

3.3 DataOps Principles Adopted on the Platform

Bringing automation into the modern data analytics workflow is a key DataOps principle. Automation helps to reduce manual labor, enables speedy delivery, and improves quality. In the context of a data platform, data pipelines are a Key ingredient of such automation.

On our Data Platform, we applied the DataOps principles to address the issues in two specific areas. Firstly, we developed data pipelines to streamline the deep learning model development work. Secondly, we implemented dataset versioning and management in our storage layer. The following section details how these approaches were implemented on our platform.

Automated Data Pipelines – A Key Contributor in DataOps: Automation is a fundamental principle of DataOps, and to align with this, we have developed automated data pipelines on our platform. These pipelines focus on three key functional areas: 1) Data Movement, 2) Data Quality Assurance, and 3) Data Visualization.

1. Data Movement Pipeline: Our storage layer is segmented into various logical sections, each representing different stages in data analysis and AI model development. The transfer of data between these sections is a repetitive task that we have optimized through a configurable Data Movement Pipeline. This pipeline provides a flexible, general-purpose solution that can accommodate various data movement scenarios. Figure 3 provides an illustrative example of how our Data Movement Pipeline manages data transfer to and from the HPC cluster.
   As shown in Fig. 3, an AI model training task on our platform retrieves input data from the Active Research Data Section, which utilizes the HPC cluster's

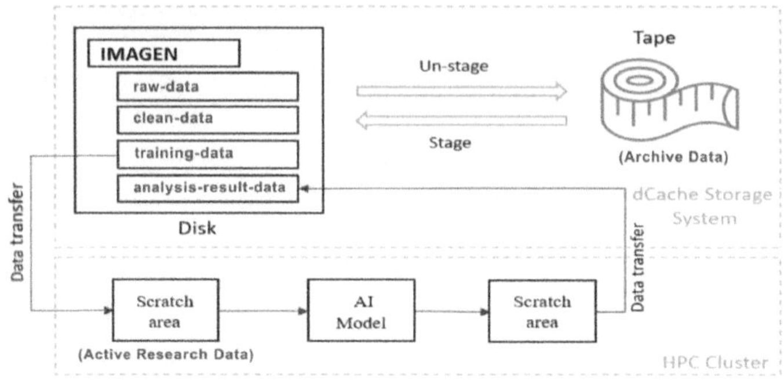

Fig. 3. Data Movement and Storage

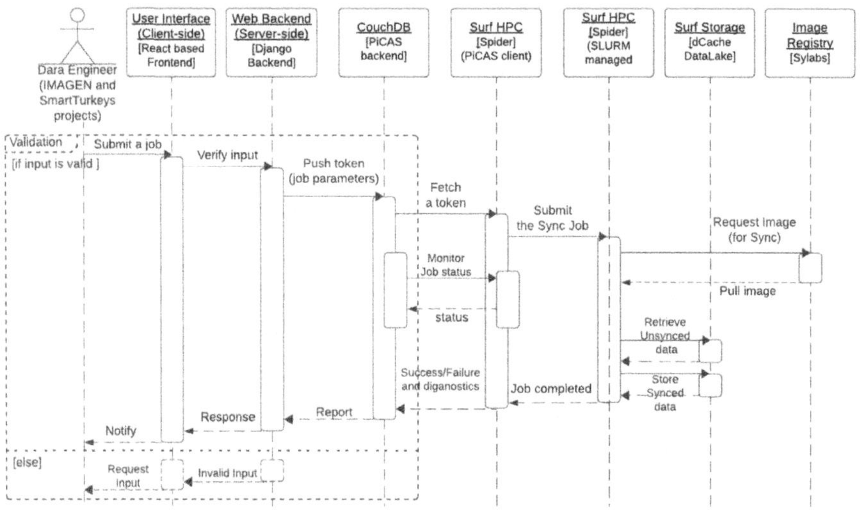

Fig. 4. Sequence diagram of frame rate synchronization data pipeline

scratch area for its high-performance I/O capabilities during model training. Transferring data from dCache to the HPC cluster's scratch area is crucial for leveraging this high-speed I/O, ensuring that the AI model has rapid access to the necessary datasets. Our Data Movement Pipeline allows users to configure both source and destination addresses as parameters, offering the flexibility to move data between any desired locations.

2. Data Quality Assurance Pipeline: Ensuring data quality is essential for the success of deep learning applications. In the IMAGEN program, cameras are the primary data sources for computer vision deep learning models. We use a multi-camera setup to record animals on the farm, combining footage from cameras installed at various points around the farm. However, an irregular frame rate across different cameras in a multi-camera setup can cause issues such as drift. Therefore, it is crucial to synchronize the data from different

cameras before feeding it into a deep learning model [12].
To address this challenge, we developed the Data Quality Assurance pipeline, which corrects irregular frame rates in a multi-camera setup. As depicted in Fig. 4:
   (a) This data pipeline reads and writes input raw videos and output synced videos from our data lake on the dCache storage system [13].
   (b) The synchronization logic is containerized using Singularity containers.
   (c) Surf's PiCAS pilot job framework is utilized for job submission and management on SLURM-based SPider HPC clusters.
3. Data Visualization Pipeline: We designed an INSIGHT [25] data visualization pipeline on the platform that caters to the needs of animal behavioral scientists. The INSIGHT is a data pipeline that analyzes the output of AI algorithms to help researchers conduct animal behavior research. The INSIGHT provides a visual interpretation of the output of AI algorithms. The pipeline generates various visualizations related to:
   (a) Trajectory analysis;
   (b) Distance traveled by each animal, e.g., it is used to calculate feeding efficiency;
   (c) Proximity encounters within a specific radius;
   (d) Basic animal state identification, e.g., Passive, Active, and Very Active;
   (e) Social network based on proximity and animal state;
   (f) Recorded states of an individual animal over time.

   A couple of example plots generated by the INSIGHT pipeline are shown in Fig. 5 and Fig. 6. The INSIGHT pipeline was successfully validated with animal behavior scientists for the purpose of downstream animal behavior research.

Implementing Dataset Management into Our Storage Layer
A good dataset management approach is important for a successful deep learning model development process. Compared to data processing and model training/serving, this aspect often gets overlooked in the deep learning model development process. A good dataset management decouples raw data collection from its consumption. This decoupling ensures model reproducibility by versioning the training datasets.

On our platform, dataset management is implemented via versioning training datasets. Our dataset management is built around the following principles:

1) Enable Model Reproducibility by versioning the datasets: Model reproducibility is achieved via versioning the training datasets. During a deep learning model development process, the versioned datasets provide the capability to run a deep learning algorithm on a certain dataset and gain the same or similar quality of results. This helps to create trust and credibility in the system that produced such a deep learning model.
2) Enable training data persistence: It is important that training datasets are stored immutably so that model reproducibility is achieved for trust and troubleshooting purposes.

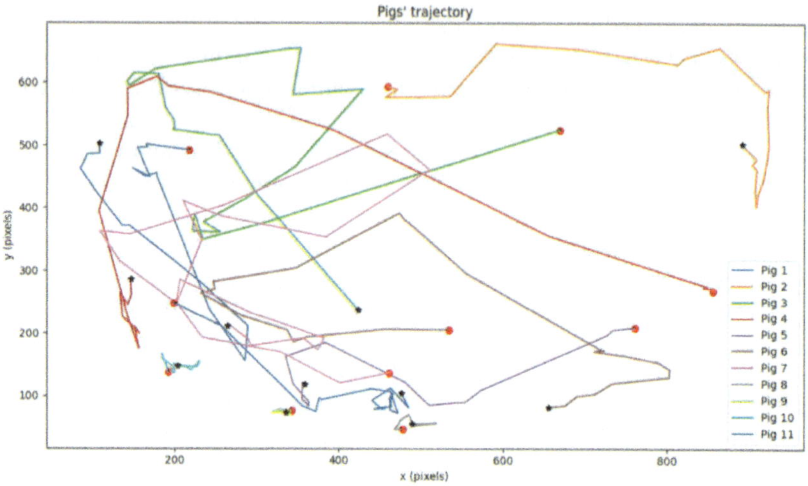

Fig. 5. Trajectory plot of pigs in a pen. Red dots are the starting points, and black stars are the ending points

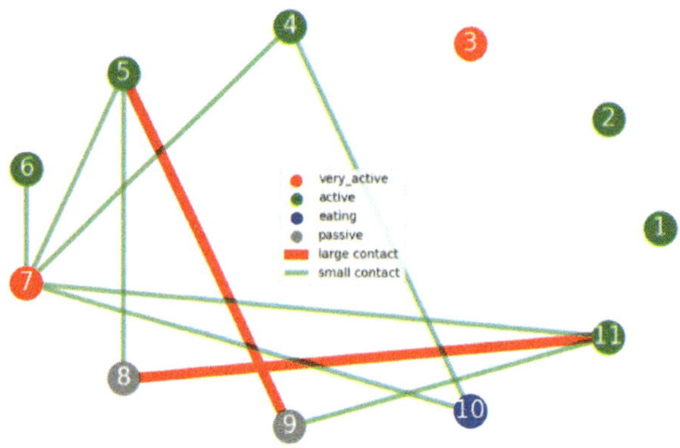

Fig. 6. Social network plot on pig data: the proximity radius is 80 pixels.

On our platform, the dataset versioning is achieved by using the open-source Data Version Control (dvc) tool [1]. We used Surf's dCache-based data lake for persisting the datasets under the "Training Data Section".

4 Conclusion and Future Work

In this chapter, we presented the key features of the IMAGEN Data Analytics Platform designed for animal phenotype detection research. We highlighted the three main components of our platform: 1) an intuitive user interface, 2) integration with a pilot-job framework, and 3) implementation of DataOps principles.

Each of these components not only supports the IMAGEN team's deep learning research but also offers design ideas that can be adopted in various other domains. Moving forward, we plan to enhance the platform's operational capabilities by recording detailed job execution statistics. Based on the analysis of these statistics, we will implement new features to further improve the platform.

Acknowledgment. This work was supported by the Nederlandse Organisatie voor Wetenschappelijk Onderzoek (NWO): AnIMAl Group sENsor - Integrating behavioral dynamics and social genetic effects to improve health, welfare and ecological footprint of livestock (IMAGEN) and Smart Turkeys, NWO Open Technology program.

References

1. Data version control. https://dvc.org/. Accessed 25 Mar 2024
2. Agarwal, M., Dovdon, E., Barge, L.R., Dajsuren, Y., de Vlieg, J.: A HPC-based data analytics platform architecture for data-driven animal phenotype detection. In: 2023 IEEE 6th International Conference on Cloud Computing and Artificial Intelligence: Technologies and Applications (CloudTech), pp. 1–6. IEEE (2023)
3. Amershi, S., et al.: Software engineering for machine learning: a case study. In: 2019 IEEE/ACM 41st International Conference on Software Engineering: Software Engineering in Practice (ICSE-SEIP), pp. 291–300. IEEE (2019)
4. Amiri-Zarandi, M., Hazrati Fard, M., Yousefinaghani, S., Kaviani, M., Dara, R.: A platform approach to smart farm information processing. Agriculture **12**(6), 838 (2022)
5. Berckmans, D.: General introduction to precision livestock farming. Anim. Front. **7**(1), 6–11 (2017)
6. Biswas, S., Wardat, M., Rajan, H.: The art and practice of data science pipelines. arXiv preprint arXiv:2112.01590 (2021)
7. Boyer, A.F., Hill, D.R., Haen, C., Stagni, F.: Pilot-job provisioning through cream computing elements on the worldwide LHC computing grid. In: 34th European Simulation and Modelling Conference (ESM) (2020)
8. Breed4Food: Smart turkeys, NWO open technology program (2023). https://www.breed4food.com/affiliate-projects/item/23-smart-turkeys-nwo-open-technology-program. Accessed 28 Sept 2023
9. Brito, L.F., Oliveira, H.R., McConn, B.R., Schinckel, A.P., Arrazola, A., Marchant-Forde, J.N., Johnson, J.S.: Large-scale phenotyping of livestock welfare in commercial production systems: a new frontier in animal breeding. Front. Genet. **11**, 793 (2020)
10. Casajus, A., Graciani, R., Paterson, S., Tsaregorodtsev, A., et al.: Dirac pilot framework and the dirac workload management system. In: Journal of Physics: Conference Series, vol. 219, p. 062049. IOP Publishing (2010)
11. Debauche, O., Mahmoudi, S., Manneback, P., Lebeau, F.: Cloud and distributed architectures for data management in agriculture 4.0: review and future trends. J. King Saud Univ.-Comput. Inf. Sci. **34**(9), 7494–7514 (2022)
12. Dovdon, E., Agarwal, M., Dajsuren, Y., de Vlieg, J.: Irregular frame rate synchronization of multi-camera videos for data-driven animal behavior detection. In: International Conference on Advances in Computing Research, pp. 97–112. Springer (2024)

13. Fuhrmann, P., Gülzow, V.: dCache, storage system for the future. In: Euro-Par 2006 Parallel Processing: 12th International Euro-Par Conference, Dresden, Germany, 28 August–1 September 2006, pp. 1106–1113. Springer (2006)
14. Garcia, R., Aguilar, J., Toro, M., Pinto, A., Rodriguez, P.: A systematic literature review on the use of machine learning in precision livestock farming. Comput. Electron. Agric. **179**, 105826 (2020)
15. Karmas, A., Tzotsos, A., Karantzalos, K.: Geospatial big data for environmental and agricultural applications. In: Big Data Concepts, Theories, and Applications, pp. 353–390 (2016)
16. Kawamura, T., Katsuragi, T., Kobayashi, A., Inatomi, M., Oshiro, M., Eguchi, H.: Development of an information research platform for data-driven agriculture. Int. J. Agric. Environ. Inf. Syst. (IJAEIS) **13**(1), 1–19 (2022)
17. Klimentov, A., et al.: Integration of panda workload management system with supercomputers for atlas and data intensive science. In: Journal of Physics: Conference Series, vol. 762, p. 012021. IOP Publishing (2016)
18. Koltes, J.E., et al.: A vision for development and utilization of high-throughput phenotyping and big data analytics in livestock. Front. Genet. **10**, 1197 (2019)
19. Krisnawijaya, N.N.K., Tekinerdogan, B., Catal, C., van der Tol, R.: Data analytics platforms for agricultural systems: a systematic literature review. Comput. Electron. Agric. **195**, 106813 (2022)
20. Merzky, A., Turilli, M., Titov, M., Al-Saadi, A., Jha, S.: Design and performance characterization of radical-pilot on leadership-class platforms. IEEE Trans. Parallel Distrib. Syst. **33**(4), 818–829 (2021)
21. Meuwissen, T., Hayes, B., Goddard, M.: Genomic selection: a paradigm shift in animal breeding. Anim. Front. **6**(1), 6–14 (2016)
22. Nederlandse Organisatie voor Wetenschappelijk Onderzoek (NWO): Animal group sensor - integrating behavioural dynamics and social genetic effects to improve health, welfare and ecological footprint of livestock (IMAGEN), pp. 18–19 (2023). https://www.nwo.nl/onderzoeksprogrammas/perspectief. Accessed 28 Sept 2023
23. Perakis, K., Lampathaki, F., Nikas, K., Georgiou, Y., Marko, O., Maselyne, J.: Cybele-fostering precision agriculture & livestock farming through secure access to large-scale HPC enabled virtual industrial experimentation environments fostering scalable big data analytics. Comput. Netw. **168**, 107035 (2020)
24. Raicu, I., Zhao, Y., Dumitrescu, C., Foster, I., Wilde, M.: Falkon: a fast and light-weight task execution framework. In: Proceedings of the 2007 ACM/IEEE Conference on Supercomputing, pp. 1–12 (2007)
25. Roma Barge, L.: The design of the insight pipeline for behavioral animal science animal breeding. Ph.D. thesis, TUE (2021). pdEng thesis
26. Sawant, M., Urkude, R., Jawale, S.: Organized data and information for efficacious agriculture using pride model. Int. Food Agribus. Manage. Rev. **19**(1030-2016-83147), 115–130 (2016)
27. Surf: Picas overview (2024). https://doc.grid.surfsara.nl/en/latest/Pages/Practices/picas/picas_overview.html
28. Turilli, M., Santcroos, M., Jha, S.: A comprehensive perspective on pilot-job systems. ACM Comput. Surv. (CSUR) **51**(2), 1–32 (2018)
29. Underwood, M.: Continuous metadata in continuous integration, stream processing and enterprise DataOps. Data Intell. **5**(1), 275–288 (2023)
30. Walsh, B.: AI Best Practice and DataOps. Springer, Cham (2022)
31. Zhu, Y., Wu, D., Lil, S.: Cloud computing and agricultural development of china: theory and practice. Int. J. Comput. Sci. Issues (IJCSI) **10**(1), 7 (2013)

A Three-Layered AI-Oriented Approach for Intelligent Generation of Standardized Privacy Policies in Smart Healthcare Environments

Ismail El Yamani[1(✉)], Souad Sadki[1], Hanan El Bakkali[1], and Driss Allaki[2]

[1] Mohammed V University in Rabat, Rabat, Morocco
ismail.elyamani@um5r.ac.ma, {souad.sadki,h.elbakkali}@um5s.net.ma
[2] INPT in Rabat, Rabat, Morocco
d.allaki@inpt.ac.ma

Abstract. The adoption of cloud computing and artificial intelligence (AI) in healthcare has revolutionized patient care, necessitating robust privacy policies to safeguard individual privacy. Yet, existing policies lack standardization due to diverse actors involved. Motivated by this, we introduce in this work an AI-oriented and multilayered framework for intelligent generation of standardized privacy policies. The proposed approach actively involves stakeholders, including healthcare providers, patients, and data custodians, in the formulation of policies. To show the comprehensiveness, accuracy, and alignment of our solution with evolving regulatory frameworks, a new conversational agent called Spp-G was implemented. This inclusive approach ensures that privacy policies, shaped by collaborative efforts and input from key stakeholders, not only comply with regulatory standards but also reflect the diverse perspectives and concerns of those impacted by data-sharing practices.

Keywords: Compliance · Privacy Policies Standardization · Data Protection · Smart Healthcare · Cloud Computing · Artificial Intelligence

1 Introduction

Smart healthcare is a game-changing development in the medical industry, integrating advances in artificial intelligence, IoT, and data analytics, among others, to provide personalized, efficient, and high-quality care on a greater scale. Despite significant investment, smart healthcare adoption has fallen short of expectations due to technological, organizational, and environmental problems [1]. Among these problems, patient privacy preservation is the most relevant [1]. Ensuring the privacy and security of the personal health data of patients remains a significant challenge for healthcare providers [2]. There is a strong emphasis on understanding local laws and regulations. Many works in the literature [3–5]

M. Zbakh et al. (Eds.): CloudTech 2023, LNNS 1220, pp. 13–28, 2024.
https://doi.org/10.1007/978-3-031-78698-3_2

have demonstrated how important it is to have a complete understanding of legal frameworks such as the General Data Protection Regulation (GDPR) and the Health Insurance Portability and Accountability Act (HIPAA). Adherence to these requirements guarantees the preservation of patients' privacy and compliance with established regulations. Legal expertise not only protects healthcare professionals from future legal lawsuits but also makes it easier to incorporate current technologies that handle and secure patient data. This is essential given the growing litigiousness of the healthcare industry [5]. Several studies in the literature have attempted to address this difficulty using various methodologies and technologies. Some works [6,7] used AI to address privacy concerns, utilizing Machine Learning and Deep Learning models to identify potential privacy issues, such as when a smartphone application accesses more data than stated in its terms of service, or when user reviews are analyzed to identify key privacy concerns.

Artificial Intelligence (AI) technologies, such as Natural Language Processing (NLP), were used in [8] to facilitate the processing and analysis of vast amounts of data to enhance the quality of life for patients with chronic pathologies. This capacity could be used for finding inconsistencies and discrepancies in policies and making sure they comply with laws and regulations. There is a need for a comprehensive and automated approach to understanding and managing policies, preferences, and compliance [9]. Merely relying on manual processes is no longer sufficient, as the sheer volume and complexity of data, coupled with the ever-evolving regulatory environment, necessitate a parallel and automated approach. This paper proposes an AI-oriented approach to a multilayered framework for the intelligent generation of standardized privacy policies. The approach relies on involving stakeholders, including healthcare providers, patients, and data custodians, in the formulation of policies through the use of a conversational agent.

The rest of the paper is structured as follows. Section 2 reviews existing research that worked in a similar direction. Section 3 states the research problem through a motivational example. Section 4 introduces the architecture behind our 3-layered approach idea and provides a detailed explanation of how it works. Section 5 presents the Spp-G conversational agent in action on a real world example. Finally, Sect. 6 concludes the paper by summarizing our findings and discussing potential future work.

2 Related Work

In the literature, many efforts have been made by researchers to tackle the issue of privacy policy formalization and translation. For example, authors in [10] developed a prototype of a privacy policy translator using NLP Techniques. This translator is capable of translating privacy policies of social networking sites from natural language to Controlled Natural Language for Data Sharing Agreements (CNL4DSA), employing techniques such as tokenization, parsing, and ontologies to automatically translate the policies. Building upon this work, the authors in [11] added a graphical user interface (GUI) to the privacy policy translator and tested it on 100 policies across five social networking sites, claiming to have achieved an accuracy rate of 93% in the translations. [12] proposes an approach to automatically detect privacy-related information in user story requirements using NLP and transfer learning, arguing that this can decrease privacy risks during agile software development. However, the study only evaluated the approach on a dataset of 1680 user stories, and the performance in other types of data or real-world settings remains unclear. Authors in [13] evaluated the use of supervised binary classification to distinguish between legitimate and rogue privacy policies, analyzing 100 policies from legitimate websites and 67 from rogue websites, and found clear statistical differences in length and adherence to privacy principles. The work in [14] used traditional and neural network models, including Convolutional Neural Networks (CNNs), Logistic Regression (LR), and support vector machines (SVMs), to classify privacy policy text into one or more privacy practice categories, demonstrating the potential of these methods in enabling downstream tasks such as filtering for more detailed data practices, extracting salient details to present to users, or summarizing privacy practices. [15] proposes PriGen, a framework that uses deep learning models and static analysis to generate privacy captions directly from an Android application's source code. The authors used a code captioning dataset to train a Neural Machine Translation (NMT) model and evaluated the quality of the generated captions. PriGen provided developers with a set of tools and methodologies to create consistent and concise privacy notices for their applications, addressing the challenge of matching privacy rationales with source code changes.

In another vein, beyond the concern of privacy policy formalization, another category of recent works emphasizes the importance of compliance with laws and regulations in protecting patients' sensitive data. Indeed, according to the authors in [2], the principal challenge for health information exchange in a healthcare context is how to share the maximum amount of data while maintaining compliance with regulations that may constrain the sharing of sensitive private data. To address this concern, the authors of [2] highlight the importance of translation guardrails to protect against problematic translations of regulatory ideas. In the same context, and considering the General Data Protection Regulation (GDPR) as an example of such regulations, the authors of [16] present a technique to formalize privacy policies and a notion of compliance in the setting of object-oriented distributed systems. In particular, they introduce a policy specification language that takes regulatory compliance into account. Additionally, they propose a high-level modeling language for distributed systems, extended with support for policies [16]. Additionally, authors of [17] developed a quantitative and systematic tool for evaluating privacy policies' compliance with the Personal Data Protection Act (PDPA) while providing recommendations for enhancing compliance practices. For this purpose, they introduced a new multi-criteria privacy policy scoring model integrated with comprehensive statistical data analyses. To prove the effectiveness of their model, this solution has been applied to 384 SMEs privacy policies from Thailand [17]. Yet, in large architectures such as IoT-based systems, the concern for compliance with legal laws and regulations in privacy policies becomes more important than ever before. Hence, as a first step towards compliance checking, it is of utmost importance to consider Privacy by Design (PbD) practices during the design phase of IoT system development [18]. In this regard, a recent work in [18] analysis Privacy by Design needs in current IoT systems and then introduces a new ontology encapsulating the existing PbD measurements and their inter relationships as a means of offering easily explainable PbD guidance [18]. Having this mind, the next challenge concerns the expression of users' (e.g. patients') privacy preferences to prevent any conflicts with service providers privacy policy . In this context, authors of [19] proposes a three-module framework, named "Ontology-Based Privacy-Preserving" (OBPP) to address the problem of heterogeneity, privacy-preserving of generated data, while ensuring a high-level services. Compared to this relevant proposition, authors of [20–22] focus of the problem of understanding patient's preferences first and how we can preserve them.[20] proposes a Privacy-Preserving architecture for mobile healthcare (PPAMH) that uses an intelligent mobile app to formulate and enforce patients' privacy preferences in the healthcare environment. However, this work did not provide a way to deal with conflicting privacy policies. This is why [21] builds on [20] by proposing a negotiation-based mechanism that uses S4P privacy defining language to detect and resolve conflicts between multiple policies. Meanwhile,[22] provides a different solution to the problem of conflicting policies by prioritizing one policy over another based on the Analytic Hierarchy Process (AHP), also using S4P for conflict discovery and resolution. The most recent of work in this chain of works would be [23] which focused on bringing these privacy-preserving mechanics to the cloud through the use case of remote healthcare monitoring. the All of these works made use of S4P's formal privacy policy language, emphasizing the importance of standardizing privacy policies to improve patient privacy and trust in mobile healthcare. This is why we propose an NLP-based approach to policy translation for comprehensive interpretation of service policies in healthcare. This will help us build on [20–22] by providing the second piece of the puzzle which is the service policies in S4P, this will allow us to use the privacy policies in conflict detection using AHP and negotiation techniques.

3 The Issue of Ambiguous, Non-Compliant, and Conflicting Privacy Policies

Privacy policies within healthcare settings are often drafted using complex legal terminology, posing significant comprehension challenges for patients. As illustrated in Fig. 1, Consider the case of Michael, a British expatriate residing in Germany for professional pursuits. Michael encounters considerable difficulties in understanding and navigating privacy policies when seeking medical treatment, which necessitates the sharing of his Personal Health Information (PHI) between his healthcare provider in the UK and the new provider in Germany. The intricate legal jargon employed in these policies, compounded by Michael's adjustment to the German healthcare system, leaves him uncertain about how his data is being handled and disseminated.

Further exacerbating this scenario is the involvement of two distinct cloud service providers (CSPs) for the storage and management of patient data, each operating under a different healthcare provider. This multi-CSP participation introduces additional complexities in privacy management, particularly in ensuring compliance with disparate regulatory frameworks such as the GDPR in Europe and any potential UK-specific regulations. Michael must navigate not only the differences in national privacy laws but also the distinct policies of each CSP, which may permit extended data retention periods or divergent protocols for data sharing with third parties. The diversity of domains of application, fixed vocabularies, and varying levels of policy abstraction contribute to the heterogeneity of these policies, leading to potentially conflicting situations that necessitate careful management and reconciliation.

Getting inspired by this scenario, recent privacy-preserving solutions have to address issues such as simplicity and clarity in privacy policies, standardization across all privacy-related documents, compliance with laws and regulations, as well as incorporating privacy-by-design principles at every stage of user privacy protection.

To address these challenges, we propose a 3-layered approach to compliance checking, conflict detection, and resolution through the use of the translator proposed in [9] and a conversational agent titled *Standard Privacy Policy-Generator (Spp-G)*. This approach not only simplifies the understanding of complex legal policies for patients but also ensures that all of the actors work together for a more effective and patient-centered compliant and conflict free privacy management.

4 Multi-layered Architecture for Intelligent Generation of Standardized Privacy Policies in Smart Healthcare Environments

In this section, we introduce a multilayered architecture that leverages the power of artificial intelligence to improve the protection of privacy in smart healthcare environments. Since smart healthcare uses multiple advanced technologies such as cloud computing, IoT, and artificial intelligence, it is imperative to take into account all of the different actors involved in providing the necessary services to patients. However, their involvement creates a considerable issue regarding the protection of private personal data. One big pillar of this protection is making sure that we are working on a patient-centric system that takes the patient's preferences into account. This was accomplished by means of earlier research [9, 20–22]. However, there was a need for a system that can check if these preferences are conflicting with the privacy practices of the various aforementioned actors, particularly the primary healthcare provider. This is the main interest of the preliminary layer of this architecture.

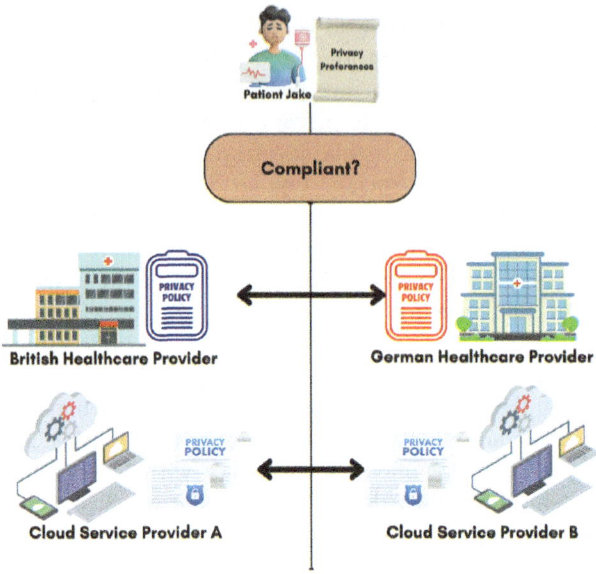

Fig. 1. Illustration of the problem

4.1 Privacy Policies and Preferences Translation Using CL2S4P

In previous work [9], we worked on developing a translator for translating privacy policies from common language to S4P (CL2S4P). S4P is a machine readable language, enabling users to define their privacy preferences as well as a framework for service providers to define their privacy policies, enabling an automated comparison with a user's expressed preferences to assess service compliance [24]. This translator leverages the capabilities of a domain-specific language to provide a standardized format for privacy-related documents, including service policies, laws, regulations, and privacy preferences. It harnesses the power of natural language processing and machine learning algorithms to decipher the complex language of privacy policies and extract the most pertinent information. Figure 2 illustrates the workflow of CL2S4P through a flowchart.

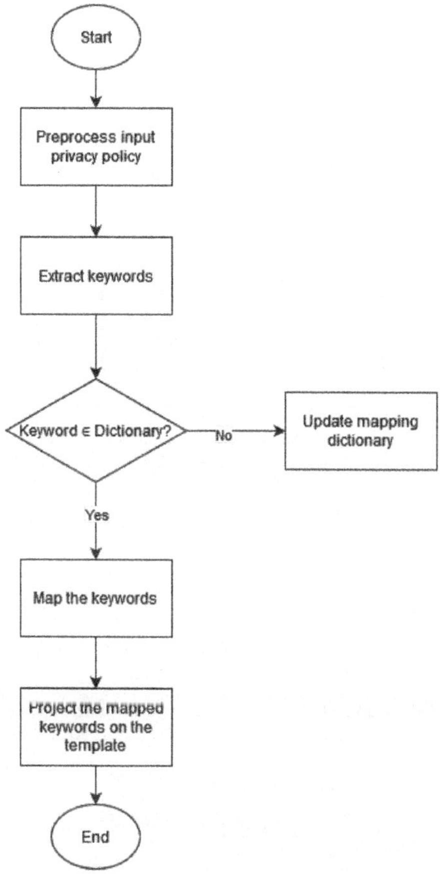

Fig. 2. Flowchart of CL2S4P

Through this preliminary layer, where the patient's preferences, the service provider's privacy policy, and relevant laws and regulations are expressed in a standardized, simplified language (S4P), our solution can benefit all actors involved. The patient will clearly understand the service provider's policies and thus be more knowledgeable about how their data is being handled. Simultaneously, the service provider will have a better understanding of the user's preferences and fine-tune their practices to avoid conflicts, leading to more trust and transparency. Furthermore, ensuring full compliance with laws and regulations through S4P could potentially save service providers from significant financial losses from lawsuits. This approach addresses the critical need for simplicity, standardization, and regulatory compliance in privacy-preserving solutions.

4.2 Overview of the Three-Layered Architecture for Preventing Privacy Conflicts and Ensuring Compliance

In order to combine this contribution with previous work we propose a 3-layered architecture for conflict prevention and privacy enhancement and standardization. As presented in Fig. 3, this architecture comprises 3 layers.

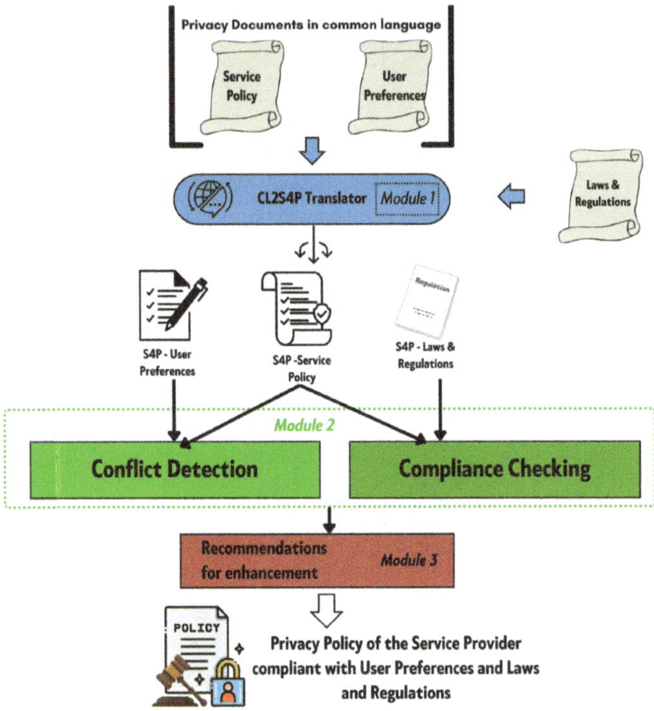

Fig. 3. The proposed 3-layered architecture

4.3 Conflict Detection and Compliance Checking Module

The objective of this layer is to make use of the S4P language and the utilities and methods associated with it in order to detect conflicts between the service provider's privacy policy and the user preferences as part of the first sub-module dedicated to Conflict Detection. Then, in parallel, it would do the same thing with the laws and regulations that apply to the service provider, such as HIPAA and GDPR, in a separate sub-module dedicated to Compliance Checking of the service policy.

4.3.1 Conflict Detection

In this sub-layer we will leverage the power of S4P for conflict detection, based on the works in [21,22] there are 2 ways we can deal with conflicts:

Negotiation Based Technique [21]**:** A negotiation-based mechanism was developed in this work, it uses S4P privacy defining language as a standard in which the privacy rules are defined. The suggested technique is divided into a five steps process ending with the establishment of a contract between the patient and the other conflicting third parties.

Prioritization Based Technique [22]**:** This technique is more advanced compared to that of [21]. It provides a different solution to the problem of conflicting privacy policies in smart healthcare contexts by prioritizing one policy over another based on the AHP (Analytic Hierarchy Process) prioritization method. This study employs S4P formal privacy policy language to aid in conflict discovery and resolution [22].

Both of these works rely on having the privacy policies in S4P in order for their detection and resolution mechanisms to work. Integrating these works into our smart framework massively improves their robustness and practicality.

4.3.2 Compliance Checking Layers

This sub-layer is very important for the service provider, privacy lawsuits could cost millions of dollars to companies handling sensitive personal information [25]. This layer assesses the compliance of the service provider's privacy policy with relevant local laws and regulations. Our conversational agent, **Spp-G**, plays a crucial role in this process. It employs a conversational approach to gather comprehensive information about the service provider's privacy practices. This conversational method proves invaluable in determining whether their practices align with the applicable laws and regulations. We can then employ the aforementioned conflict detection and resolution mechanisms to solve any conflicts that may affect the user and service provider. At this stage the agent will have gathered enough information to go to the next layer.

4.4 Recommendation Layer

In the recommendation layer, we will use every piece of information gathered in previous layers and employ AI to facilitate communication with the service provider in order to gather more information about them and eventually propose the changes that they must make to their privacy policy for it to be compliant with the laws and regulations as well as the privacy preferences of the user. This approach will minimize future conflicts and ensure that both the patient and the service provider are protected.

5 The Role of the Conversational Agent Spp-G (Standard Privacy Policy Generator) in Policy Generation

To facilitate a collaborative approach to privacy policy enhancement or generation, we introduce a conversational agent called Standard Privacy for Service Providers (Spp-G). This agent collaboratively analyzes the current privacy practices of service providers and provides recommendations to ensure their policies comply with relevant laws and regulations. Additionally, it suggests ways to improve the clarity around how service providers handle user/patient data. By adopting this collaborative approach, we aim to enhance transparency regarding how service providers manage user/patient data.

5.1 Introduction to RASA

Spp-G is based on RASA, which is an accessible, flexible, robust, and transparent Python based, machine learning framework for building smart conversational agents [26]. RASA employs Conversation Driven Development (CDD), to achieve rapid development [26]. RASA benefits, among others, include modular development, adaptability and data control [27]. RASA comprises a micro-services architecture that gives it a lot of flexibility to work in any scenario, it is used for a lot of applications including smart healthcare, for example, it was employed in [28] to build an app to accompany elderly patients and remind them to do their daily medical routine as well give recommendations for their diet based on their condition.

Rasa comprises two important modules, Rasa Natural Language Understanding(NLU) and Rasa Core [28], responsible for understanding the user utterances and performing dialogue management, respectively [29]. Figure 4 shows the complete data pipeline for the conversational agent module.

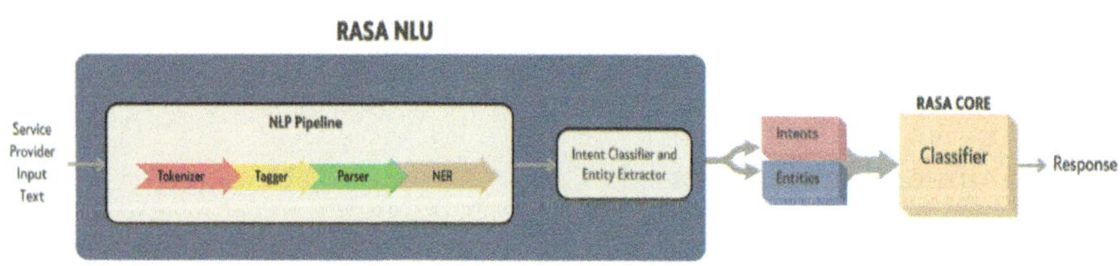

Fig. 4. RASA Data pipeline Architecture

Rasa Natural Language Understanding (NLU) comprises two components:

Natural Language Processing (NLP) Component: The input text is passed through a pipeline of NLP operations, including Tokenization, Part of Speech (POS) Tagging, Parsing, and Named Entity Recognition. The tokenizer splits the input text into tokens, which can then be labeled by the POS tagger into grammatical types such as nouns, verbs, and adjectives. The parser then analyzes the grammatical structure of the text and establishes relationships between tokens, as well the syntactic structure of the sentence. Finally, the NER is used to identify and classify the entities of the text into categories such as actors, organizations, dates, and more.

Intent Classifier and Entity Extractor: The processed query from the NLP component is passed on to the intent classifier and the entity extractor module. This component determines the user's intention (what they are trying to achieve with their message) and extracts the important entities. Entities in RASA are pieces of information that are relevant to the user's intent [29]. The module maps the user query to a pre-defined intent and extracts important entities from it. RASA uses a multi-task transformer architecture that handles both intent classification and entity recognition together as part of its NLU component [30].

RASA Core Component: After the Natural Language Understanding (NLU) component extracts the intents and entities from the user's input, this information is forwarded to the second component of the framework, known as RASA Core Dialogue Management. RASA Core employs a classifier as a response selector, which is responsible for identifying and providing the most appropriate response to the user's query. In situations where RASA cannot match the query to a predefined intent, a fallback policy is triggered. This fallback policy generates a fallback response, which is sent to the user, requesting them to rephrase their query in a different manner.

5.2 Spp-G Architecture

Standard Privacy Policy Generator (Spp-G) is a multi-layered smart conversational agent designed to assist service providers in enhancing their privacy policies, ensuring compliance with relevant laws and regulations, and providing users with clear and transparent privacy policies. Leveraging RASA's open-source framework and its ability to integrate custom modules and functionalities, Spp-G seamlessly navigates through its various layers. Algorithm 1 describes the workflow of the conversational agent, which can transition seamlessly between layers and fall back to other layers as necessary.

Here is a breakdown of each layer of Spp-G:

- **Initiation Layer:** This layer establishes the initial interaction between the service provider and the conversational agent. It aims to create a friendly tone and address any preliminary concerns.
- **Statement of Intent Layer:** Here, the agent seeks to understand the service provider's objectives for the conversation to determine whether it can provide assistance and how best to proceed. This layer is crucial in shaping the direction of the subsequent interaction.
- **Policy Definition Layer:** As the first step in data collection, the agent gathers primary information from the privacy policy, such as the service provider's geographical location and operating field.

Algorithm 1. Conversational Agent for Privacy Policy Analysis

```
Begin
% Initiation Layer
initializeChatSession()
displayMessage("Welcome! I'm here to help you assess and improve your privacy policy. Let's get started!")
% Statement of Intent Layer
userResponse ← getUserResponse("Could you tell me more about your service and what you hope to achieve today?")
policyInfo["intent"] ← extractIntent(userResponse)
if not policyInfo["intent"] then
   displayMessage("It seems like I might not be the right fit for your needs.")
   endChatSession()
end if
% Policy Definition Layer
userResponse ← getUserResponse("Where is your company located, and what is your field of operation?...")
policyInfo["location"], policyInfo["field"], ... ← extractGeneralInfo(userResponse)
% Data Layer
userResponse ← getUserResponse("Can you provide details about how you collect, use, and retain user data?")
policyInfo["data handling"] ← extractDataHandlingDetails(userResponse)
% User-Centric Layer
userResponse ← getUserResponse("How do you ensure that your privacy policy is user-centric?")
policyInfo["user-centricity"] ← assessUserCentricity(userResponse)
% Security and Compliance Layer
userResponse ← getUserResponse("What measures do you have in place to comply with local data protection laws?")
policyInfo["compliance"] ← checkCompliance(userResponse)
% Recommendation Layer
recommendations ← generateRecommendations(policyInfo)
displayRecommendations(recommendations)
newPolicy ← generateNewPrivacyPolicy(policyInfo, recommendations)
displayMessage("Based on our discussion and the following recommendations, we have drafted a new privacy policy for you:")
displayNewPolicy(newPolicy)
userResponse ← getUserResponse("Please review the updated policy. Do you have any modifications or further concerns?")
if userResponse indicates modifications needed then
   modifyPolicy(newPolicy, userResponse)
   displayNewPolicy(newPolicy)
   goto the previous step for further modifications if needed
end if
displayMessage("Thank you for using our service. Have a great day!")
endChatSession()
End
```

- **Data Layer:** In this layer, the agent collects comprehensive information about data collection, usage, retention, and other relevant aspects. The gathered information will prove invaluable for providing highly valuable recommendations in later stages.
- **User-Centric Layer:** Aligning with our previous work [9,20–23], we maintain a user/patient-centric approach. This layer focuses on assessing the user-centricity of the privacy policy, determining the degree of control users have over their data.
- **Security and Compliance Layer:** This critical layer addresses compliance with local laws and regulations. The conversation at this stage is influenced by the data gathered in previous layers, ensuring a comprehensive evaluation.
- **Recommendation Layer:** Finally, this layer provides tailored advice and industry best practices for data and privacy protection. This guidance helps service providers maintain compliance, avoid legal issues, and contribute to more transparent data practices, ultimately benefiting users by ensuring their protection.

5.2.1 The Exchange Process

SPP-G uses a layered architecture to ensure it goes through everything it needs to discuss with the service provider. It starts off with the initiation layer where it states its purpose and how it's going to help the service provider, then it makes sure that this interaction is going to be beneficial for the service provider by inquiring about what they are looking for from their interaction with it. Then it quickly moves on to gathering information about the service provider as shown in Fig. 5. It starts off by gathering general information about the content of the service provider's privacy policy to understand their practices in general.

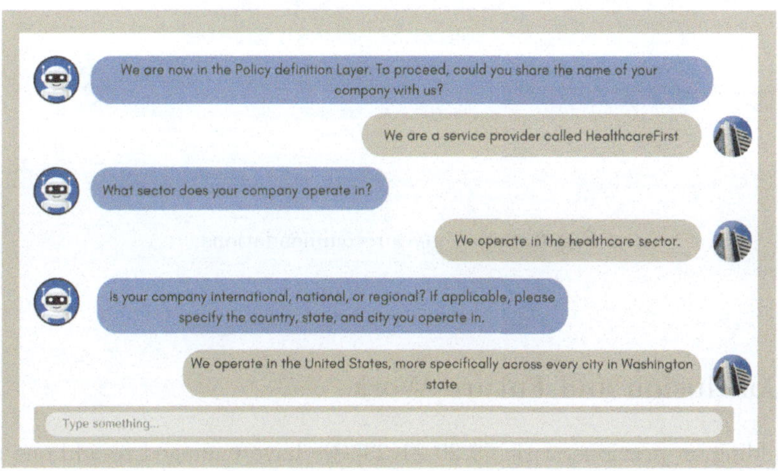

Fig. 5. Screenshot of Spp-G

Then the conversational agent tries to get deeper in the data layer to understand thoroughly what kind of data is being collected, how it's being collected and for what purpose. It also tries to look for special cases such as the handling of minors' data which has special privacy regulations [31]. Then it moves on to discuss the rights the privacy policy gives to the user, to check the degree of control given to the user to perform actions such as modifying and deleting their own data. Based on the information gathered so far, the agent will then select the relevant laws and regulations that apply to the concerned service provider. For example, if they are a healthcare provider in the EU, then they must abide by GDPR [4].

Once the data gathering is done it moves on to giving recommendations based on what it gathered, these recommendations are tailored based on a lot of factors such as the geographical location of the Service provider, what kind of data they handle, and more. As seen in Fig. 6, Spp-G provides general and specific recommendations, such as the use of basic policies of data minimization, regular updates to terms and services, transparency in data practices. The more data the Service provider is willing to give, the better the recommendations will be.

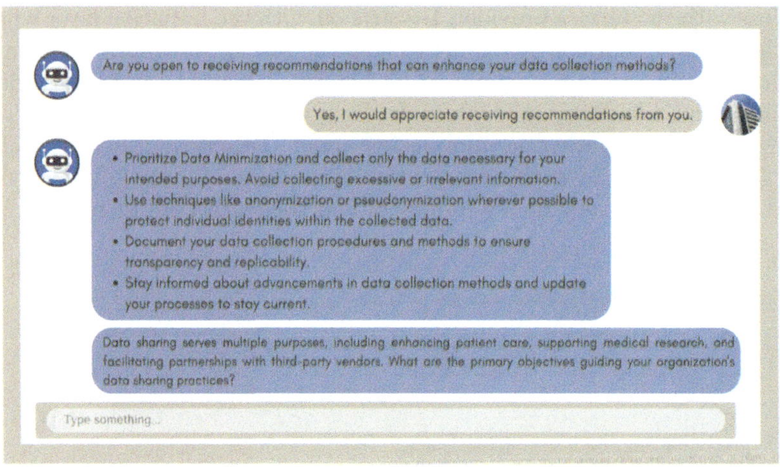

Fig. 6. Spp-G gives recommendations

6 Conclusion and Future Work

By building on previous works [9,20,21,23], we have managed to add one more piece to the overall patient-centric privacy-oriented smart healthcare architecture. This work focuses on improving privacy preservation by adopting privacy by design principles to avoid conflicts that may jeopardize the protection of a patient's or user's private data. We have successfully developed a conversational agent that takes a collaborative approach to compliance checking, conflict detection, and resolution. There is definitely more to be done on this subject. In future work, we will also conduct a systematic literature review to study the structure of privacy policies and explore how we can further leverage them to enhance privacy protection in smart healthcare. We will also analyze a database of privacy policies from multiple organisations across various domains with a focus on smart healthcare. This will help us automate the process of conflicts detection and compliance checking.

References

1. Renukappa, S., Mudiyi, P., Suresh, S., Abdalla, W., Subbarao, C.: Evaluation of challenges for adoption of smart healthcare strategies. Smart Health **26**, 100330 (2022)
2. Anderson, C., Baskerville, R., Kaul, M.: Managing compliance with privacy regulations through translation guardrails: a health information exchange case study. Inf. Organ. **33**(1), 100455 (2023)
3. Nichols, A., Trimble, P., Stebbings, A.: The use of inter-professional education (IPE) healthcare law and ethics scenario based learning sessions amongst nursing, midwifery and law students: a qualitative investigation. Nurse Educ. Today **109**, 105254 (2022)
4. Häuselmann, A., Custers, B.: Substantive fairness in the GDPR: fairness elements for article 5.1a GDPR. Comput. Law Secur. Rev. **52**, 105942 (2024)
5. Mia, M.R., et al.: A comparative study on HIPAA technical safeguards assessment of android mHealth applications. Smart Health **26**, 100349 (2022)
6. Nema, P., Anthonysamy, P., Taft, N., Peddinti, S.T.: Analyzing user perspectives on mobile app privacy at scale. In: 2022 IEEE/ACM 44th International Conference on Software Engineering (ICSE), pp. 112–124 (2022)
7. Rahman, M.S., et al.: PermPress: machine learning-based pipeline to evaluate permissions in app privacy policies. IEEE Access **10**, 89248–89269 (2022)
8. Lázaro, E., et al.: Efficiency of natural language processing as a tool for analysing quality of life in patients with chronic diseases. A systematic review. Comput. Hum. Behav. Rep. **14**, 100407 (2024)
9. Yamani, I., Sadki, S., El Bakkali, H., Allaki, D.: CL2S4P: a privacy policy translator for comprehensive interpretation in smart healthcare systems, pp. 01–07 (2023)
10. Tanoli, I.K., Petrocchi, M., De Nicola, R.: Towards automatic translation of social network policies into controlled natural language. In: 2018 12th International Conference on Research Challenges in Information Science (RCIS), Nantes, pp. 1–12. IEEE (2018)
11. Tanoli, I.K., Amin, I., Junejo, F., Yusoff, N.: Systematic machine translation of social network data privacy policies. Appl. Sci. **12**, 10499 (2022)
12. Casillo, F., Deufemia, V., Gravino, C.: Detecting privacy requirements from user stories with NLP transfer learning models. Inf. Softw. Technol. **146**, 106853 (2022)
13. Adhikari, A., Das, S., Dewri, R.: Privacy policy analysis with sentence classification. In: 2022 19th Annual International Conference on Privacy, Security & Trust (PST), pp. 1–10 (2022)
14. Liu, F., Wilson, S., Story, P., Zimmeck, S., Sadeh, N.M.: Towards automatic classification of privacy policy text (2017)

15. Jain, V., Gupta, S.D., Ghanavati, S., Peddinti, S.T.: PriGen: towards automated translation of android applications' code to privacy captions. In: Cherfi, S., Perini, A., Nurcan, S. (eds.) RCIS 2021. LNBIP, vol. 415, pp. 142–151. Springer, Cham (2021). https://doi.org/10.1007/978-3-030-75018-3_9
16. Tokas, S., Owe, O., Ramezanifarkhani, T.: Static checking of GDPR-related privacy compliance for object-oriented distributed systems. J. Log. Algebraic Methods Program. **125**, 100733 (2022)
17. Chatsuwan, P., Phromma, T., Surasvadi, N., Thajchayapong, S.: Personal data protection compliance assessment: a privacy policy scoring approach and empirical evidence from thailand's smes. Heliyon **9**(10), e20648 (2023)
18. Alkhariji, L., De, S., Rana, O., Perera, C.: Semantics-based privacy by design for internet of things applications. Futur. Gener. Comput. Syst. **138**, 280–295 (2023)
19. Gheisari, M., et al.: OBPP: an ontology-based framework for privacy-preserving in iot-based smart city. Futur. Gener. Comput. Syst. **123**, 1–13 (2021)
20. Sadki, S., Bakkali, H.E.: PPAMH: a novel privacy-preserving approach for mobile healthcare. In: The 9th International Conference for Internet Technology and Secured Transactions (ICITST-2014), London, United Kingdom, pp. 209–214. IEEE (2014)
21. Sadki, S., El Bakkali, H.: An approach for privacy policies negotiation in mobile health-Cloud environments. In: 2015 International Conference on Cloud Technologies and Applications (CloudTech), pp. 1–6, IEEE (2015)
22. Sadki, S., El Bakkali, H.: Resolving conflicting privacy policies in M-health based on prioritization. SCPE **17**, 207–226 (2016)
23. El Yamani, I., El Mrabet, Y., Sadki, S., El Bakkali, H.: A privacy-preserving and cloud-based framework for medical emergencies detection in smart systems. In: El Bhiri, B., Saidi, R., Essaaidi, M., Kaabouch, N. (eds) ICATH 2022. Advances in Science, Technology & Innovation, pp. 105–112. Springer, Cham (2024). https://doi.org/10.1007/978-3-031-46849-0_12
24. Becker, M., Malkis, A., Bussard, L.: S4P: a generic language for specifying privacy preferences and policies. Technical report MSR-TR-2010-32 (2010)
25. Online, C.: The biggest data breach fines, penalties, and settlements so far (2023)
26. Malamas, N., Symeonidis, A.: Embedding rasa in edge devices: capabilities and limitations. Procedia Comput. Sci. **192**, 109–118 (2021). Knowledge-Based and Intelligent Information & Engineering Systems: Proceedings of the 25th International Conference KES2021
27. Braun, D., Hernandez Mendez, A., Matthes, F., Langen, M.: Evaluating natural language understanding services for conversational question answering systems. In: Jokinen, K., Stede, M., DeVault, D., Louis, A. (eds.) Proceedings of the 18th Annual SIGdial Meeting on Discourse and Dialogue, Saarbrücken, Germany, pp. 174–185. Association for Computational Linguistics (2017)
28. Kumar, A., Haider, G., Khan, M., Khan, R.Z., Raza, S.S.: SAATHI: an Urdu virtual assistant for elderly aging in place. In: Aloulou, H., Abdulrazak, B., de Marassé-Enouf, A., Mokhtari, M. (eds.) ICOST 2022. LNCS, vol. 13287, pp. 73–85. Springer, Cham (2022). https://doi.org/10.1007/978-3-031-09593-1_6
29. Bocklisch, T., Faulker, J., Pawlowski, N., Nichol, A.: Rasa: Open source language understanding and dialogue management (2017)
30. Bunk, T., Varshneya, D., Vlasov, V., Nichol, A.: Diet: lightweight language understanding for dialogue systems (2020)
31. DataGuidance. International handling of children's personal data (2022)

CAReNet: A Promising AI Architecture for Low Data Regime Mixing Convolutions and Attention

Aurélie Cools[1(✉)], Mohammed Amin Belarbi[2], and Sidi Ahmed Mahmoudi[1]

[1] University of Mons, 7000 Mons, Belgium
{aurelie.cools,sidi.mahmoudi}@umons.ac.be
[2] AminTechs, 7000 Mons, Belgium

Abstract. Modern deep learning architectures often face significant challenges in terms of computational and memory resource consumption, as well as an increasing reliance on extensive annotated datasets. This issue is particularly problematic in areas where acquiring labeled data is costly or challenging. In this context, we introduce CAReNet *(Convolutional Attention Residual Network)*, an innovative architecture designed to effectively address these challenges. By skillfully merging convolutional layers with attention mechanisms and residual connections, CAReNet facilitates feature extraction and representation learning, optimizing model efficiency and compactness. CAReNet is particularly advantageous in the current landscape where minimizing dependence on large datasets and reducing memory consumption are priorities. It demonstrates an improvement in accuracy compared to other convolutional networks, with up to a 2.61% increase. CAReNet not only stands out for its performance but also for its size, which is nearly half that of others networks. This efficiency makes CAReNet ideal for applications that require less resources without compromising the result quality. This blend of compactness and increased efficiency underscores CAReNet's potential in meeting the demands of current deep learning architectures. It highlights the need for adopting architecture-specific training strategies to enhance efficiency and performance in deep learning, especially in scenarios where minimizing data and memory consumption is crucial.

Keywords: Deep Learning · Computer Vision · Efficiency · Compactness

1 Introduction

In the landscape of artificial intelligence and machine learning, supervised learning techniques have demonstrated remarkable success in solving a multitude of complex tasks, ranging from computer vision to natural language processing. Despite their efficiency, these approaches come with a set of inherent limitations

M. Zbakh et al. (Eds.): CloudTech 2023, LNNS 1220, pp. 29–39, 2024.
https://doi.org/10.1007/978-3-031-78698-3_3

that cause significant challenges for broad adoption. Foremost among these is the insatiable appetite for labeled data, which not only necessitates considerable human effort but often requires domain expertise for accurate annotation. The need for vast amounts of training data creates bottlenecks in data acquisition and preparation, limiting the applicability of supervised methods in low-resource settings or specialized domains where labeled examples are scarce. Even when an adequate dataset is assembled, the computational cost of training sophisticated models is another critical concern. The required hardware for high-quality model training, typically comprising multiple GPUs or TPUs, presents a substantial financial burden. Furthermore, the time investment for model training and hyperparameter tuning is significant, making it a resource-intensive endeavor that could span days or even weeks for particularly complex models. Thus, while supervised learning methods continue to push the envelope in predictive performance, the trade-offs in terms of data, computational resources, and time raise important questions about their practicality and scalability in real-world applications.

In this paper, we propose a novel architecture designed to mitigate these inherent difficulties. Uniquely, our model is not pre-trained on a large, external dataset but is instead tailored to perform exceptionally well using only the dataset in focus. By eschewing the need for expansive labeled datasets and high-computational training environments, our approach aims to democratize the application of machine learning models. This not only reduces the time and financial costs associated with training but also enhances the model's applicability in specialized or low-resource settings. In doing so, we present a feasible alternative to the traditional paradigms of supervised learning, emphasizing efficiency and accessibility without compromising on performance.

In the domain of image classification, Convolutional Neural Networks (CNNs) have long been considered the gold standard, achieving state-of-the-art results in a variety of benchmarks [12,21]. Their hierarchical structure naturally lends itself to the progressive abstraction of features, from edges to complex shapes, making them exceptionally well-suited for tasks that require spatial awareness [16]. However, the recent surge of Vision Transformers (ViTs) has opened new avenues for exploration [9]. These architectures leverage self-attention mechanisms to capture long-range dependencies and have shown comparable, if not superior, performance to CNNs on several classification benchmarks [24]. However, ViTs often require large amounts of data and computational resources for effective training, making them less accessible for low-resource settings [2]. Despite these limitations, the flexibility of Vision Transformers in handling different modalities and their capability for better interpretability make them an attractive alternative to traditional CNNs [23].

Despite their data-intensive nature, ViTs have shown remarkable performance even in low-data regimes when equipped with appropriate regularization and transfer learning techniques [19]. Their ability to capture both local and global features makes them a compelling choice for image classification tasks, often outperforming state-of-the-art CNNs under similar conditions.

The fusion of Convolutional Neural Networks (CNNs) and Vision Transformers (ViTs) [13] presents a groundbreaking approach to image classification, capitalizing on the unique strengths of each architecture. On the one hand, CNNs excel in parameter efficiency and local feature extraction, making them well-suited for tasks that require a nuanced understanding of spatial hierarchies. On the other hand, ViTs are adept at capturing global contextual information through self-attention mechanisms, providing a more holistic understanding of the image. By integrating these two architectures, a hybrid model can leverage the local feature recognition of CNNs and the global contextual awareness of ViTs, creating a more robust and versatile system for image classification.

The remainder of this paper is organized as follows: Sect. 2 provides a comprehensive review of existing literature and precedents in the domain. In Sect. 3, we elaborate on our proposed architectural design, discussing the datasets employed and elaborating on the specific training parameters. Our experiments' outcomes and findings are showcased in Sect. 4. The Sect. 5 concludes with a summary of our contributions, their potential implications, and future perspectives.

2 Related Works

In the area of computer vision, Convolutional Neural Networks have long been the cornerstone for solving a myriad of tasks, ranging from simple image classification to complex object detection. However, the recent emergence of Vision Transformers [13] has ushered in a new era, challenging the traditional paradigms and offering fresh perspectives on how vision problems can be approached. While these Transformer-based models have shown remarkable promise, they are hampered by their computational efficiency and the hard ability to scale. Various innovative solutions have been proposed to mitigate these issues, such as introducing sparse attention mechanisms and developing hybrid models that marry the best of both worlds-ConvNets and Transformers. In this context, Zihang Dai and al. in [6], proposed a novel hybrid architecture, CoAtNet, that combines the strengths of Convolutional Neural Networks (CNNs) and Transformers. While Transformers excel in terms of model capacity, they suffer from poor generalization due to the absence of the right inductive bias. On the other hand, CNNs possess this inductive bias but may lack the model capacity for certain tasks requiring complex spatial transformations and understanding hierarchical structures in images, such as 3D object recognition and scene understanding.

In [5], Dagli introduced a novel method designed for learning from limited data to classify galaxy morphologies. The approach employs a hybrid architecture that combines Transformer and Convolutional Neural Networks (CNNs). To enhance the model's performance, data augmentation and regularization techniques are also incorporated. They achieves state-of-the-art results in predicting galaxy morphologies and demonstrates strong performance on other datasets like CIFAR-100 [14] and Tiny ImageNet [25]. The paper also delves into the challenges of using large models and training methods in low-data regimes. It explores semi-supervised learning approaches that utilize both labeled and unlabeled data from different datasets. Despite promising results reported for the

Astroformer architecture, the absence of publicly available code prevented us from conducting our own experiments to challenge these findings.

Xinyu and al. in [18] addressed the computational inefficiencies of Vision Transformers (ViTs) by proposing a new family of models called EfficientViT. They identify memory access, computation redundancy, and parameter usage as the three main bottlenecks affecting the speed of ViTs. They introduce a new building block with a "sandwich layout", which places a single memory-bound Multi-Head Self-Attention (MHSA) layer between efficient Feed-Forward Network (FFN) layers. This design improves memory efficiency and enhances channel communication. The paper also introduces a Cascaded Group Attention (CGA) module that feeds different splits of the full feature to different attention heads, reducing computational redundancy and improving attention diversity. EfficientViT models outperform existing efficient models in terms of both speed and accuracy. For instance, EfficientViT-M5 surpasses MobileNetV3-Large by 1.9% in accuracy while achieving higher throughput on both GPU and CPU.

In [22], Zhengzhong and al. presented a new type of Transformer module, called multi-axis self-attention (MaxVit), that capably serves as a basic architecture component which can perform both local and global spatial interactions in a single block. Compared to full self-attention, MaxVit enjoys greater flexibility and efficiency, i.e., naturally adaptive to different input lengths with linear complexity; in contrast to (shifted) window/local attention, MaxVit allows for stronger model capacity by proposing a global receptive field.

In [1], Barhoumi and al. proposed a hybrid architecture that leverages the strengths of both Convolutional Neural Networks and Vision Transformers. By utilizing multiple CNNs for feature extraction, they enrich the input that the Vision Transformer attends to, thereby enhancing its ability to capture relevant features. This approach not only improves the performance of the Vision Transformer but also introduces a degree of modularity and scalability, making it adaptable for various medical imaging tasks.

All the previous recent works focused on fusion ViT and CNN. Our approach aligns with the same paradigm, albeit with an emphasis on more effectively and flexibly integrating the two components with a lower data dependency. Specifically, our contributions aim to develop a model with fewer parameters, resulting in a smaller model size that can be trained effectively with limited data. This contrasts with current practices in the literature, where networks are often pre-trained on large datasets like ImageNet and consist of several tens of millions of parameters, making them less accessible for widespread use. We will elaborate on our architecture in the subsequent section.

3 Methodology

3.1 The Proposed Architecture

As depicted in Fig. 1, our architecture starts with an introductory convolutional block, tailored to extract elementary feature maps from the input image. Subsequent to this, we introduce a duo of parallelized, distinct blocks. The first block within this duo integrates a bottleneck block, composed of multiple convolutional layers, offering enhanced feature extraction capabilities. Simultaneously, the secondary block introduces a residual unit, explicitly designed to bolster information flow and empower the model to learn profounder representations. The resultant feature maps from these parallel constructs are seamlessly merged using an element-wise addition operation, which is then succeeded by a Max pooling layer to further refine the spatial hierarchy of the feature maps.

This modular assembly is systematically reiterated thrice, employing varying channel dimensions to adeptly capture multi-scale feature intricacies. The terminal segment of our architecture encompasses a CareNet block, which is structured with a bottleneck block followed by an attention mechanism operating in parallel across both grid and window patterns. The outputs from these attention mechanisms are subsequently averaged to obtain a consolidated feature representation. Running in tandem with the CareNet block is an additional residual unit, ensuring consistent feature propagation.

To conclude the architectural design, we employ a Max pooling layer, which is then followed by a flatten operation to reshape the feature maps into a singular vector. The final component of our architecture is a linear classification layer, dedicated to producing the predictive outcomes. In the following sections, we delineate the advantages of the various layers employed in our architecture.

Convolutional Layers. Convolutional layers serve as the building blocks for many state-of-the-art computer vision architectures, including VGG [20] and ResNet in [10]. These layers are particularly effective in capturing local features like edges, corners, and textures in an image and preserving the spatial relationships between pixels. The virtues of Convolutional Neural Networks (CNNs) are manifold. Initially, their architecture is inherently well-suited for capturing hierarchical patterns in the data, enabling them to excel particularly in tasks that necessitate an understanding of spatial hierarchies, such as object recognition and segmentation. Subsequently, their parameter-sharing mechanism enhances computational efficiency, thereby permitting the training of deeper, more complex models without a commensurate increase in computational resources.

Max Pooling. Max-pooling layers have become an integral component in numerous convolutional neural network architectures due to their numerous advantages. Primarily, max-pooling reduces the spatial dimensions of the feature maps, leading to a decrease in the amount of parameters and computations and thus avoid the overfitting in the network. This reduction helps in combating the curse of dimensionality, enabling the network to focus on the most salient features and thus enhancing its generalization capability. Notably, max-pooling layers also play a pivotal role in hierarchically capturing abstract representations in deep CNNs. In sum, the inclusion of max-pooling layers in CNN architectures offers a systematic approach to achieving computational efficiency, robustness, and a hierarchical representation of features, making them indispensable in modern deep learning paradigms.

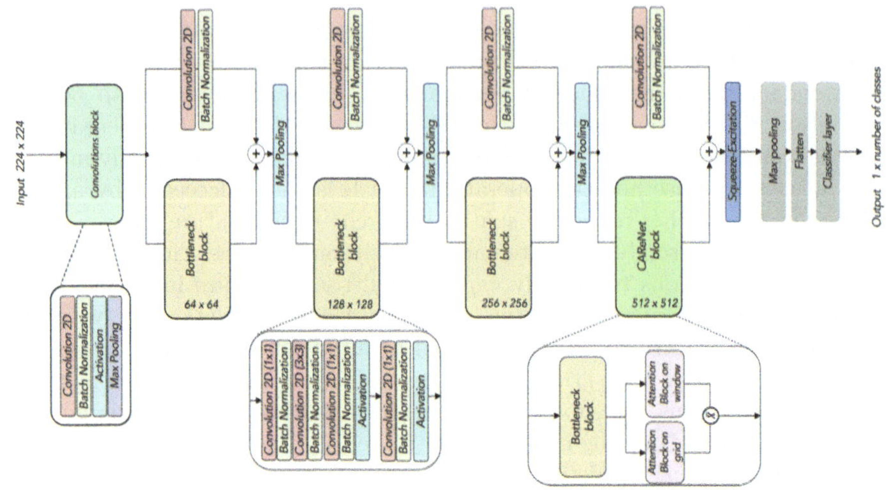

Fig. 1. Illustration of our architecture

Attention Mechanisms. Attention mechanisms have gained prominence primarily in the field of natural language processing [3] with architectures like Transformers [23]. However, their application in computer vision, as seen in models like ViT in [9], has shown significant promise. Attention layers weigh the importance of different regions in the image, allowing the model to focus on more relevant parts. They capture long-range dependencies without requiring higher computational costs.

Squeeze-and-Excitation Layers. Authors in [11] presents the Squeeze-and-Excitation (SE) layers which offer a mechanism to adaptively recalibrate channel-wise feature responses. SE layers dynamically adjust the importance of each channel based on the global average pooling. These layers add minimal computational overhead while offering performance gains. Finally, by allowing the model to focus on more informative channels, SE layers improve the representational power of the network.

Residual Connections. Introduced by He et al. in ResNet [10], residual connections have revolutionized the training of deep networks. In our architecture, Squeeze-and-Excitation (SE) layers are incorporated for a variety of reasons, notably to enhance the network's capacity for feature recalibration, thereby leading to a more adaptive representation of input data. Additionally, SE layers contribute to the overall interpretability of the model by providing a mechanism to weigh the importance of different features.

3.2 Experimentations

Methodology. In the methodology adopted for this study, we utilize several benchmark datasets, each of which is elaborated upon in the *Data* section. To ensure a fair and unbiased comparison, we commit to training a variety of models from scratch, or "from zero", without the influence of any pre-existing weights or pre-trained parameters. This approach ensures that each model's performance is a direct result of its architecture and training regime, devoid of any external influences. All experiments conducted herein are grounded in our original efforts, employing the bare codebases of the models without any pre-loaded weights, ensuring the purity and repeatability of our results.

Data. In this study, we leverage a diverse set of benchmark datasets to rigorously evaluate the performance and robustness of our proposed models. The Tiny-ImageNet dataset [15], a downscaled variant of the much larger ImageNet [7], comprises 200 classes with each class having 500 training and 50 validation images of 64×64 resolution. It provides a challenging yet computationally accessible platform for fine-grained classification tasks. For the Tiny-ImageNet dataset, we opted for a reduced set of 450 training images per class, rather than the full 500, in an effort to investigate the model's performance under a more constrained training scenario. Additionally, we used 50 images from each class for validation to evaluate the generalization capabilities of our approach.

We also utilize the Fashion-MNIST dataset [26], an alternative to the original MNIST dataset [17] of handwritten digits. Fashion-MNIST contains grayscale images of 10 different clothing types, such as shirts, trousers, and footwear, each normalized to a 28×28 pixel resolution. This offers a more intricate classification landscape compared to MNIST, which features single-channel, 28×28 grayscale images of numerals 0–9 and serves as a cornerstone for basic machine learning algorithms.

Finally, we incorporate the STL-10 dataset [4], designed for unsupervised and transfer learning. STL-10 contains 10 classes of 96×96 color images, with 500 labeled training instances and 800 test instances per class, in addition to an unlabeled set of 100,000 images. The selection of these datasets provides a broad testing ground to validate the versatility and scalability of our methods across different domains and levels of complexity.

Training Parameters. During the training phase, we used an RTX 3090 graphics card and opted for a maximum of 100 epochs to ensure comprehensive model training. Nevertheless, understanding the risk of overfitting and to capitalize on computational efficiency, we employed an early stopping mechanism. This mechanism monitors the validation loss during training. If the validation loss fails to improve over a span of 10 consecutive epochs, referred to as 'patience', the early stopping procedure is triggered, and the training is halted. This approach not only aids in preventing overfitting but also ensures that the model training is efficient, by eliminating the redundancy of epochs that do not contribute to model improvement. In our experimentation phase, meticulous attention was given to the selection of hyperparameters to optimize model performance across various image datasets. A learning rate of 1×10^{-4} was universally applied to all datasets, with the notable exception of STL10, for which a learning rate of 5×10^{-4} was chosen. These parameters were not arbitrarily set; instead, they emerged as optimal from a series of rigorous experiments. Similarly, the optimizer's selection was grounded in empirical evaluations. After juxtaposing several candidates, the AdamW optimizer was identified as the most efficacious for our task and subsequently adopted.

4 Results

In this section, we present a comprehensive evaluation of our proposed architecture, CAReNet, across multiple benchmark datasets. The primary focus of this evaluation is on the Top-1 validation accuracy and the model size, as detailed in Table 1. This comparison is crucial for illustrating the efficiency and effectiveness of CAReNet, particularly when trained from scratch.

4.1 Performance Across Datasets

The datasets used for this evaluation include STL10 [4], Tiny ImageNet, Fashion-MNIST [26] and MNIST [8]. These datasets were chosen to demonstrate the architecture's versatility and performance in diverse scenarios.

STL10 Dataset. For the STL10 dataset, CAReNet achieved a Top-1 accuracy of 80.26%, outperforming other architectures. This was accomplished with nearly half the number of parameters used in some networks, demonstrating CAReNet's efficiency in parameter utilization and its effectiveness in balancing accuracy with model compactness.

Tiny ImageNet. On the Tiny ImageNet dataset, CAReNet recorded a Top-1 accuracy of 54.4%. While this is slightly lower than MaxVit's 58.28% accuracy, it is important to note that CAReNet achieved this with a significantly smaller model size, highlighting its capability to handle complex, high-resolution image data effectively.

Table 1. Top-1 accuracy and model size comparison of neural network architectures trained from scratch across multiple datasets.

Architecture	Number of parameters (Million)	Model size (MB)	Top1 Validation Accuracy (%)			
			STL10	Fashion Mnist	Mnist	Tiny Imagenet (ours)
ResNet50	23.53	91	78.36	95.15	99.35	51.79
ResNet34	21.29	82	76.72	95.11	99.43	56.19
ResNet18	11.18	44	75.33	94.37	99.35	54.35
CoatNet_0	17.03	67	74.75	94.99	99.17	54.56
MaxVit	24.68	95	76.95	93.48	99.42	58.28
VGG16	134.30	513	10.00	94.91	99.33	52.17
CAReNet (ours)	**12.75**	**50**	**80.26**	**94.94**	**99.4**	**54.4**

Fashion-MNIST and MNIST. CAReNet demonstrates robust performance on Fashion-MNIST and MNIST, with accuracies nearing 95% and over 99%, respectively. These results are indicative of the model's adaptability to different types of image data and complexities, further emphasizing its practical utility.

4.2 Model Size and Efficiency

One of the key strengths of CAReNet is its compactness combined with high performance. The architecture's ability to maintain high accuracy with a reduced parameter count and smaller model size makes it a promising candidate for applications where computational resources are limited, such as mobile and edge computing.

4.3 Comparative Analysis

The evaluation also included a comparative analysis with other well-established models like ResNet50, ResNet18, VGG16, and emerging architectures like MaxVit and CoatNet. The results show that CAReNet is either equivalent or superior in terms of Top-1 accuracy across the evaluated datasets, while maintaining a smaller model size and fewer parameters.

4.4 Implications

The findings from our experiments suggest that CAReNet is not only an efficient and effective architecture for image classification tasks but also a viable solution in scenarios demanding computational frugality. Its ability to deliver high accuracy with a compact model size is pivotal for extending the applicability of deep learning models to a broader range of real-world scenarios, particularly where computational and data resources are scarce.

The presented results furnish an in-depth comparison among several cutting-edge neural network architectures, focusing on their Top-1 validation accuracy across diverse datasets while highlighting their model sizes and parameter counts of the architectures evaluated, ResNet50, with its 23,528,522 parameters, led the pack in the Mnist dataset with an accuracy of 99.35%. However, its performance on STL10, at 78.36%, though commendable, is surpassed by our proposed CAReNet, which achieves an impressive 80.26% while only utilizing 12,745,930 parameters. This is a testament to CAReNet's optimization, effectively leveraging a reduced parameter set to outperform a renowned architecture like ResNet50 on the STL10 dataset.

In summary, the deep learning landscape often poses the challenge of balancing model size, computational efficiency, and performance. However, CAReNet emerges as a sterling example, demonstrating that with judicious architecture design and parameter optimization, it's possible to craft models that are compact yet rival, if not surpass, the performance of their larger counterparts. Our findings model CAReNet's potential as a frontrunner for scenarios demanding computational frugality without compromising on performance.

5 Conclusion

In conclusion, the proposed CAReNet architecture shows promise in achieving competitive, if not superior, performance with a relatively lower parameter count, making it an efficient and robust model choice. On the one hand, our CAReNet model showed an improvement in top-1 accuracy, being up to 5.51% higher compared to other architectures. On the other hand, in terms of model complexity, our model requires fewer parameters with a factor ranging from 12.3% more to 46.04% less compared to other architectures. Its consistent performance across various datasets indicates its potential for broad applications in diverse computer vision tasks. Further studies and refinements on this approach could lead to even more efficient and high-performing models suitable for real-world deployment.

As future work, the CAReNet architecture will be adapted for semi-supervised and self-supervised learning paradigms. The ultimate goal is to push the envelope further in terms of achieving state-of-the-art performance with minimal labeled data. By leveraging unlabeled data, which is abundantly available but significantly underutilized, the aim is to explore the efficacy of CAReNet in settings where acquiring labeled data is either too expensive or impractical. The challenge lies in fine-tuning the architecture and loss functions to maximize the utility of both labeled and unlabeled data, thereby mitigating the data scarcity issue that often hampers the deployment of machine learning models in real-world applications.

References

1. Barhoumi, Y., Rasool, G.: ScopeFormer: N-CNN-ViT hybrid model for intracranial hemorrhage classification. arXiv preprint arXiv:2107.04575 (2021)
2. Brown, T., et al.: Language models are few-shot learners. Adv. Neural. Inf. Process. Syst. **33**, 1877–1901 (2020)
3. Chowdhary, K., Chowdhary, K.R.: Natural language processing. Fundam. Artif. Intell. 603–649 (2020)
4. Coates, A., Ng, A., Lee, H.: An analysis of single-layer networks in unsupervised feature learning. In: Gordon, G., Dunson, D., Dudík, M. (eds.) Proceedings of the Fourteenth International Conference on Artificial Intelligence and Statistics. Proceedings of Machine Learning Research, vol. 15, pp. 215–223. PMLR, Fort Lauderdale (2011). https://proceedings.mlr.press/v15/coates11a.html
5. Dagli, R.: AstroFormer: more data might not be all you need for classification. arXiv preprint arXiv:2304.05350 (2023)
6. Dai, Z., Liu, H., Le, Q.V., Tan, M.: CoatNet: marrying convolution and attention for all data sizes. Adv. Neural. Inf. Process. Syst. **34**, 3965–3977 (2021)
7. Deng, J., Dong, W., Socher, R., Li, L.J., Li, K., Fei-Fei, L.: ImageNet: a large-scale hierarchical image database. In: 2009 IEEE Conference on Computer Vision and Pattern Recognition, pp. 248–255. IEEE (2009)
8. Deng, L.: The MNIST database of handwritten digit images for machine learning research. IEEE Signal Process. Mag. **29**(6), 141–142 (2012)

9. Dosovitskiy, A., et al.: An image is worth 16×16 words: transformers for image recognition at scale. arXiv preprint arXiv:2010.11929 (2020)
10. He, K., Zhang, X., Ren, S., Sun, J.: Deep residual learning for image recognition. In: Proceedings of the IEEE Conference on Computer Vision and Pattern Recognition, pp. 770–778 (2016)
11. Hu, J., Shen, L., Sun, G.: Squeeze-and-excitation networks. In: Proceedings of the IEEE Conference on Computer Vision and Pattern Recognition, pp. 7132–7141 (2018)
12. Johnson, R., Zhang, T.: Deep pyramid convolutional neural networks for text categorization. In: Proceedings of the 55th Annual Meeting of the Association for Computational Linguistics (Volume 1: Long Papers), pp. 562–570 (2017)
13. Khan, S., Naseer, M., Hayat, M., Zamir, S.W., Khan, F.S., Shah, M.: Transformers in vision: a survey. ACM Comput. Surv. (CSUR) **54**(10s), 1–41 (2022)
14. Krizhevsky, A., Hinton, G., et al.: Learning multiple layers of features from tiny images (2009)
15. Le, Y., Yang, X.: Tiny ImageNet visual recognition challenge. CS 231N **7**(7), 3 (2015)
16. LeCun, Y., Bottou, L., Bengio, Y., Haffner, P.: Gradient-based learning applied to document recognition. Proc. IEEE **86**(11), 2278–2324 (1998)
17. LeCun, Y., Cortes, C.: MNIST handwritten digit database (2010). http://yann.lecun.com/exdb/mnist/
18. Liu, X., Peng, H., Zheng, N., Yang, Y., Hu, H., Yuan, Y.: EfficientViT: memory efficient vision transformer with cascaded group attention. In: Proceedings of the IEEE/CVF Conference on Computer Vision and Pattern Recognition, pp. 14420–14430 (2023)
19. Parvaiz, A., Khalid, M.A., Zafar, R., Ameer, H., Ali, M., Fraz, M.M.: Vision transformers in medical computer vision-a contemplative retrospection. Eng. Appl. Artif. Intell. **122**, 106126 (2023)
20. Simonyan, K., Zisserman, A.: Very deep convolutional networks for large-scale image recognition. arXiv preprint arXiv:1409.1556 (2014)
21. Smith, L.N., Topin, N.: Deep convolutional neural network design patterns. arXiv preprint arXiv:1611.00847 (2016)
22. Tu, Z., et al.: MaxViT: multi-axis vision transformer. In: European Conference on Computer Vision, pp. 459–479. Springer (2022)
23. Vaswani, A., et al.: Attention is all you need. Adv. Neural Inf. Process. Syst. **30** (2017)
24. Wang, W., et al.: Pyramid vision transformer: a versatile backbone for dense prediction without convolutions. In: Proceedings of the IEEE/CVF International Conference on Computer Vision, pp. 568–578 (2021)
25. Wu, J., Zhang, Q., Xu, G.: Tiny ImageNet challenge. Technical report (2017)
26. Xiao, H., Rasul, K., Vollgraf, R.: Fashion-MNIST: a novel image dataset for benchmarking machine learning algorithms. arXiv preprint arXiv:1708.07747 (2017)

Designing Converged Middleware for HPC, AI, and Big Data: Challenges and Opportunities

Dhableswar K. Panda(✉), Hari Subramoni, Mustafa Abduljabbar, Aamir Shafi, Nawras Alnaasan, and Shulei Xu

The Ohio State University, Columbus, OH 43210, USA
{panda,subramon}@cse.ohio-state.edu,
{abduljabbar.1,shafi.16,alnaasan.1,xu.2452}@osu.edu

Abstract. The field of computing has been evolving over the years with the need for High-Performance Computing (HPC), Deep Learning (DL), and Machine Learning (ML) on heterogeneous architectures. These modern computing environments are spread over edges to clouds/HPC centers. This chapter focuses on challenges and opportunities in designing HPC and Artificial Intelligence (AI) middleware on these systems with both scale-up and scale-out strategies. The first part of the chapter focuses on the broad approaches/solutions to address these challenges. Next, we present an overview of the High-Performance MVAPICH Message Passing Interface (MPI) library, its features, and sample solutions on current generation CPU and GPU on-premise systems. Next, we present an overview of the MPI-driven framework to provide high-performance training and inference for emerging AI workloads. An overview of the MPI-driven approach towards Big Data stacks (Spark and Dask) is presented next. Finally, we demonstrate how these solutions are being transitioned into public cloud environments. The results demonstrate that an MPI-driven converged middleware stack can be used on current and emerging on-premise and cloud systems for diverse HPC, AI, and Big Data workloads and frameworks.

Keywords: High-Performance Computing (HPC) · Cloud · Message Passing Interface (MPI) · AI · Big Data · Data Science

1 Introduction

The computing industry has undergone significant transformations over the last few decades, which are marked by distinct phases reshaping technology and society. The 1980s witnessed the advent of personal computers bringing computing power into homes and small businesses. The early versions of modern supercomputers—including Cray X-MP and Cray-2 [1]—were also introduced around the same time and led to the birth of Scientific Computing or traditional High Performance Computing (HPC). The second phase started in the early

M. Zbakh et al. (Eds.): CloudTech 2023, LNNS 1220, pp. 40–63, 2024.
https://doi.org/10.1007/978-3-031-78698-3_4

2000s, fueled by the rise of the Internet connecting the world in unprecedented ways, saw the proliferation of large amounts of data that was typically processed by parallel and distributed computing resources—this is known as Data Analytics or Big Data processing [2]. The third phase of computing, starting in the 2010s, saw the growth of cloud computing and the emergence of AI [3] enabling innovation through vast data storage, processing, and intelligent automation. Each phase in this evolution reflects the relentless pace of technological advancement and its profound impact on every aspect of modern life.

Traditional applications used to use HPC, AI, or Big Data in an isolated manner. As indicated in Fig. 1, many modern application workflows involve HPC, AI, and Big Data/Data Science in an interleaved manner through multiple phases. In some of these workflows, these phases also happen concurrently in different parts of the system. Thus, these applications require *converged middleware with high performance and scalability* from the underlying systems.

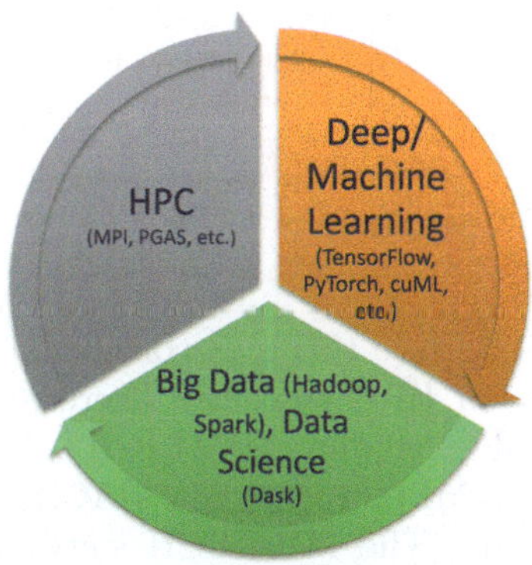

Fig. 1. High-level overview of modern workflows using HPC, AI (Deep/Machine Learning), Big Data, and Data Science in a converged manner.

1.1 Challenges in Designing Converged Middleware

The challenges faced in designing a converged communication middleware are depicted in Fig. 2. The design goals for the middleware runtime include high performance (low-latency and high-throughput), scalability on large systems, and resilience to failures. Also, the converged communication middleware needs to support three distinct workloads: traditional HPC, data analytics or Big Data, and modern AI or deep/machine learning (DL/ML) applications. Such a unified runtime reduces the operating costs of data centers as well as makes it easier for the maintenance staff. We also envisage that the runtime will support a range of programming models ranging from Message Passing Interface (MPI), Partitioned Global Address Space (PGAS) libraries (UPC, OpenSHMEM), Big Data stacks (Spark, Dask), and Deep Learning (PyTorch, JAX, DeepSpeed). Apart from the support of point-to-point and collective communication routines, the runtime aspires to support energy-efficiency and fault tolerance. Lastly, a range of emerging interconnects (Slingshot, InfiniBand, Omni-Path Express, etc.), processors (Grace, ARM, Intel Rapids Sapphire, etc.), accelerators (from NVIDIA, AMD, Intel, etc.), and Data Processing Units (NVIDIA BlueField) will need to be supported by the communication middleware.

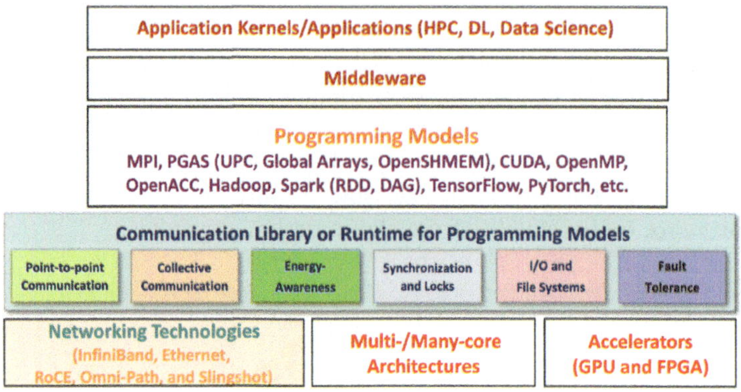

Fig. 2. Overview of the framework for designing converged middleware

The Message Passing Interface (MPI) is considered a de facto standard for writing parallel programs on HPC systems. There are many MPI standard-compliant libraries available including MVAPICH2, MPICH, and Open MPI. The idea is to extend our MPI library, MVAPICH, to meet the challenges of the converged middleware. The MVAPICH software [4] provides point-to-point and collective communication support [5] [6] [7] for multi- and many-core CPUs and GPUs on a range of high-performance interconnects including InfiniBand [8], Omni-Path [9,10], Ethernet/iWARP, RoCE(v1/v2) [11], Cray Slingshot 10/11 [12], and Rockport Networks [13].

1.2 High-Level Overview

This book chapter describes the efforts at the Network-based Computing Laboratory (NOWLAB) to achieve a converged software stack for HPC, Big Data, and AI. Our approach is particularly appealing to audiences interested in both on-premise and cloud computing systems and the associated applications involving high-performance and data-intensive tasks.

A key component of this effort is the MVAPICH project highlighted in Sect. 2. The MVAPICH MPI library provides high-performance, scalability, and fault tolerance for various on-premise computing systems and interconnects including those adopted in the public cloud. The project is also distributed by many vendors as part of their software packages, demonstrating its critical role in modern HPC environments.

Building on the foundation provided by MVAPICH, HiDL, the High-Performance Deep Learning project featured in Sect. 3 plays a pivotal role in this initiative. HiDL focuses on developing scalable and efficient solutions for deep learning workloads, adaptable to both on-premise and cloud environments. By leveraging advanced HPC hardware, high-speed interconnects, and optimized software stacks, HiDL aims to accelerate deep/machine learning applications involving both training and inference activities.

The HiDL project addresses key challenges in deep/machine learning, such as optimizing communication and computation overlap, improving scalability across distributed systems, and enhancing fault tolerance. HiDL achieves significant performance improvements and scalability as we share in Sect. 3.3.

In addition to HiDL, the High-Performance Big Data (HiBD) project described in Sect. 4 focuses on optimizing Big Data and Data Science applications by utilizing HPC hardware and specialized processors within on-premise and cloud environments. Key components of HiBD include MPI4Dask and MPI4Spark, which integrate HPC capabilities into popular Big Data frameworks. MPI4Dask modifies the Dask Distributed library to utilize the MVAPICH MPI library for GPU-aware MPI communication, addressing challenges like communication isolation and dynamic connectivity encountered in cloud infrastructures. Similarly, MPI4Spark accelerates Apache Spark's shuffle engine by incorporating an MPI-based communication backend. This integration results in higher performance for communication-intensive operations, with substantial reductions in execution times and improvements in communication efficiency, as seen in evaluations on platforms like TACC Frontera as we show in the results in Sect. 4.3.

Both HiBD and HiDL projects exemplify how the convergence of HPC, AI, Big Data, and Data Science can be effectively achieved within on-premise and cloud environments as these stacks are developed on top of the MVAPICH library. In Sect. 5, we show our successful exploitation of cloud infrastructures that not only enhance the scalability and flexibility of these applications but also ensure efficient handling of diverse and demanding workloads. The optimizations made for cloud deployments, such as improving intra-node and inter-node communication latency and bandwidth, further reinforce the feasibility and advantages of using a converged communication runtime for a wide range of applications.

2 MVAPICH MPI Library Project

The MVAPICH MPI library [4] project is led by the Network-Based Computing Laboratory (NBCL) of The Ohio State University. The MVAPICH software delivers the best performance, scalability, and fault tolerance for high-end computing systems and servers using InfiniBand, Omni-Path, Ethernet/iWARP, RoCE(v1/v2), Cray Slingshot, and Rockport Networks networking technologies. Figure 3 depicts an overview of the software architecture of the MVAPICH MPI library.

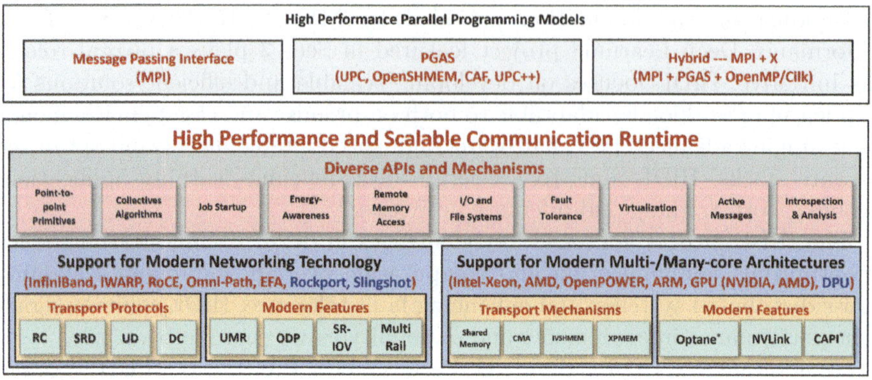

Fig. 3. Software Architecture Overview of the MVAPICH MPI Library [4]

2.1 Motivation and Challenges

The main motivation of the MVAPICH project is to provide high-performance MPI implementation support with scalability and fault tolerance for various emerging computing, accelerators, and networking technologies. The MPI standard has many APIs. Thus, the primary challenge of the MVAPICH project has been to achieve novel MPI-level designs for these APIs with good performance (low latency and high throughput) while exploiting maximum overlap between computation and communication. The designs exploit the features of the underlying computing, accelerator, and networking technologies to the fullest extent.

2.2 Highlights of Some Designs and Optimizations Used in MVAPICH

2.2.1 MVAPICH-3.0

MVAPICH-3.0 is the latest MVAPICH MPI library version that was released in November 2023. Based on MPICH-3.4.3, the ch4:ucx and ch4:ofi devices are supported in MVAPICH-3.0 with optimized collective designs inherited from MVAPICH2. The Infiniband and RoCE inter-node communication are supported by IB verbs over Unified Communication X (UCX). Similarly, the support for Slingshot and Omni-Path is provided through the Open Fabrics Interface (OFI). As shown in Fig. 4, MVAPICH-3.0 provides competitive MPI point-to-point communication latency performance compared to the point-to-point communication latency of the previous MVAPICH2 series. The series also includes optimized collective communication performance.

2.2.2 MVAPICH-GPU

The rapid development in big data, AI, and large-scale high-performance GPU systems drive the increasing need for high-performance MPI library support. The MVAPICH-GPU library provides programming model support with the

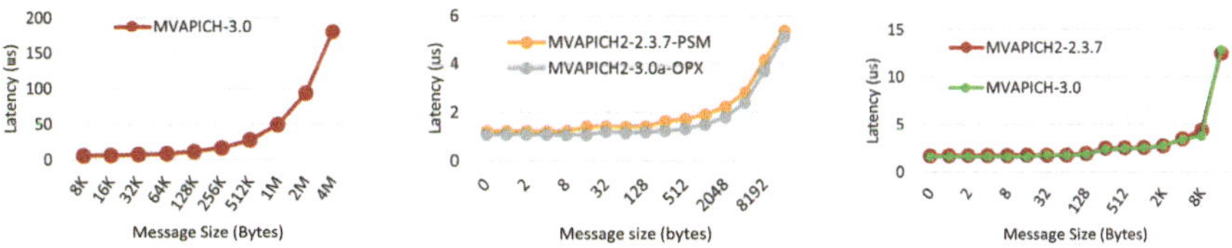

(a) Slingshot11 (b) OPX (c) InfiniBand

Fig. 4. MVAPICH-3.0 Point-to-point communication performance over different interconnects

standard MPI interfaces for unified data movement between CPU and GPU. By taking advantage of the Unified Virtual Addressing (UVA), which is supported in CUDA 4.0 and later versions, MVAPICH-GPU overlaps data movement from GPU with RDMA transfers.

Based on the GPU-Aware design of MVAPICH-GPU, we provide high-performance and scalable MPI support for GPUs from various vendors including Nvidia and AMD. For example, "On-the-fly" message compression optimization [14] takes advantage of the existing GPU compression algorithms to accelerate MPI point-to-point communication performance of transferring large GPU-to-GPU data. Figure 5 shows the performance improvements of the message compression design with a popular HPC application AWP ODC. We observe up to 35% GPU higher computing flops, and 26% faster runtime per timestep in ZFP-OPT for two of the GPU configurations.

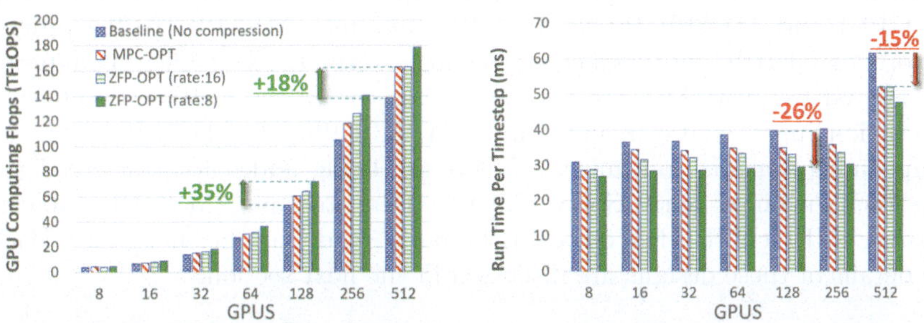

(a) GPU Computing Flops (TFlops) (b) Run Time Per Timestep (ms)

Fig. 5. Performance improvements of "On-the-fly" compression support in MVAPICH-GPU, running with weak-scaling of HPC application AWP-ODC on Lassen cluster (Nvidia v100 GPU) [14]

The ROCm-aware MVAPICH-GPU support also delivers significant performance improvements on AMD GPUs. Figure 6 presents the performance comparison of MVAPICH2-GDR with ROCm-aware support and OpenMPI with

UCX. With the enhanced AMD GPU support (including CUDA-aware point-to-point and collective operations), MVAPICH-GPU can achieve over 60× lower MPI_Allreduce latency and 3× lower MPI_Bcast latency in the scale of 8 nodes and 8 GPUs per node.

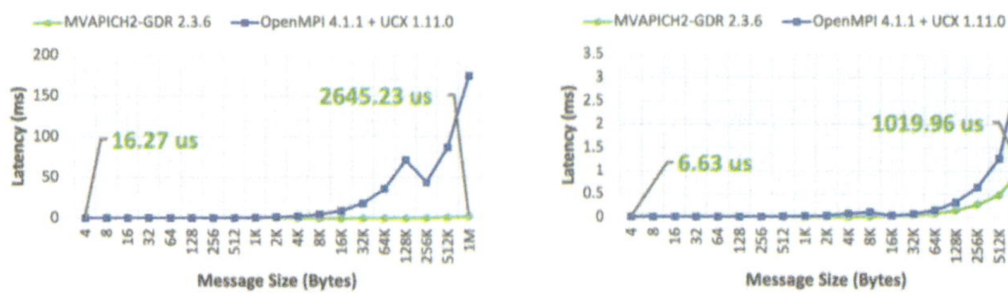

(a) Allreduce (b) Bcast

Fig. 6. Performance Evaluation of ROCm-aware MVAPICH-GPU with optimized support for AMD GPUs, running on the system configured with 8 nodes and 8 GPU per node (64 GPUs) [15]

2.2.3 MPI-Driven Converged Software Stack: MVAPICH-Plus

To address the need for a unified parallel programming model for an increasing variety of high-performance computing devices and diverse use cases like AI and Big Data, the MVAPICH-Plus project was introduced as an MPI-driven converged software stack. The latest version of MVAPICH-Plus 3.0 was released in March 2024. The MVAPICH-Plus project unifies the existing MVAPICH2-GDR (for GPU) and MVAPICH2-X (for CPU) features. MVAPICH-Plus provides support for all HPC interconnect technologies over UCX or OFI, as appropriate. The GPU-Aware MPI support for Nvidia and AMG GPUs, introduced in the previous part, is also included in MVAPICH-Plus. In MVAPICH-Plus and its upcoming versions, optimized designs are being added for converged software stack with next-generation workflows – Deep Learning (MPI4DL), Machine Learning (MPI4cuML), Big data (MPI4Spark), and Data Science (MPI4Dask). The details of these designs are discussed in the next section.

2.3 Discussion

The MVAPICH MPI library software family has become an active participator and an important collaborator in many research projects in High-Performance Computing and other fields. For example, the MVAPICH2 MPI library lived at the core of the NASA-DART mission, which is researching ways to prevent near-earth asteroids from causing global catastrophes. The MVAPICH2 library enabled scalability underneath large-scale hydrodynamical and gravitational simulations required to compute the impact such as Spheral models. In the Lawerence Livermore National Laboratory (LLNL) Nuclear Fusion Research project, the MVAPICH MPI library also has been used as the default MPI library on the LLNL systems, enabling the thousands of simulation jobs that have led to the nuclear fusion research achievements.

3 High-Performance Deep Learning and Machine Learning

Recent advancements in AI have been propelled by the resurgence of state-of-the-art Machine Learning (ML) and Deep Learning (DL) models, driving a revolution in both classical and emerging application areas such as Image Recognition [16], Natural Language Processing [17], and Autonomous Vehicle systems [18]. The popularity of DL as an approach for automation spans various sectors including business operations, identity verification, advertisements, and marketing. This rapid growth in DL is fueled by several factors: the public availability of extensive datasets like ImageNet [16] and large language corpora, the widespread adoption of data-parallel hardware such as GPUs and specialized accelerators for Deep Neural Network (DNN) training, and the development of more efficient and effective DNN architectures like GPT [19], DALL·E, LLaMA [20], BERT, ViT, and ResNet [21]. As models become deeper and larger, the need for HPC systems has become paramount in achieving significant improvements in training time and accuracy. In this context, HPC systems with high-bandwidth and low-latency interconnects have become essential for efficiently scaling out DNN training and inference across thousands of nodes.

3.1 Motivation and Challenges

In the fields of machine learning (ML) and deep learning (DL), HPC plays a crucial role in training cutting-edge models. It provides the computational power needed to handle large datasets and complex models. By utilizing the exceptional speed and efficiency of HPC systems, researchers and practitioners can push the boundaries of what's possible with AI, leading to innovations that could potentially transform society. However, fully utilizing the power of HPC for DL/ML presents significant challenges that need to be addressed to achieve optimal performance and efficiency. These challenges include:

1. Computational Demands of Training State-of-the-Art DNNs: Training cutting-edge DNNs is inherently computationally intensive, necessitating the development of sophisticated parallelization strategies. These strategies must effectively utilize HPC resources while minimizing communication and synchronization overhead, which can otherwise significantly hinder performance.
2. Memory Constraints with Large DNNs: Some DNNs are so large that they cannot fit within the memory of a single GPU. This necessitates splitting the model or layers across multiple GPUs and ensuring proper synchronization between them. This is particularly crucial when training on high-resolution images, where memory demands are exceedingly high.

3. Underutilization of Modern GPUs with Lightweight DNNs: Advanced GPUs like the NVIDIA A100 are powerhouses in HPC environments, yet they are often not fully utilized by DNN training workloads. This is especially true for smaller to medium-sized DL models and input sizes, where the computational capabilities of these GPUs are not fully leveraged.
4. Variability in Communication Runtime for Distributed DL: Effective DL parallelism requires a blend of collective and point-to-point communication operations across a wide range of message sizes and scales. The performance of communication libraries such as MPI and NVIDIA Collective Communication Library (NCCL) can vary dramatically depending on the type of communication operation, its scale, and the message size involved.
5. Inference Challenges with Autoregressive Models: Inference on autoregressive models involves a temporal dependency, where the probability distribution of the current token is conditioned on all preceding tokens. This creates significant computational inefficiencies during inference, as generating each token requires loading the entire model's weights. This issue is exacerbated in real-world deployments where inference requests arrive unpredictably and with varying generation lengths, making the process heavily memory-bound.
6. MPI-Based Communication Support for GPU-Accelerated ML: Providing robust MPI-based communication support for GPU-accelerated machine learning applications presents unique challenges. There is a pressing need for a Python API that can leverage MPI-based communications, enabling seamless integration and optimization of ML applications within HPC environments.

3.2 Design and Implementation

In addressing the multifaceted challenges of using HPC for DL and ML, a comprehensive design and implementation strategy is essential. This section delves into the methodologies and solutions developed to overcome the critical issues identified, ranging from the computational demands of training state-of-the-art DNNs to the complexities of efficient parallelism and communication. We explore advanced parallelization strategies that maximize resource utilization while minimizing communication and memory overheads. We provide an overview of several such designs and implementations below:

3.2.1 MVAPICH (MPI)-Driven Infrastructure for ML/DL Training

In Data Parallelism (DP), DNN is replicated across multiple Processing Elements (PEs) like GPUs. An Allreduce operation is used to synchronize the DNN's weights across multiple replicas by reducing the gradients from all replicas and pushing the result to all the replicas. Figure 7 shows our MVAPICH2 [22] (MPI)-driven Infrastructure for ML/DL Training [23]. MVAPICH2 provides an optimized Allreduce operation to accelerate DNN training on a large number of PEs/GPUs. Horovod [24] is a distributed deep learning training framework, which supports popular deep learning frameworks like TensorFlow and PyTorch.

Horovod with MVAPICH2 provides scalable distributed DNN training solutions for both CPUs and GPUs. Similar integration has also been done for Torch distributed and DeepSpeed frameworks.

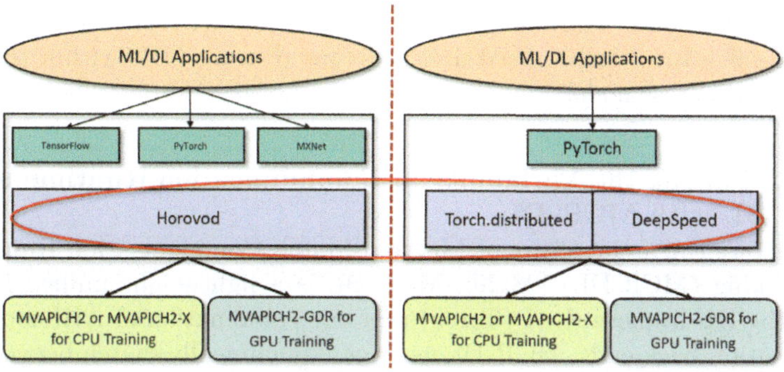

Fig. 7. MVAPICH2 (MPI)-driven Infrastructure for ML/DL Distributed Training with Scalable Performance

3.2.2 Exploiting 5D Parallelism for Large Models and Inputs with MPI4DL

MPI4DL [25,26] efficiently utilizes distributed training for dense convolutional neural networks and very high-resolution images that appear in real-world applications such as medical images. Figure 8 shows the integration of spatial, layer, data, bi-directional, and pipeline parallelism strategies (as discussed in [25,26]), to solve the memory and performance challenges.

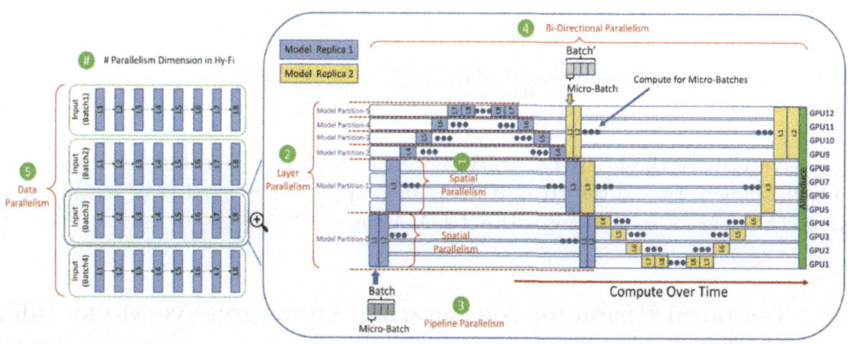

Fig. 8. Workflow of 5D Parallelism combining spatial, bidirectional, data, pipeline, and layer parallelism to enable efficient out-of-core training. [25,26]

3.2.3 Alleviating the Underutilization of Modern GPUs for Lightweight DNNs with AccDP

To improve the underutilization of GPUs with lightweight DNNs, we leverage MPI and NVIDIA MPS features to place multiple model replicas on the same GPU. AccDP [27] initializes an MPS server on each GPU to execute multiple processes with spatial sharing of resources. Each model replica produces local gradients which are synchronized and aggregated using MPI within the GPU, the node, and across nodes.

3.2.4 Stabilizing the Variability in Communication Runtime for Distributed DL with MCR-DL

Figure 9 shows the architecture of Mix-and-Match Communication Runtime for Deep Learning (MCR-DL) [28,29]. MCR-DL is a lightweight unified interface between the DL framework (PyTorch) and any combination of ABI-compatible communication back-ends. MCR-DL users can dynamically switch between communication back-ends during distributed DL training. MCR-DL supports many existing communication back-ends (by implementing them as a high-level back-end class) and provides an extensible design to enable new communication back-ends and performance optimizations.

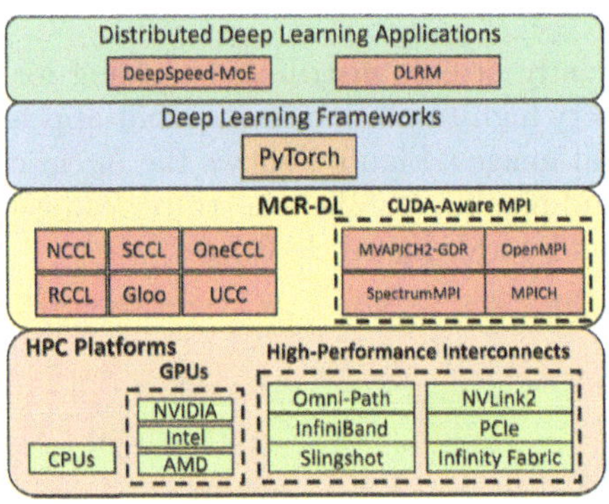

Fig. 9. Overview of the MCR-DL Software Stack. MCR-DL is a thin layer between a target DL framework and the HPC system, and supports any number of stream-aware communication back-ends along with CUDA-Aware MPI. [28,29]

3.2.5 Temporal Fusion for Accelerating Autoregressive Model Inference with Flover

Flover [30,31] is a temporal fusion framework tailored to the context of parallel inference in autoregressive models. The main contribution of Flover is to promptly process incoming requests, eliminating the need for batching or time window allocation, while not triggering the launch of redundant model instances or kernel calls. Figure 10 shows how Flover maintains one main computing stream to handle any number of requests throughout the lifecycle of inference, largely

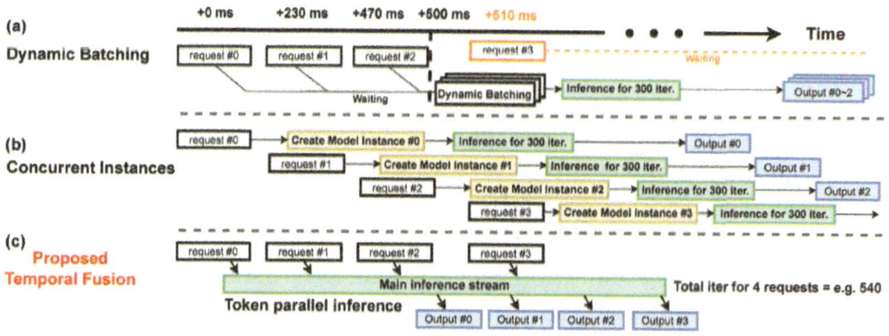

Fig. 10. Workflow comparison on dynamic batching, concurrent model instances, and Temporal fusion (Flover). [31]

reducing the overhead in numerous separated kernel calls and scheduling redundant collective communicators.

3.2.6 MPI-Based Communication Support for GPU-Accelerated ML with MPI4cuML

Figure 11 shows the architecture of MPI4cuML [32], which is a custom version of the cuML machine learning library that contains high-performance MPI communication backend for cuML. The MPI back-end in MPI4cuML uses mpi4py over the MVAPICH2-GDR library and targets modern HPC clusters built with GPUs. MPI4cuML provides point-to-point asynchronous I/O communication co-routines, which are non-blocking concurrent operations.

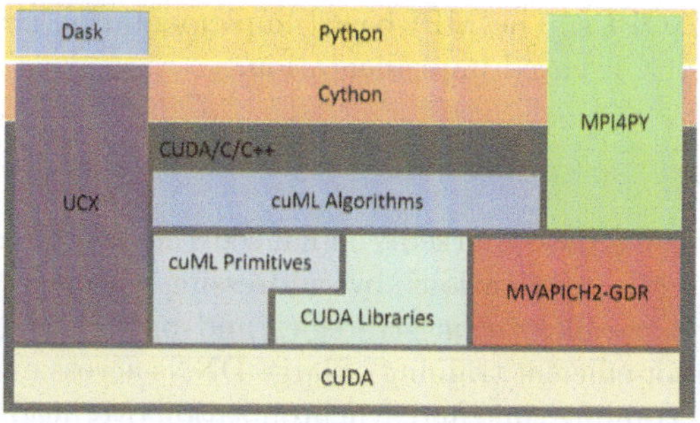

Fig. 11. Overview of the cuML software stack with MPI-based communication (MPI4cuML). [32]

3.3 Performance Evaluation

Figure 12 presents a comprehensive analysis of the implemented strategies, examining their impact on computational efficiency, resource utilization, and overall performance. We provide further details for each sub-figure below.

Figure 12(a) demonstrates the scalability of training with the MPI-driven MVAPICH2-GDR infrastructure for distributed DNN training on GPU systems. This experiment was conducted on up to 1,536 NVIDIA GPUs, achieving a throughput of 0.42 million images per second and a training time of 3 s per epoch on the entirety of the ImageNet-1k dataset. With increase in the number of GPUs, scalable performance is observed.

Figure 12(b) illustrates the ability to train large DNN models such as AmobeNet-f214 on large-resolution pathology images (8,192 X 8,192 pixels). This experiment was conducted on up to 257 NVIDIA GPUs, showing a speedup of 2.26x by using the 5D parallelism scheme offered by the MPI4DL framework over other state-of-the-art methods.

Figure 12(c) compares regular data parallelism and the enhanced AccDP parallelism for different lightweight DNN models on the specified hardware with the specified clients per GPU. This evaluation indicates up to a 58% improvement over regular data parallelism.

Figure 12(d) demonstrates the throughput improvements on the DS-MoE model using the MCR-DL framework over other communication backends on up to 256 NVIDIA GPUs. This evaluation shows a consistent speedup as the number of GPUs increases, with up to a 1.45x speedup compared to the best baseline.

Figure 12(e) compares the performance of Flover and FasterTransformer frameworks on the specified hardware with different numbers of parallel inference requests for the Llama 65B model. Flover shows a significant speedup as the number of parallel requests increases, with up to a 16.7x speedup with 32 parallel requests.

Finally, Fig. 12(f) shows the performance and speedup of MPI4cuML, which enables MPI-based distributed ML implementations over pure NCCL communication on up to 32 GPUs. The MPI-based implementation shows up to a 1.6x speedup over the NCCL-based implementation.

3.4 Discussion

Our proposed solutions have markedly enhanced the performance of HPC systems for DL and ML applications by addressing the critical challenges of Sect. 3.1. Advanced parallelization strategies and optimized memory management have allowed for efficient training of large DNNs across multiple GPUs, significantly reducing training time and synchronization overhead [23]. The full utilization of modern GPUs and the optimization of communication libraries (MPI and NCCL) have minimized latency and bandwidth issues, ensuring smoother and faster computations. Inference optimizations for autoregressive models have reduced memory overhead and improved efficiency in real-world deployment scenarios. The development of a Python API for MPI-based communication has facilitated seamless integration of GPU-accelerated ML applications within HPC environments. Overall, these solutions collectively have contributed to a substantial improvement in the scalability, efficiency, and performance of DL and ML workloads on HPC systems.

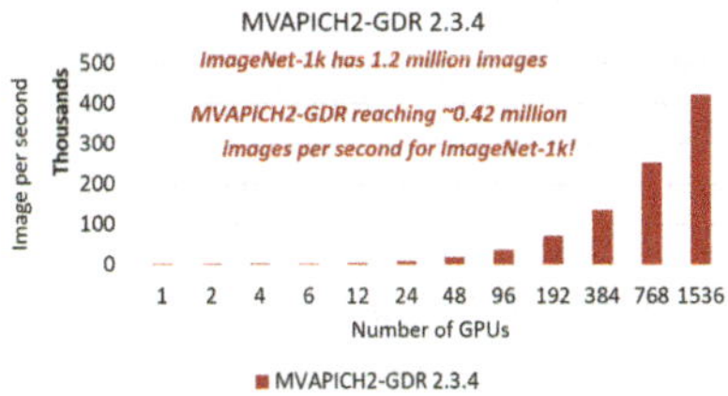

(a) MVAPICH2 Scalability for DL Training on the ImageNet dataset with ResNet-50.

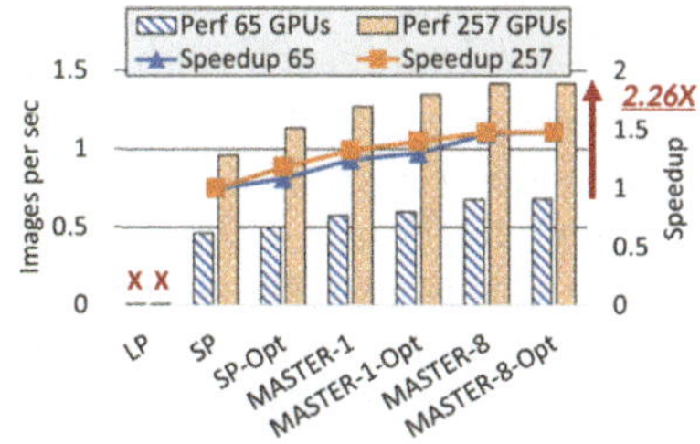

(b) Enabling training on very high-resolution images (8,192 X 8,192 pixels) on the AmobeNet-f214 model. [26]

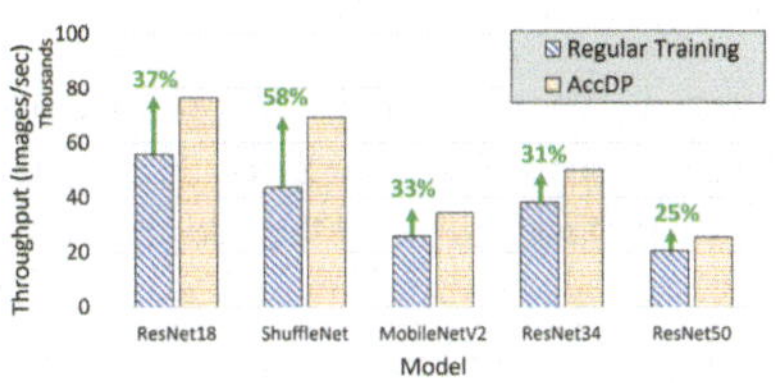

(c) Comparison between regular data parallelism and AccDP for different lightweight DNN models on A100 GPU with 4 MPS clients. [27]

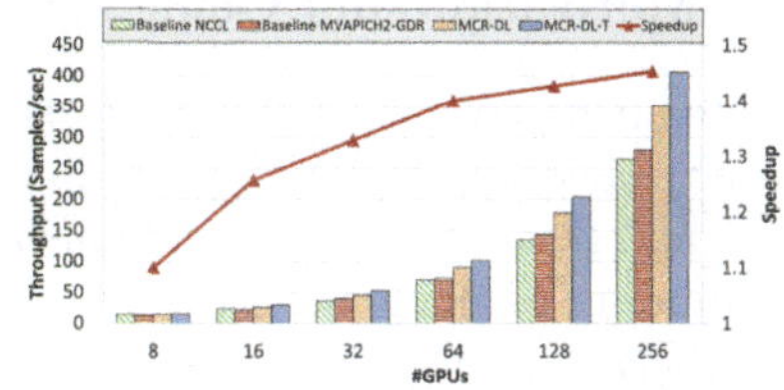

(d) Throughput improvements for DS-MoE with pure MVAPICH2-GDR, pure NCCL, and mixed-backends with MCR-DL. [29]

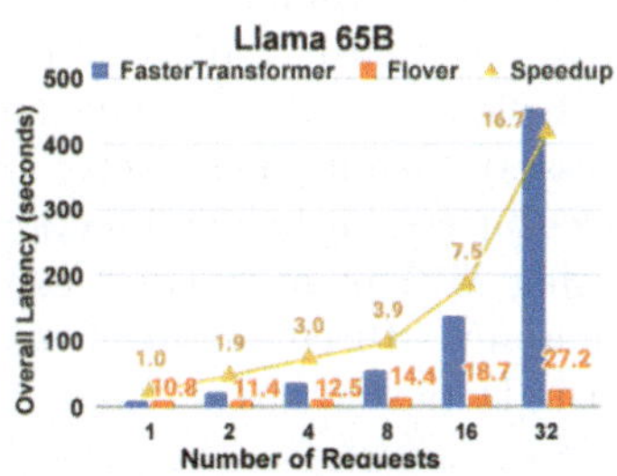

(e) Speedup using Flover for Parallel inference latency on the Llama 65B model with different number of parallel requests. [31]

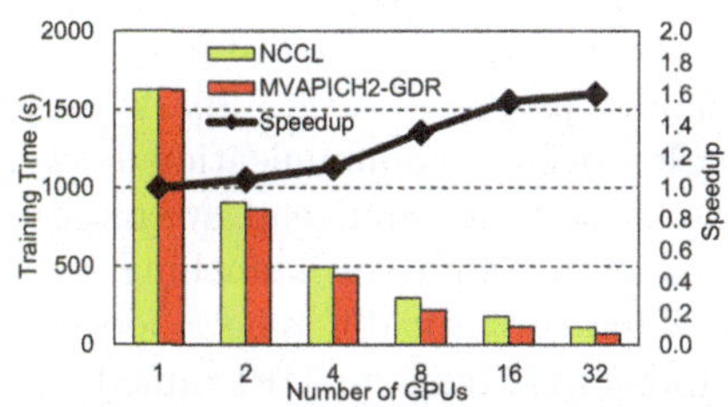

(f) MPI4cuML improvements in the training time of the distributed K-Means algorithm on multi-GPU system. [32]

Fig. 12. Evaluations of State-of-the-art High-Performance DL and ML solutions with the MPI-driven converged software stack.

4 High-Performance Big Data

The High-Performance Big Data (HiBD) project [33] is an umbrella of projects that focus on achieving a converged software stack for HPC, Big Data, and Data Science. The main goal of the HiBD project is to enhance the performance and scalability of Big Data and Data Science applications. This is done by utilizing HPC hardware, such as, high-speed interconnects (e.g., InfiniBand [8], Intel Omni-Path [9], and HPE Slingshot [12]) and specialized processors (e.g., GPUs, CPUs designed for heavy computations) through the MVAPICH MPI library.

Recent efforts under the HiBD project such as MPI4Dask [34] and MPI4Spark [35] achieve this goal. MPI4Dask is a custom version of the Dask [36] Distributed library that relies on the MVAPICH MPI library. MPI4Spark is a version of Apache Spark [37] designed with a high-performance MPI-based Netty [38] communication back-end for accelerating Spark's communication-intensive shuffle phase.

4.1 Motivation and Challenges

The main motivation behind the HiBD project is to accelerate Big Data and Data Science applications, providing a convergent software stack for HPC, Big Data, and Data Science. Recent efforts such as MPI4Dask and MPI4Spark presented a number of challenges.

MPI4Dask addresses two main challenges. The first challenge revolves around maintaining communication isolation and support for dynamic connectivity between Dask components including the scheduler, workers, and client.

The second challenge entails sustaining the Dask Distributed asynchronous communication pattern. This means that any point- to-point communication operations—implemented via MPI—must be integrated as communication co-routines that typically have performance penalties over regular Python functions.

The MPI4Spark effort also addresses several challenges, namely, launching dynamic processes (e.g., Spark Executor processes), maintaining Netty's—Spark's default communication back-end—connection establishment functionality while write and read events are handled using MPI point-to-point communication primitives, launching a Spark cluster in an MPI environment, and matching semantic differences between MPI and Big Data stacks (e.g., mapping endpoints/channels to MPI ranks).

4.2 Design and Implementation

4.2.1 MPI4Dask

The Dask Distributed library provides TCP and UCX-Py [39] backends represented in Fig. 13(b) in the "Comm Layer" layer by tcp.py and ucx.py. MPI4Dask is implemented as part of the Dask Distributed "Comm Layer" as a multi-layered software sitting over `mpi4py`, which in turn exploits MVAPICH2-GDR [22] for GPU-aware MPI communication.

The MPI4Dask effort uses the `mpirun_rsh` utility from MVAPICH2-GDR to launch the Dask scheduler, client, and workers. MPI4Dask also uses non-blocking communication to support asynchronous co-routine execution for data communication. Large message sizes are handled by chunking them into 1 GB-sized chunks.

MPI Sub-communicators are used in MPI4Dask for providing communication isolation between control and data messages between process pairs. Dynamic connectivity is handled in MPI4Dask by duplicating sub-communicators with `Comm.Dup()` for each new dynamic connection.

4.2.2 MPI4Spark

The MPI4Spark package is aimed at accelerating Spark's shuffle engine using the MVAPICH MPI library. The design was implemented at the Netty layer. Netty is the default communication backend that Spark uses that allows for adding new transport layers. Since Netty is Java-based, MPI4Spark was implemented using Java bindings that allow for calling native MPI functions through Java.

The MPI4Spark design, as shown in Fig. 13(a), intercepts shuffle messages sent out through Netty's non-blocking IO (NIO) transport and sends their contents using MPI point-to-point primitives. Since Spark needs to be launched in an MPI environment to utilize MPI communication, a Java wrapper program was used to launch Apache Spark with MPI. The wrapper program uses the launcher scripts that Spark provides to launch the Spark cluster.

Spark also requires that new processes be dynamically launched, MPI4Spark relies on MPI's Dynamic Process Management (DPM) [40] feature to meet this requirement by launching processes dynamically through MPI.

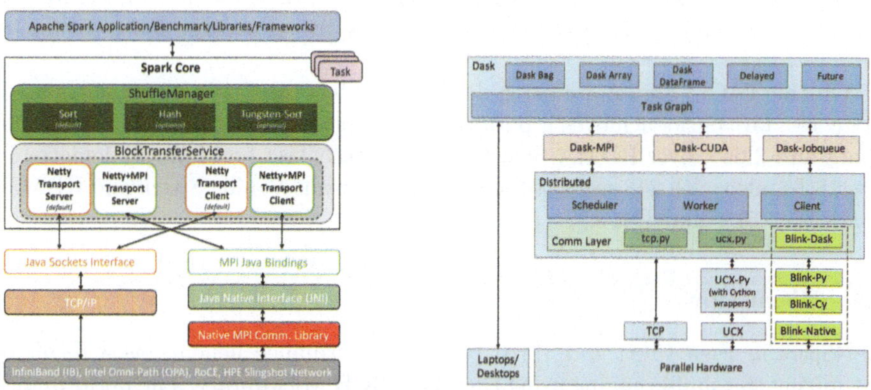

(a) MPI4Spark Design (b) MPI4Dask Design

Fig. 13. Overview of the MPI4Spark [35] and MPI4Dask [34] designs.

4.3 Performance Evaluation

In this subsection, the performance evaluation of MPI4Dask and MPI4Spark are presented. The evaluation was conducted on two systems: An in-house cluster (RI2) at The Ohio State University, and the Texas Advanced Computing Center (TACC) Frontera system.

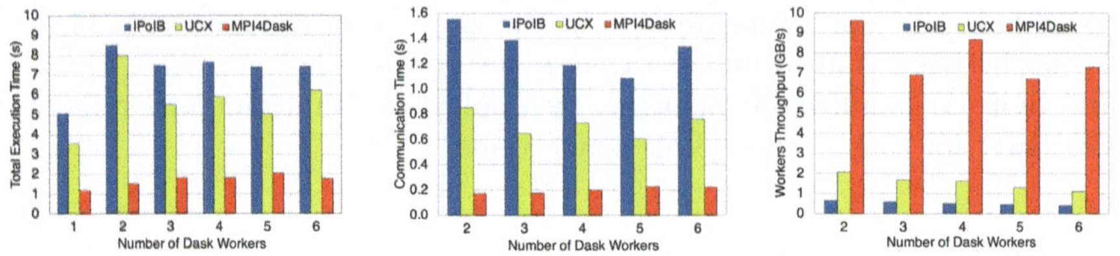

(a) Execution Time Comparison (b) Communication Time Comparison (c) Aggregate Throughput Between Workers

Fig. 14. Strong-scaling performance results produced by the sum of a cuPy array and its Transpose (Dimensions: 16K × 16K, Chunk size: 4K, Partitions: 16). Carried out on the RI2 cluster comparing IPoIB, UCX, and MPI4Dask . Each Dask worker was configured to use 28 threads representing the total number of cores on single node. [34]

On the RI2 cluster, each node is equipped with a single GPU. Therefore, we run a single-worker process with 28 worker threads to fully utilize the available CPU cores. Figure 14(a) shows execution times, where it can be observed that MPI4Dask achieves an average speedup of 3.47× for 2–6 Dask workers. Figure 14(b) presents the communication time for the benchmark run, demonstrating that MPI4Dask outperforms UCX by an average factor of 6.92× for 2–6 workers. Average throughput comparison for Dask workers is illustrated in Fig. 14(c), showing that MPI4Dask exceeds UCX by an average factor of 5.17× for 2–6 workers.

On the TACC Frontera cluster, looking at GroupByTest in Fig. 15(a), with 448 cores, MPI4Spark achieves a performance improvement of 3.72× over Vanilla Spark and 2.06× over RDMA-Spark. In the SortByTest benchmark, depicted in Fig. 15(b), also with 448 cores, MPI4Spark outperforms Vanilla Spark and RDMA-Spark by 3.51× and 1.41×, respectively.

4.4 Discussion

MPI4Spark in Figs. 15(a) and 15(b) reveals performance benefits as a result of optimizing the shuffle engine using the MVAPICH MPI library. The performance benefits seen compared to RDMA-Spark are a result of MPI outperforming the unified communication runtime (UCR) that the RDMA-Spark design relies on.

The Dask Distributed library was extended with a new communication device called MPI4Dask that is based on the MPI standard. MPI4Dask exploits the `mpi4py` python bindings on top of the `MVAPICH2-GDR` library. This design offers a high-performance alternative to the existing TCP and UCX communication backends. The performance benefits seen are due to the implementation of efficient coroutines that use non-blocking point-to-point MPI calls.

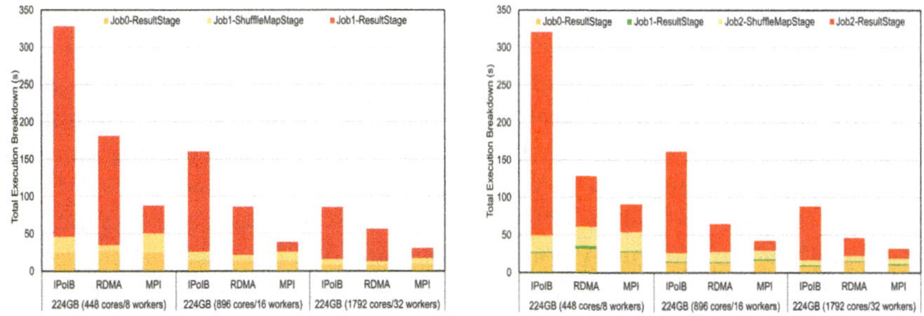

(a) GroupByTest Execution Breakdown Evaluation

(b) SortByTest Execution Breakdown Evaluation

Fig. 15. Strong scaling OSU HiBD Benchmark (OHB) numbers comparing RDMA (RDMA-Spark), IPoIB (Vanilla Spark) and MPI-based designs (MPI4Spark). The numbers were collected on the TACC Frontera cluster using 224 GB. Job1-ResultStage in (a) and Job2-ResultStage in (b) refer to the shuffle read time. [35]

5 Optimizations and Deployments in Public Cloud

To extend the MVAPICH MPI library support on cloud systems, we have been making efforts to characterize different cloud system architectures systematically and propose optimization designs for these cloud systems, respectively.

5.1 Motivation and Challenges

Over the last decade, cloud-based systems have been trying to meet the performance of bare-metal hardware while keeping the benefits of the computing-as-service model intact. Today's HPC cloud offerings such as Microsoft Azure H-series and Amazon EC2 provide near-native performance, high customizability, resource provisioning and isolation, and elasticity to scale applications dynamically.

The first challenge of moving an MPI software stack to the cloud is efficient support for Virtual Machines(VM). Reducing overhead over network virtualization is crucial for competitive MPI performance. In some cloud system virtualization designs, a set of CPU cores need to be avoided for system virtualization-related tasks. The other challenge is to provide various performance optimization approaches by leveraging the interconnect and CPU/GPU architectures of each cloud system. For example, Microsoft Azure uses the Infiniband Network adapter, while AWS uses the Elastic Fabric Adapter. In the last few years, the MVAPICH team has collaborated with Microsoft Azure, AWS, and Oracle OCI to provide easy-to-deploy and high-performance MPI support on their HPC cloud systems.

5.2 Design and Implementation

In this section, we briefly introduce the MVAPICH optimization designs on Microsoft Azure, AWS, and OCI.

Microsoft Azure configures the HPC VM systems with InfiniBand adapters. The MVAPICH software family, encompassing MVAPICH2, MVAPICH-GPU, HiBD, and HiDL libraries, benefits from extensive years of performance optimization efforts. These solutions operate seamlessly on Azure HPC VMs equipped with InfiniBand network infrastructure. Moreover, we have been working with the Azure team to optimize the MVAPICH software on different HPC VM types including HB, HC, and ND (GPU system) series. Performance enhancement efforts, including comprehensive point-to-point and collective algorithm selection, and optimized XPMEM support for intra-node communication have been made to optimize the MVAPICH performance further. The MVAPICH software family has also been integrated into Azure's official CentOS HPC Images, which allows users to launch HPC VMs with optimized MVAPICH library support through one-click operation.

AWS configures the HPC instances with its own designed network adapter – Elastic Fabric Adatper (EFA). The RDMA communication on EFA is supported by either libfabric or EFA verbs. Different from IB adapters, EFA supports an Unreliable Datagram (UD) and a newly defined Scalable Reliable Datagram (SRD) transport protocol. However, it does not support the most commonly used Reliable Connection (RC) transport protocol. As introduced in Sect. 2, MVAPICH-3.0 and MVAPICH-Plus can both run on AWS HPC EFA instances through OFI support. On the other hand, to address some of the EFA's performance limitations, such as higher latency, no RC protocol support, and more likely to have out-of-order received UD or SRD data, we have proposed EFA-verbs-level support in MVAPICH with many performance optimization efforts on different AWS HPC instance types. Details of the designs are discussed in [41,42]. The latest MVAPICH MPI library version with EFA optimization is available as MVAPICH2-X-AWS-2.3.7.

OCI configures its bare-metal HPC VMs with RDMA over Converged Ethernet (RoCE) network adapters. The RoCE network supports IB verbs, which allows the MVAPICH library to run on OCI HPC VM systems seamlessly. Through the collaboration with OCI team, we have further enhanced the MVAPICH MPI performance over RoCE systems by comprehensive point-to-point and collective tuning. The detailed results are available from [43].

5.3 Performance Evaluation

5.3.1 Microsoft Azure

Figure 16 shows the point-to-point performance of Azure GPUs with various message sizes. MVAPICH2-GDR achieved a maximum of 23.04GB/s for GPU to GPU communication.

Figure 17 compares the performance of MVAPICH2 and MVAPICH2-X-XPMEM on Azure's HBv2 resources. When comparing the performance of the WRF application with various numbers of processes, XPMEM-enabled MVAPICH demonstrates higher performance and scalability.

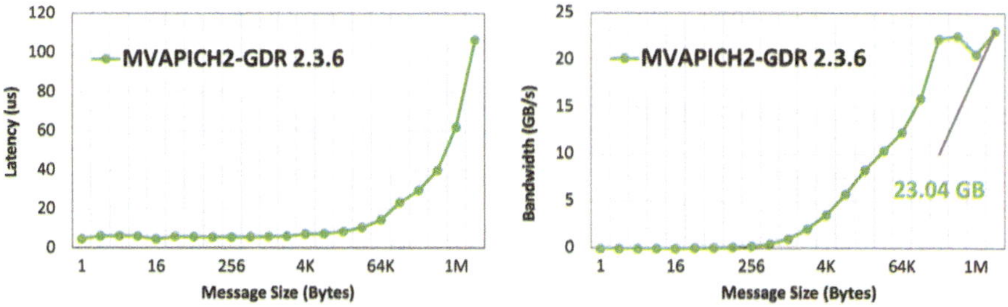

(a) Inter-node Latency (b) Inter-node Bandwidth

Fig. 16. MVAPICH2-GDR Point-to-point Performance on Azure NDv4 GPU VMs [44]

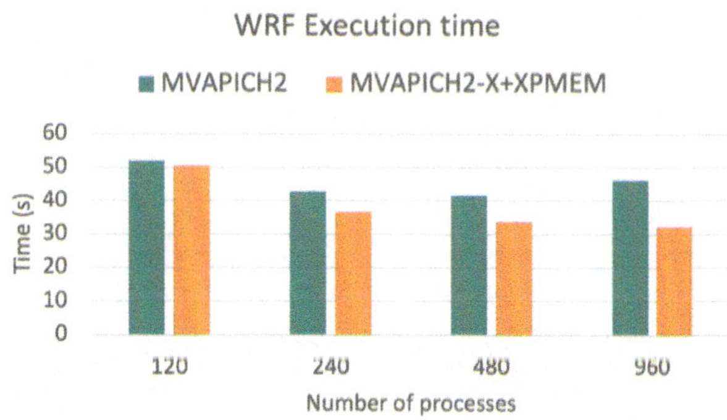

Fig. 17. Comparison of WRF Execution Time with MVAPICH2 and MVAPICH2-X-XPMEM [42]

5.3.2 AWS

Figures 18(a) and 18(b) show the inter-node point-to-point performance on c6gn instances equipped with AWS's Graviton2 processor. MVAPICH2-X-AWS demonstrates low latency for both small and large messages.

Figures 18(c) and 18(d) illustrate the collective performance and scalability of MVAPICH2-X-AWS on the same c6gn instances. The experimental results for representative collective algorithms, Allgather and Allreduce, confirm excellent communication performance across various message sizes.

Figure 19 compares the execution time of WRF and miniAMR on AWS with MVAPICH2, MVAPICH2-X-Optimized, and OpenMPI. When comparing the completion time on 32 nodes, the MVAPICH2-X-Optimized version shows 31% faster on WRF and 19% faster on miniAMR compared to OpenMPI's completion time. Overall trending of completion time shows optimized MVAPICH demonstrates much more stable and scalable performance than other MPI libraries.

These results indicate that the MVAPICH2 library can also be effectively used on cloud systems, in addition to on-premise systems, to get good performance and scalability compared to other MPI libraries.

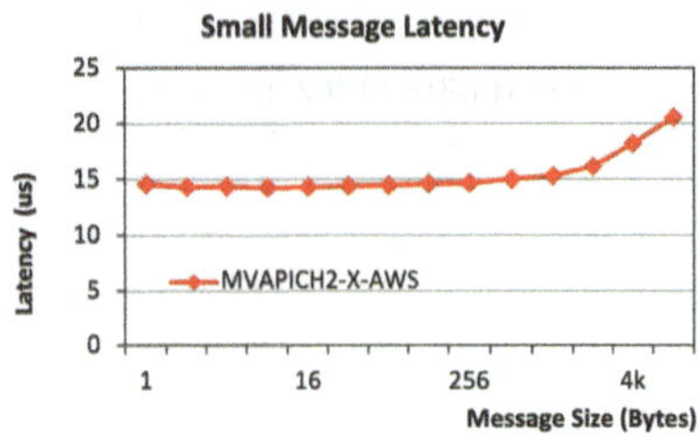

(a) Small Message Latency

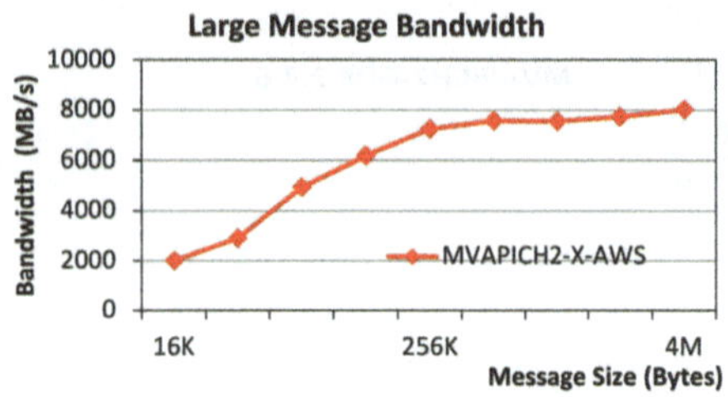

(b) Large Message Bandwidth

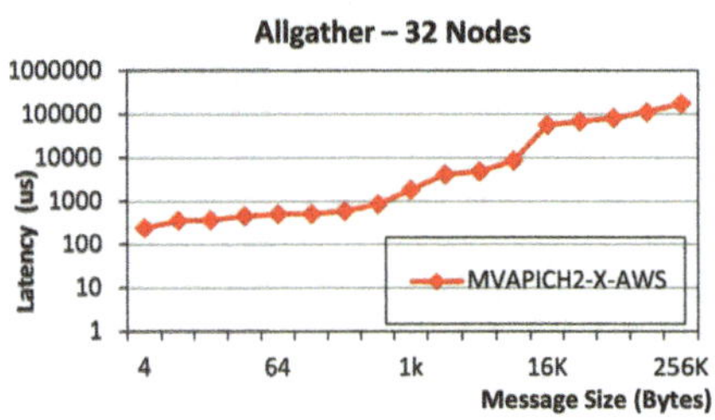

(c) Allgather

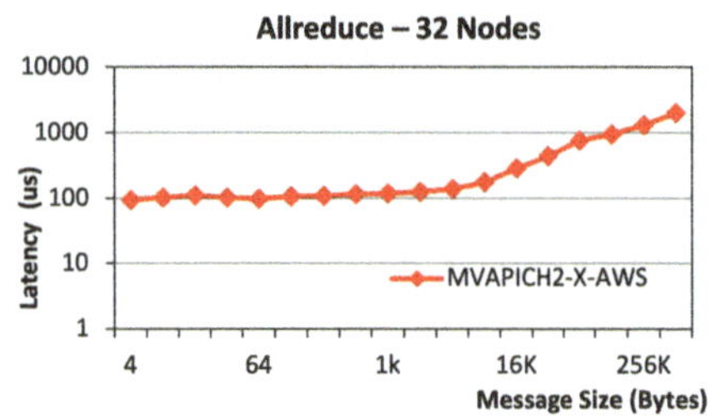

(d) Allreduce

Fig. 18. MVAPICH2-X-AWS Point-to-point and collective Performance on AWS c6gn instances with EFA [42]

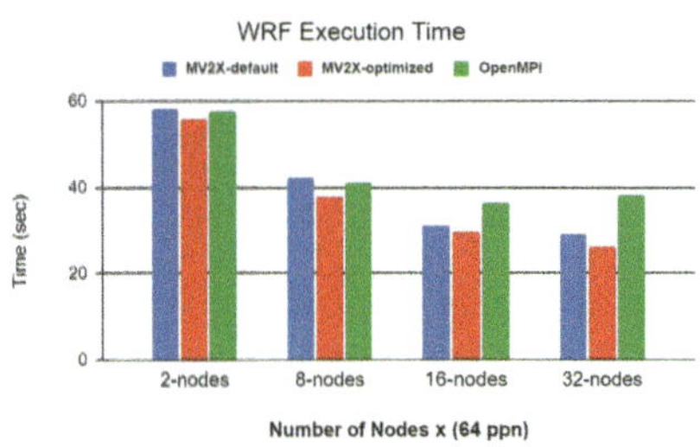

(a) WRF

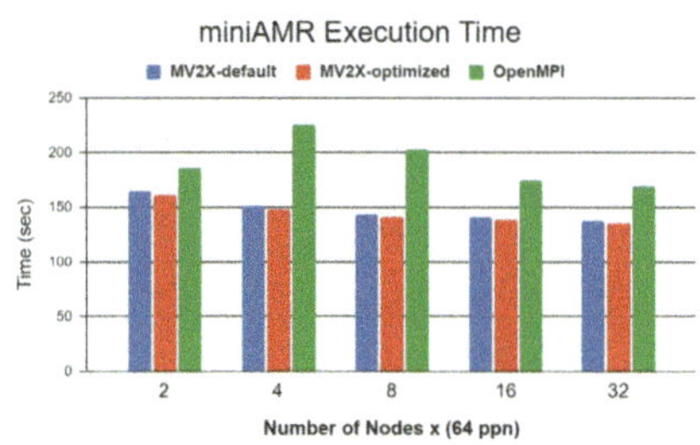

(b) MiniAMR

Fig. 19. Comparison of HPC Application Performance with Various MPIs on AWS [42]

6 Conclusion

The development of converged middleware for HPC and AI aims to ensure that DL, ML, and Big Data applications work in a scalable and efficient manner with various network technologies, computing architectures, and accelerators. The MVAPICH MPI library is a representative MPI library that facilitates efficient communication not only in the traditional on-premise HPC systems but also in cloud environments. This chapter demonstrates how the MVAPICH-driven components for AI (MPI4DL, MPI4cuML, and MCR-DL) and Big Data (MPI4Dask and MPI4Spark) can be effectively designed to provide the overall converged middleware. The chapter also provides performance evaluation results for a range of HPC, AI, and Big Data applications on on-premise and cloud systems using the converged middleware. Converged middleware for HPC, AI, and Big Data requires continuous development and optimization with the emergence of new computing, accelerator, and networking devices, used on both on-premise and rapidly changing cloud environments. The MVAPICH-driven converged middleware project is aimed towards satisfying this objective and helping the HPC, AI, and Big Data communities to take advantage of the modern HPC systems to accelerate their workloads.

Acknowledgment. The authors would like to acknowledge contributions made by many teams members on the various designs and their corresponding software releases, as included in this chapter: Kinan Al Attar, Quentin Anthony, Chen Chun Chen, Sungjae Lee, Nat Shineman, Jinghan Yao, and Qinghua Zhou.

References

1. Simmons, M.L., Wasserman, H.J.: Performance comparison of the CRAY-2 and CRAY X-MP/416. In: Supercomputing 1988: Proceedings of the 1988 ACM/IEEE Conference on Supercomputing, vol. I, pp. 288–295 (1988)
2. Dean, J., Ghemawat, S.: MapReduce: simplified data processing on large clusters. Commun. ACM **51**(1), 107–113 (2008)
3. Coates, A., Huval, B., Wang, T., Wu, D.J., Ng, A.Y., Catanzaro, B.: Deep learning with COTS HPC systems. In: Proceedings of the 30th International Conference on International Conference on Machine Learning, ICML 2013, vol. 28, pp. III-1337–III-1345. JMLR.org (2013)
4. Panda, D.K., Subramoni, H., Chu, C.-H., Bayatpour, M.: The MVAPICH project: transforming research into high-performance MPI library for HPC community. J. Comput. Sci. **52**, 101208 (2021). Case Studies in Translational Computer Science
5. Ramesh, B., Zhou, Q., Shafi, A., Abduljabbar, M., Subramoni, H., Panda, D.K.: Designing efficient pipelined communication schemes using compression in MPI libraries. In: 2022 IEEE 29th International Conference on High Performance Computing, Data, and Analytics (HiPC), pp. 95–99 (2022)
6. Khorassani, K., Chen, C., Subramoni, H., Panda, D.K.: Designing and optimizing GPU-aware nonblocking MPI neighborhood collective communication for PETSc*. In: 2023 IEEE International Parallel and Distributed Processing Symposium (IPDPS), Los Alamitos, CA, USA, pp. 646–656. IEEE Computer Society (2023)
7. Suresh, K.K., et al.: Efficient personalized and non-personalized alltoall communication for modern multi-HCA GPU-based clusters. In: 2022 IEEE 29th International Conference on High Performance Computing, Data, and Analytics (HiPC), pp. 100–104 (2022)
8. NVIDIA. NVIDIA Infiniband platform. https://www.nvidia.com/en-us/networking/products/infiniband/. Accessed 21 Nov 2024
9. Intel. Omni-path. https://www.intel.com/content/www/us/en/products/network-io/high-performance-fabrics.html. Accessed 21 Nov 2024
10. Cornelis Networks. Omni-path express (OPX). https://www.cornelisnetworks.com/. Accessed 21 Nov 2024

11. NVIDIA. NVIDIA RoCE v1/v2. https://enterprise-support.nvidia.com/s/article/roce-v2-considerations. Accessed 21 Nov 2024
12. HPE. Slingshot interconnect. https://www.hpe.com/sa/en/compute/hpc/slingshot-interconnect.html. Accessed 21 Nov 2024
13. Rockport Networks. Switchless scale fabrics. https://rockportnetworks.com/. Accessed 21 Nov 2024
14. Zhou, Q., et al.: Accelerating MPI all-to-all communication with online compression on modern GPU clusters. In: Varbanescu, A.-L., Bhatele, A., Luszczek, P., Marc, B. (eds.) High Performance Computing, pp. 3–25. Springer, Cham (2022)
15. Khorassani, K.S., Hashmi, J., Chu, C.-H., Chen, C.-C., Subramoni, H., Panda, D.K.: Designing a ROCm-aware MPI library for AMD GPUs: early experiences. In: Chamberlain, B.L., Varbanescu, A.-L., Ltaief, H., Luszczek, P. (eds.) High Performance Computing, pp. 118–136. Springer, Cham (2021)
16. Russakovsky, O., et al.: Imagenet large scale visual recognition challenge (2015)
17. Patwardhan, N., Marrone, S., Sansone, C.: Transformers in the real world: a survey on NLP applications. Information **14**(4) (2023)
18. Chen, L., Wu, P., Chitta, K., Jaeger, B., Geiger, A., Li, H.: End-to-end autonomous driving: challenges and frontiers (2024)
19. Yenduri, G., et al.: Generative pre-trained transformer: a comprehensive review on enabling technologies, potential applications, emerging challenges, and future directions (2023)
20. Touvron, H., et al.: Llama: open and efficient foundation language models (2023)
21. He, K., Zhang, X., Ren, S., Sun, J.: Deep residual learning for image recognition (2015)
22. MVAPICH2-GDR Development Team. MVAPICH2-GDR User Guide. http://mvapich.cse.ohio-state.edu/userguide/gdr
23. The High-Performance Deep Learning Project. http://hibd.cse.ohio-state.edu/
24. Sergeev, A., Del Balso, M.: Horovod: fast and easy distributed deep learning in tensorflow (2018)
25. MPI4DL. https://github.com/OSU-Nowlab/MPI4DL
26. Jain, A., Shafi, A., Anthony, Q., Kousha, P., Subramoni, H., Panda, D.K.: Hy-fi: hybrid five-dimensional parallel DNN training on high-performance GPU clusters, pp. 109–130. Springer, Heidelberg (2022)
27. Alnaasan, N., Jain, A., Shafi, A., Subramoni, H., Panda, D.K.: ACCDP: accelerated data-parallel distributed DNN training for modern GPU-based HPC clusters. In: 2022 IEEE 29th International Conference on High Performance Computing, Data, and Analytics (HiPC), pp. 32–41 (2022)
28. MCR-DL. https://github.com/OSU-Nowlab/MCR-DL
29. Anthony, Q., et al.: MCR-DL: mix-and-match communication runtime for deep learning (2023)
30. ParaInfer-X. https://github.com/OSU-Nowlab/Flover
31. Yao, J., Alnaasan, N., Chen, T., Shafi, A., Subramoni, H., Panda, D.: Flover: a temporal fusion framework for efficient autoregressive model parallel inference (2023)
32. Ghazimirsaeed, S.M., Anthony, Q., Shafi, A., Subramoni, H., Panda, D.K.D.: Accelerating GPU-based machine learning in python using MPI library: a case study with mvapich2-GDR. In: 2020 IEEE/ACM Workshop on Machine Learning in High Performance Computing Environments (MLHPC) and Workshop on Artificial Intelligence and Machine Learning for Scientific Applications (AI4S), pp. 1–12 (2020)

33. HiBD. Nowlab HiBD. https://hibd.cse.ohio-state.edu/
34. Shafi, A., Hashmi, J.M., Subramoni, H., Panda, D.K.: Efficient MPI-based communication for GPU-accelerated dask applications (2021)
35. Al-Attar, K., Shafi, A., Abduljabbar, M., Subramoni, H., Panda, D.K.: Spark meets MPI: towards high-performance communication framework for spark using MPI. In: 2022 IEEE International Conference on Cluster Computing (CLUSTER), pp. 71–81 (2022)
36. Dask Development Team. Dask: Library for dynamic task scheduling (2016)
37. Zaharia, M., et al.: Apache spark: a unified engine for big data processing. Commun. ACM **59**(11), 56–65 (2016)
38. The Netty Project (2024). https://netty.io/
39. UCX. UCX-Py Github. https://github.com/rapidsai/ucx-py
40. Gropp, W., Lusk, E.: Dynamic process management in an MPI setting, pp. 530–533 (1995)
41. Subramoni, H., Panda, D., Chakraborty, S., Xu, S.: Designing scalable and high-performance MPI libraries on amazon elastic fabric adapter (2019)
42. Hashmi, J., Subramoni, H., Panda, D., Xu, S., Ghazimirsaeed, M.: MPI meets cloud: case study with Amazon EC2 and Microsoft azure (2020)
43. Basu, S.: Evaluation report of MVAPICH2-X on Oracle Cloud HPC shapes. https://blogs.oracle.com/cloud-infrastructure/post/evaluation-report-of-mvapich2-x-on-oracle-cloud-hpc-shapes
44. Microsoft. Azure ND A100 v4-series. https://docs.microsoft.com/en-us/azure/virtual-machines/nda100-v4-series

Key Mechanisms and Emerging Issues in Cloud Identity Systems

Khalid Maidine[1(✉)] and Ahmed El-Yahyaoui[2]

[1] Mathematics, Computer Science, and Information Security Laboratory (LABMIA-SI), Faculty of Sciences, Mohammed V University in Rabat, Rabat, Morocco
khalid_maidine@um5.ac.ma

[2] Intelligent Processing and Security of Systems (IPSS), Faculty of Sciences, Mohammed V University in Rabat, Rabat, Morocco
a.elyahyaoui@um5r.ac.ma

Abstract. The rise in popularity and success of services based on cloud technology is evident. However, this growth has not been without its security challenges, which Cloud Service Providers (CSPs) must urgently address to earn the trust of their users. A critical area of focus is the management of user identities, ensuring that access to services is restricted to authorized individuals. This paper provides an accessible discussion on the strengths, weaknesses, and security concerns associated with prevalent identity management strategies employed by CSPs. We examine various approaches, including password-based systems, biometric verification (like fingerprint scanning), Multi-Factor Authentication (MFA), Enterprise Single Sign-On (SSO), OAuth 2.0, OpenID Connect, and SAML. Our aim is to demystify these technologies and highlight their role in securing cloud services while maintaining user convenience.

Keywords: Cloud computing · Identity · SSO · OAuth 2.0 · OIDC · SAML

1 Introduction

In cloud computing contexts, identity management incorporates various methods that leverage unique characteristics or attributes of entities, such as names, passwords, signatures, and biometrics. The frequent occurrence of security breaches has highlighted the critical necessity for enhanced security in cloud computing environments, especially within public cloud services. In response to these security concerns, Cloud Service Providers (CSPs) utilize a variety of protocols in their identity management frameworks. These include Lightweight Directory Access Protocol (LDAP), Secure Shell (SSH) keys, Multi-Factor Authentication (MFA), Enterprise Single Sign-On (SSO), OAuth 2.0, OpenID Connect, and Security Assertion Markup Language (SAML) 2.0 [1]. This paper presents an operational prototype of these protocols within cloud environments. This study seeks to shed light on the strengths and weaknesses of each protocol, presenting a side-by-side comparison while also focusing on their individual security challenges and the strategies for addressing them. Additionally, the paper underscores

M. Zbakh et al. (Eds.): CloudTech 2023, LNNS 1220, pp. 64–88, 2024.
https://doi.org/10.1007/978-3-031-78698-3_5

the importance of authentication, a fundamental process in establishing the true identity of individuals or entities, confirming their claimed identities. Various authentication methods are employed in cloud computing, ranging from login credentials and multifactor authentication to third-party authentication systems, passwords (including graphical ones), biometric authentication, and digital device verification [2]. In cloud computing, access to resources is commonly controlled through an identity management system that employs one or more of these authentication methods, categorically divided into physical and digital types. Further, this paper is structured into five sections. Section 2 examines various physical authentication mechanisms, highlighting their advantages and limitations. In Sect. 3, the paper delves into digital authentication mechanisms, analyzing their effectiveness and vulnerabilities. Section 4 focuses on a comparison of the key features of the most widely used protocols by CSPs: OAuth 2.0, OpenID Connect, and SAML 2.0. Section 5 examines the main security concerns associated with these protocols.

2 Physical Mechanisms

In the realm of Cloud Data Centers (CDCs), the consolidation of servers and networking resources into a single, centralized location facilitates uninterrupted access for users. To maintain secure access and mitigate the risk of data leakage, CDCs employ a variety of authentication mechanisms. These mechanisms range from access cards to advanced biometric systems, which include technologies for scanning the iris or retina, as well as recognizing fingerprints, faces, or palm prints. The repertoire of security measures also extends to the use of smart cards, security tokens, key fobs, physical keys, or specially designed security badges. Primarily, these physical methods of authentication are deployed to control access to the physical infrastructure of the CDCs.

Nevertheless, these physical security measures are not devoid of challenges. Key concerns include the possibility of misplacement or theft of these authentication devices and their vulnerability to physical wear and tear. Additionally, the practicality of managing such devices can be cumbersome. Scalability also emerges as a concern, particularly in how these methods adapt to accommodate expanding user bases or facilities. Other significant issues involve delays in the authentication process (latency), the financial implications of setting up and maintaining these systems, concerns about the privacy of biometric data [3], and the risk of erroneous identifications (false positives or negatives). Moreover, these physical security systems face various security threats, such as spoofing attacks where an impostor mimics a legitimate user, skimming which involves illegal data capture, social engineering tactics that deceive individuals into bypassing security measures, and the direct theft of physical authentication devices. These myriad challenges underscore the need for a comprehensive, multi-faceted approach to security within cloud data center environments, blending physical safeguards with advanced digital security strategies [4] (Fig. 1).

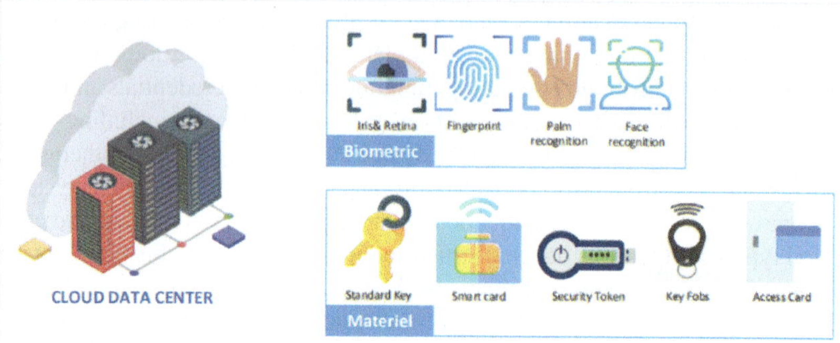

Fig. 1. Common physical access used in cloud context.

3 Digital Mechanisms

3.1 Credentials

3.1.1 Dictionary

Lightweight Directory Access Protocol (LDAP) and Active Directory (AD) are the most widely used protocol as dictionary mechanisms. These systems are fundamental in the administration of access credentials. Envision them as the vigilant custodians in the digital domain, rigorously regulating who gains access to various resources within the cloud. Their proficiency lies in centralizing critical aspects such as user accounts, permissions, and directory information, thereby facilitating streamlined management and control over access to cloud resources. Moreover, LDAP and AD are equipped with an array of security features, including robust encryption to safeguard data, comprehensive password policies to enforce strong passwords, and access control lists (ACLs) to meticulously manage entry permissions. However, integrating LDAP or AD across varied applications, particularly in extensive, geographically dispersed cloud environments with a vast user base, presents considerable complexities. Administrators tasked with managing and maintaining these systems often face an escalated workload. A notable challenge with LDAP is its original design limitation, not being primed for secure credential transmission over public networks, akin to sending a confidential message in a transparent envelope. Furthermore, the journey is not without turbulence for Active Directory and LDAP. They confront numerous cunning cyber threats, ranging from Distributed Denial of Service (DoS) attacks, which aim to inundate and incapacitate the system, to brute force attacks aggressively attempting to crack passwords [1]. *Additionally, there are Man-in-the-Middle (MitM) attacks, where data is covertly intercepted and altered during transmission, and LDAP injection attacks, where vulnerabilities are exploited for unauthorized access. Thus, while LDAP and AD are invaluable in the cloud computing sphere, they navigate a landscape replete with challenges and dynamic security threats. It resembles navigating a digital wilderness, where* maintaining acute awareness and continuously adapting security strategies are imperative for enduring success and safety (Fig. 2).

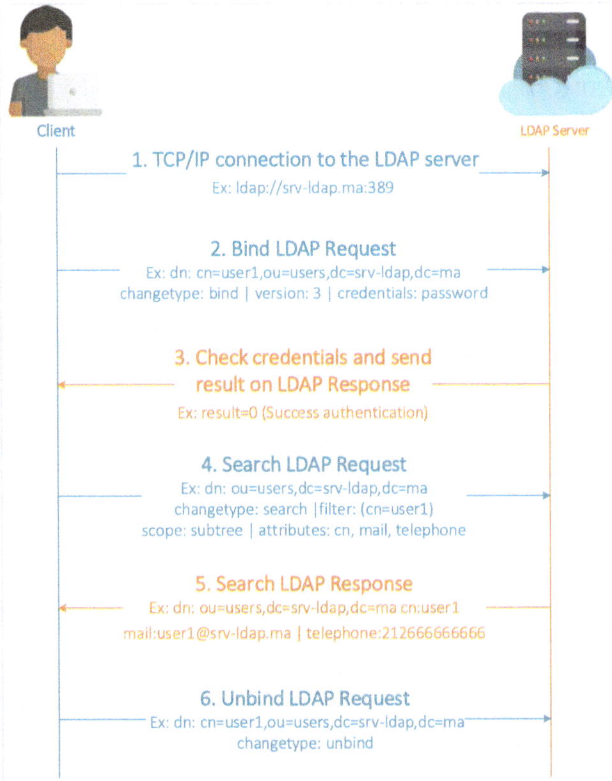

Fig. 2. Common LDAP flow exchanges.

3.1.2 Secure Shell Keys

SSH keys are employed as a reliable mechanism for secure authentication in environments involving remote connections. This system comprises two key elements: the private key, which is strictly kept confidential by the user, and the public key. The private key's role is to decode data that has been locked (encrypted) using the public key. Conversely, the public key, once distributed to a remote server or device, functions to lock (encrypt) data in a manner that only its corresponding private key can unlock (decrypt). By leveraging the concept of asymmetric cryptography, SSH keys significantly bolster security. This is achieved by making it exceptionally difficult for any unauthorized entity to infer the private key based on the public key [5]. *In practical application, the user needs to place their public key into the 'authorized_keys' directory on the remote server or gadget. Then, the private key is employed to confirm the user's identity during attempts to connect to the server. Encrypting information with these paired keys* substantially reduces the chance of external entities breaching or intercepting the communication. Maintaining the integrity and secrecy of the private key is paramount, necessitating its protection with a strong passphrase to fend off brute force attacks. While SSH is acclaimed for its efficacy in securing connections, the intricacies involved in its setup and administration cannot be overlooked. Moreover, it's imperative to remain vigilant about potential security and performance weaknesses inherent in SSH, necessitating ongoing monitoring and careful management [6] (Fig. 3).

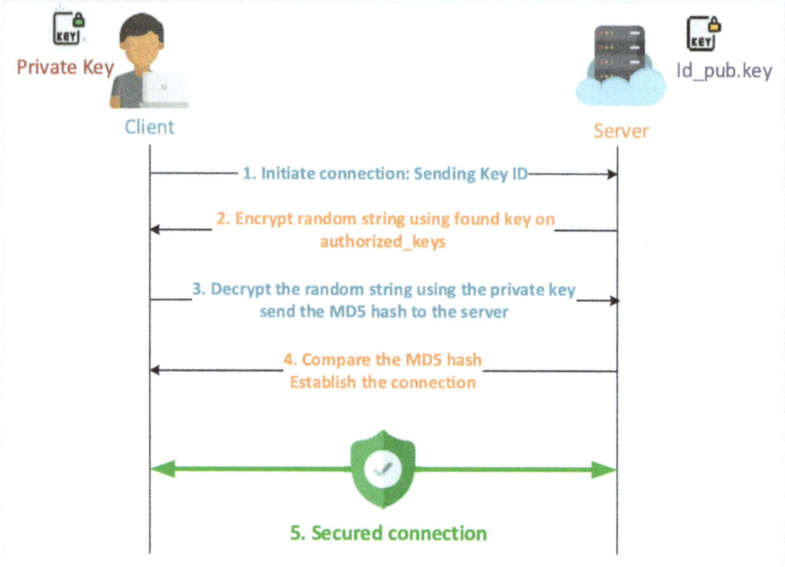

Fig. 3. Flow of SSH key connection establishing

3.2 Multifactor Authentication

Multifactor authentication (MFA) operates as an auxiliary safeguard, markedly diminishing the potential for illegitimate access to accounts and networked systems. This protocol necessitates that users substantiate their identity by furnishing multiple evidentiary elements, colloquially termed as "factors." These elements encompass knowledge-based factors such as PINs or security queries; possession-based factors like smartphones or security tokens; and inherence-based factors, exemplified by biometric traits. In the realm of financial transactions, One-time passcodes (OTPs) are prevalently employed within MFA frameworks. This system entails the generation and transmission of passcodes to a user's telecommunication device or electronic mail. The user must promptly enter this passcode within a prescribed temporal window to effectuate the authentication sequence. Contemporary trends show an inclination towards mobile application-based multifactor authentication, introducing diverse operational modalities. Furthermore, methodologies such as pattern recognition and captcha systems are being integrated as ancillary authentication modalities. However, these methodologies are not devoid of limitations. Challenges include dependence on telecommunication or email accessibility, restrictions on permissible authentication attempts leading to potential account lockout, susceptibility to fraudulent interception of email or SMS-based codes, risks related to the misplacement or theft of devices, and the menace of malware installation on user devices.

Attack vectors such as phishing, man-in-the-middle schemes, malware proliferation, and social engineering tactics represent prevalent threats undermining the robustness of MFA mechanisms.

3.3 SSO and Federation

In cloud computing paradigms, traditional methods of authentication typically necessitate that users submit their access identifiers, such as usernames and passwords, to access applications and services hosted on the cloud. This procedure can be cumbersome and susceptible to inaccuracies, especially as users are required to recall a multitude of login credentials for diverse applications and services. When this situation involves a multitude of users and an array of applications, the complexity in managing and overseeing these credentials intensifies significantly. To avoid these challenges, CSPs have adopted single sign-on (SSO) mechanism. This approach is enabled through the adoption of specialized protocols like Security Assertion Markup Language (SAML), OAuth, or OpenID, which effectively simplify and streamline the authentication process across various platforms and services.

3.3.1 Enterprise SSO

Enterprise Single Sign-On (SSO) architecture is devised to curtail the repetitiveness of users entering their authentication details for Service Provider (SP) access [7]. *It streamlines access to web applications by authenticating users via an encrypted cookie. When a user* endeavors to access a web application, the application inspects the user's device for this cookie. If present, it is sent to the SSO server for verification. Absent the cookie, users are prompted to enter and submit their encrypted login details to the SSO server. Upon receiving these details, the SSO server decrypts and assesses their validity. If deemed valid, the server generates a new encrypted cookie containing the user's identifier and necessary authorization information, transmits it for browser storage, and notifies the application of successful user authentication. If the credentials are invalid, the SSO server conveys an error message to the application [6]. *Future attempts by the user to access the application trigger a reevaluation for the encrypted cookie, enabling access if it* remains valid. An invalid or expired cookie, however, necessitates reentry of login credentials. While Enterprise SSO offers advantages like improved user experience and more streamlined identity management, it also carries potential disadvantages. These include being a singular failure point, its inherent complexity, and susceptibility to various cyber-attacks, especially in shared user environments or when inadequately configured. Threats such as phishing, Man-In-The-Middle attacks, and security breaches risking exposure of user credentials are prevalent concerns. To enhance the security of Enterprise SSO, implementing measures like ensuring high availability and integrating multi-factor authentication are pivotal (Fig. 4).

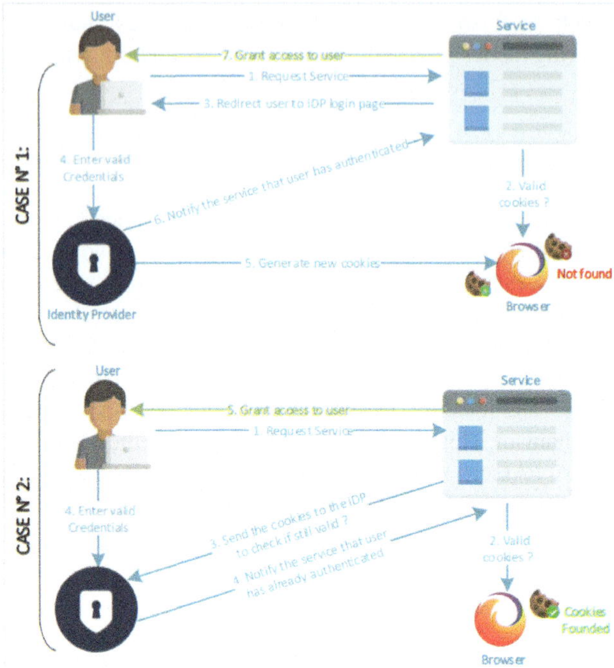

Fig. 4. Enterprise SSO cases flow.

3.3.2 OAuth 2.0

OAuth 2.0, recognized as a prevalent protocol in the realm of authorization, facilitates the secure exchange of user resources stored on a specified server with third-party entities, while simultaneously preserving the confidentiality of user credentials. This protocol hinges on the issuance of tokens by the authorization server, following the resource owner's approval for access. These tokens are subsequently employed by third-party entities to execute actions authorized by the user. In its latest update, OAuth 2.0 introduces a suite of refinements and advancements, including the transition from HTTP to HTTPS to enhance secure communication. A significant simplification in this version is the removal of signature requirements for token exchanges [8], *thereby easing the implementation process. Contrasting with OAuth 1.0, which lacks refresh tokens, leading to perpetual validity of access tokens, OAuth 2.0 delineates diverse strategies for obtaining access tokens. The primary flows in OAuth 2.0 encompass the Authorization Code and Implicit Code flows. The former* utilizes an Authorization Code, exchangeable by the third-party for an access token, while the latter involves the direct exchange of the access token by the authorization server with the third-party, a process that potentially exposes it to browser-level security risks. The Authorization Code flow further authenticates the third-party entity using a client secret [9]. *Tokens produced under this protocol are time-limited in terms of validity and are subject to revocation in cases of suspected compromise. Despite its advantages, OAuth 2.0's implementation complexity, particularly concerning the secure handling and storage of client secrets and access tokens,* remains a challenge. Inadequate configuration of OAuth 2.0 can expose it to assorted cyber threats, such as Cross-Site Request Forgery (CSRF), Access Token Injection, Identity Provider Mix-Up attacks, Authorization Token Injection, clickjacking, and token leakage. The OAuth 2.1 draft, released in March 2023, proposes pivotal changes, notably the elimination of Implicit and Resource Owner Password flows and the stipulation for refresh tokens to be either sender-constrained or utilized once. An additional significant revision is the mandatory application of Proof Key for Code Exchange (PKCE) within the Authorization Code flow, involving the creation of PKCE from a hash of a randomly generated value (code verifier) prior to initiating the authorization request [10] (Fig. 5).

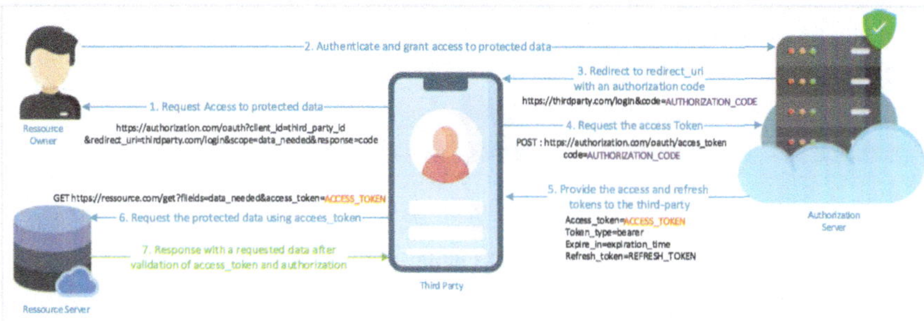

Fig. 5. OAuth 2.0 authorization code flow.

3.3.3 OpenID Connect

OpenID Connect (OIDC) operates as an advanced open standard, constructed atop the OAuth 2.0 protocol, to streamline identity verification via RESTful APIs [11]. *As the latest iteration in the OpenID series, it introduces the ID Token* [12], *encapsulating user authentication details in a JWT (JSON Web Token) format. This token, authenticated by the identity provider, is designed for independent readability and verification. The structure of the JWT within OIDC comprises three distinct parts: The Header, detailing the token's type and the algorithm used for its digital signature. The Payload,* containing critical data such as the issuer's identity, the end-user's details, client information, and the timestamps for the token's issuance and expiry [13]. *The Signature, integral for* validating the token's origin and ensuring it emanates from a trustworthy source. OIDC integrates both authentication and authorization processes into a singular, streamlined flow. Its primary aim is to simplify user authentication with an Identity Provider (IdP) while enabling Relying Parties (RP) to confirm user identity and obtain permission for accessing resources. It supports various flows, including authorization code, implicit, and hybrid, with the specific flow determined by the 'response_type' parameter in the RP's authentication request. The 'openid' scope parameter is crucial to differentiate an OIDC request from a standard OAuth 2.0 request. Predominantly, the Authorization Code flow is favored in OIDC for its enhanced security features: it conceals the code exchange in a server-to-server back channel and requires applications to authenticate via 'client_id' and 'client_secret' (Fig. 6).

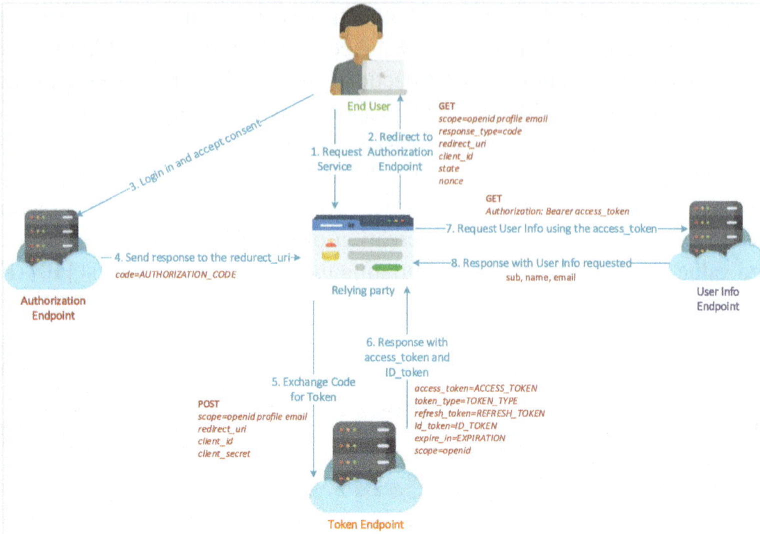

Fig. 6. Openid Connect authorization code flow.

OIDC delineates five principal roles in its authentication and authorization scheme: the End User, the Relying Party (RP), the Authorization Endpoint (AE), the Token Endpoint (TE), and the User Info Endpoint. RPs engage in a discovery process, known as WebFinger, to ascertain the configuration details of the OIDC identity provider before commencing the authentication flow. OIDC identity providers facilitate metadata discovery, allowing them to present configuration details at a standardized endpoint ('/.well-known/openid-configuration'). This feature enables RPs to automatically register with the IdP, eliminating the need for manual configurations. An exemplar of this is Google Identity's openid-configuration endpoint. Since its release in 2014, OIDC has seen widespread adoption by leading corporations like Facebook, Google, and Microsoft's Office 365 for their SSO implementations. However, incorrect configuration of OIDC can lead to a spectrum of security vulnerabilities, such as Cross-Site Request Forgery, Mix-Up attacks, Passive web threats, DDoS, and Phishing attacks. To bolster security, RPs and OIDC Providers are encouraged to enforce the use and verification of the 'state' variable and adopt secure HTTPS connections.

3.3.4 SAML

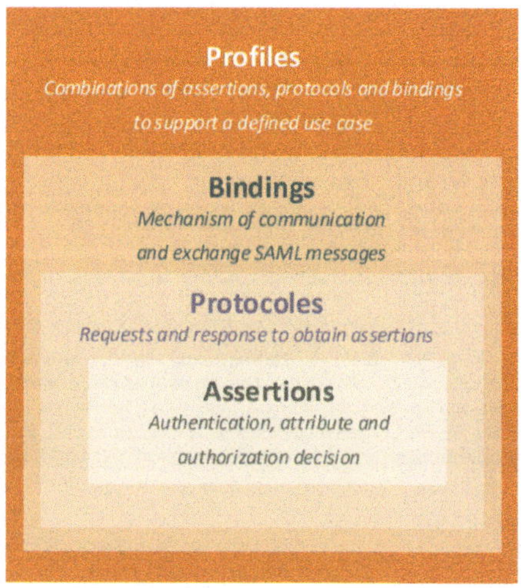

Fig. 7. SAML components.

Security Assertion Markup Language (SAML) is an open standard that enables to transmit authentication and authorization data between identity providers (IdP) and service providers (SP) using Extensible Markup Language (XML) for standardized communications [14]. *The last version of SAML is 2.0 approved by OASIS Consortium in 2005. SAML 2.0 comes with major changes that making it incompatible with the version 1.1* [15]. *One of the main changes was the introduce of new protocols including assertion queries and requests, authentication request, artifact resolution, name identifier management, name identifier mapping, and single logout protocols. The SAML architecture* contains of components that collectively serve several use cases [16] (Fig. 7).

Assertions: *SAML authorities can generate three assertion statements*

- *Authentication*: Subject was authenticated by a certain mechanism at a given time.
- *Attribute*: Attributes that describe the subject.
- *Authorization* Decision: Grant or reject access to a specified resource for the subject (Fig. 8).

Fig. 8. SAML Assertion example.

Protocols: *SAML defines several request/response protocols including Assertion Query and Request Protocol, Authentication Request Protocol, Artifact Protocol, Name Identifier Management Protocol, Single Logout Protocol and Name Identifier Mapping Protocol.*

- **Assertion Query and Request**: Assertion query is used by SP to query an attribute of user to the Identity provider.
- **Authentication Request Protocol**: Used to request authentication of a user by an Identity Provider (IdP) and receive response containing one or more assertions pertaining to the authenticated user.
- **Artifact Protocol**: Help SP to get a previously generated assertion on IdP using a reference "artifact".

- **Name Identifier Management Protocol:** Used to change the value or format of the name of subject.
- **Single Logout Protocol:** Allow user to log out of all active SP and IdP.
- **Name Identifier Mapping Protocol:** Provides a flexible way to map identifiers between different formats, making it easier for users to access multiple services using a single identity.

Binding: *SAML bindings define the mechanism used for SAML messages exchange between IdPs and SPs. The main types of binding used on SAML 2.0 protocol are:*

- **HTTP Redirect Binding**: This binding utilizes URL parameters for the conveyance of SAML messages, capitalizing on an HTTP user agent. It stands out for its simplicity and directness, especially in scenarios requiring minimal data transmission.
- **HTTP POST Binding**: Here, SAML messages are transmitted via an HTML form, encoded in base64. This method is particularly adept at handling substantial message sizes, thereby offering a robust alternative to the HTTP Redirect approach.
- **HTTP Artifact Binding**: This unique binding transmits SAML messages through a specific identifier known as an artifact. It is especially advantageous in cases where direct message transmission is impractical.
- **SOAP Binding**: Characterized by its encapsulation of SAML messages within a SOAP envelope, this binding enables direct exchanges between SP and IdP without the mediation of a user agent, thereby streamlining the process.

- **Profiles:** *SAML Profiles describes how a set of assertions, protocols and bindings might be implemented to solving a particular scenario* [17]. *SAML 2.0 comes with numerous profiles including:*
- **Web Browser Single Sign-On (SSO)**: This profile is particularly noteworthy for its facilitation of seamless user authentication across various SPs. After initial authentication at one SP, it negates the need for further credentials at others. This not only enhances user experience by reducing repetitive authentication processes but also improves overall system efficiency. It accomplishes this through the communication of SAML assertion tokens, leading to necessary access permissions. Notably, it supports both SP-initiated and IdP-initiated SSO flows, showcasing its versatility.
- **Additional Profiles:** These include Enhanced Client or Proxy (ECP), Identity Provider Discovery, Single Logout, Name Identifier Management, Artifact Resolution, and Name Identifier Mapping Profile. Each profile is tailored to cater to specific facets of digital identity management, thus enriching the SAML 2.0 framework's capability to handle a broad spectrum of authentication and identity management requirements (Figs. 9 and 10).

Metadata

In the Security Assertion Markup Language (SAML) ecosystem, metadata assumes a crucial role, functioning as an XML-formulated instrument that is indispensable for the establishment of trust between Identity Providers (IdPs) and Service Providers (SPs). This metadata is a repository of several vital elements:

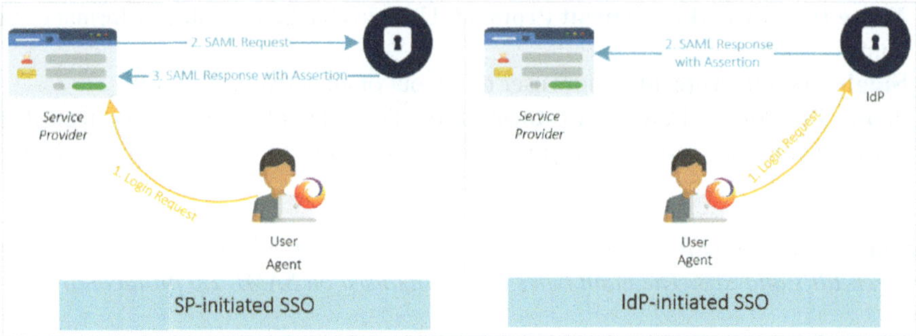

Fig. 9. SAML SP-initiated SSO vs IdP-initiated SSO

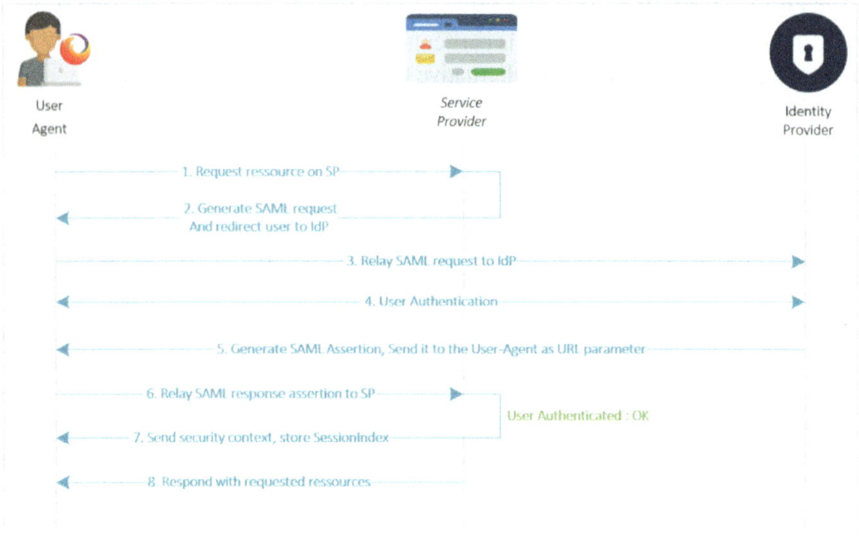

Fig. 10. SAML SP-initiated SSO flow.

- **Entity Designation**: This aspect of the metadata serves as a unique identifier for each participant in the SAML process, ensuring unequivocal recognition of entities.
- **Communication Endpoints**: It precisely delineates the designated locations for the exchange of SAML messages, pivotal for guiding the communication trajectory.
- **Public Key Data**: Central to the metadata's function, this information is key to authenticating and corroborating the integrity of the exchanged messages, thereby fortifying security.
- **Encryption and Signature Protocols**: The metadata explicitly details the cryptographic algorithms employed, a critical factor for safeguarding message confidentiality and ensuring compatibility across different SAML participants.
- **Protocol and Binding Compatibilities**: This section of the metadata specifies the range of SAML profiles and bindings that an entity can support, a fundamental aspect for ensuring interoperability and the fluid execution of SAML processes (Fig. 11).

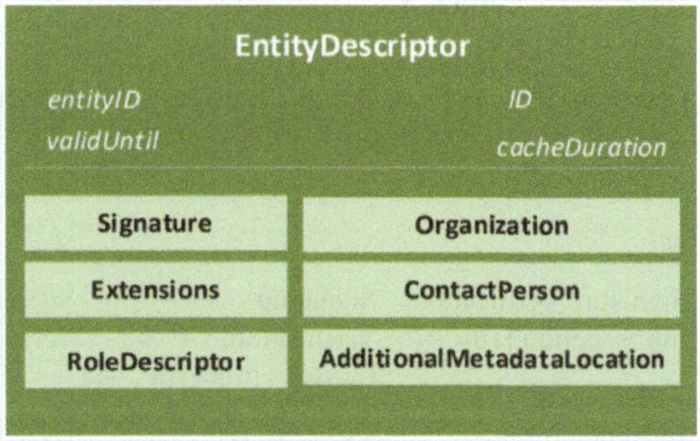

Fig. 11. SAML Metadata architecture.

4 Comparison

Within the domain of cloud services, the implementation of Identity Management systems by Cloud Service Providers (CSPs) prominently involves the deployment of protocols such as Security Assertion Markup Language (SAML), OAuth 2.0, and OpenID Connect (OIDC). These protocols, designed with precision and specificity, play a critical role in ensuring robust and effective authentication and authorization frameworks. Each protocol exhibits distinct advantages and functional nuances. The following table offers a comparative of the key features and capabilities of these widely adopted protocols (Table 1).

Table 1. OAuth 2.0, OIDC, SAML features comparison

Feature	OAuth 2.0	OpenID Connect	SAML
Standardization	IETF, OAuth Working Group	OpenID Foundation, OAuth Working Group	OASIS, Security Services Technical Committee (SSTC)
Introduction Year	2012	2014	2001
Current Version	RFC 6749 (2012)	OIDC Core 1.0 (2014)	SAML 2.0 (2005)
Protocol Type	Authorization	Authentication and Authorization	Authentication and Authorization
Protocols used	XML, HTTP, REST [18]	JSON, HTTP, REST [18]	XML, HTTP, SOAP [18]
Token Type	Access Token	ID Token	Assertion
Token Format	JWT, JSON, XML	JWT, JSON	XML

(*continued*)

Table 1. (*continued*)

Feature	OAuth 2.0	OpenID Connect	SAML
Revocation Support	Yes	Yes	Yes
Federation Support	No	Yes	Yes
Token Validation Method	Signature Verification, Introspection [19]	Signature Verification, Introspection [19]	Signature Verification[19]
Security Considerations	Designed for Delegation	Designed for Federated Identity and Single Sign-On (SSO)	Designed for Federated Identity and Single Sign-On (SSO)
Identity Provider Discovery	No	Discovery Endpoint	Metadata
Mobile App Support	Yes	Yes	No
User Consent	Yes	Yes	No
Use Cases	Mobile and Web Applications, Social Login	Single Sign-On (SSO), User Profile Exchange, Multi-Factor Authentication	Single Sign-On (SSO), Identity Federation

The table compares OAuth 2.0, OpenID Connect, and SAML, three protocols for identity management in cloud computing. OAuth 2.0 focuses on authorization using XML/HTTP, OpenID Connect extends OAuth 2.0 for authentication with JSON/HTTP, and SAML is an older protocol using XML for authentication and authorization. While OAuth 2.0 lacks native federation support, OpenID Connect and SAML include it. All support token revocation and secure transactions, with specific uses in web and mobile applications, and SSO functionalities.

5 Security Issues

Although the Security Assertion Markup Language (SAML), OpenID Connect (OIDC), and OAuth 2.0 are prevalent protocols for authentication and authorization in cloud contexts, they are not impervious to security vulnerabilities. Improper implementation of these protocols can engender significant risks, potentially allowing malicious entities to access confidential data or disrupt cloud service operations. Consequently, it is imperative to thoroughly understand the security challenges inherent to each protocol and implement effective countermeasures. This section aims to delineate common security pitfalls associated with SAML, OIDC, and OAuth 2.0 in cloud environments and provide recommendations for their mitigation.

5.1 OAuth 2.0

In this section, we intend to examine the vulnerabilities commonly found in OAuth 2.0, largely due to misconfiguration. Echoing Eran Hammer's observations from hueniverse.com on July 26, 2012, while OAuth 2.0 can be securely implemented by developers with an extensive understanding of web security, it often leads to less secure outcomes when handled by the average developer. Our goal here is to identify these vulnerabilities and suggest practical best practices to address them, enhancing the security effectiveness of OAuth 2.0 applications [20].

5.1.1 307 Redirect

Uncovered by Fett et al. in 2016, this particular attack manifests during a specific phase of the OAuth process, where an Identity Provider (IdP) opts for the HTTP 307 status code to redirect authentication results to the user agent, as delineated in step 3 (Fig. 5) [1]. *The HTTP 307 status code, designed for redirecting a user agent along with all data from* a previous request, presents an inherent security flaw in this context. Given that OAuth permits any form of HTTP redirection, this becomes a viable attack vector. Exploiting this vulnerability, a nefarious Resource Provider (RP) could potentially intercept the credentials of a user authenticated by the IdP, redirected via the 307-status code, and subsequently impersonate the user. To mitigate this attack, it is advisable for Identity Providers (IdPs) to employ HTTP status code 303 for user agent redirection, rather than the 307 code. The significance of using HTTP 303 lies in its explicit definition to discard the body of any preceding HTTP POST request. This characteristic of the 303 redirections effectively prevents the transmission of credentials to a potentially malicious Resource Provider (RP), thereby offering a robust defense against this form of attack. Adopting HTTP 303 is a strategic approach to enhance the security of the OAuth process by ensuring that sensitive authentication details are not inadvertently redirected to unauthorized entities.

5.1.2 Cross Site Request Forgery

This attack involves a malevolent website prompting the user's web browser to execute a specific request directed at a target website. The mechanism of the attack includes embedding a malicious GET request within an HTML image element, such as <IMG SRC = ATTACK_URL> [21]. *Executed silently, this request capitalizes on the user's existing authenticated state with the target system. Unbeknownst to the user, their browser automatically sends authentication headers and cookies during this process. The attacker's strategy involves intercepting their own code or access token, halting the redirection response that was initially meant for the Resource Provider (RP), and* retaining the response's URL. Subsequently, the attacker deceives the victims into clicking on this saved URL. Once clicked, the RP processes the URL, inadvertently retrieving data from the Identity Provider (IdP) and associating the victim's session with resources controlled by the attacker. The severity of this attack varies based on the nature of data handled by the RP. In a critical scenario, the end-user might inadvertently upload confidential files to an account under the attacker's control, thereby compromising sensitive information. To effectively counteract Cross-Site Request Forgery (CSRF) attacks within the OAuth framework, it is strongly advised to incorporate the usage of a 'State' parameter. This parameter should be uniquely generated and associated with the initial authorization request initiated by the Resource Providers (RPs). Upon receiving this request, Identity Providers (IdPs) are expected to include this 'State' parameter in their authorization response. The critical step in this mitigation strategy involves the RPs verifying the 'State' parameter received from the IdP against the value they initially stored. This verification acts as a safeguard, ensuring that the response is indeed corresponding to the request originally sent. An optimal approach for generating the 'State' parameter is to use an unpredictable value, ideally derived from a hash of a session-specific component, created at the onset of the OAuth flow. This method ensures that the 'State' parameter is not only unique but also securely tied to the user's session, significantly reducing the risk of successful CSRF exploits [22] (Fig. 12).

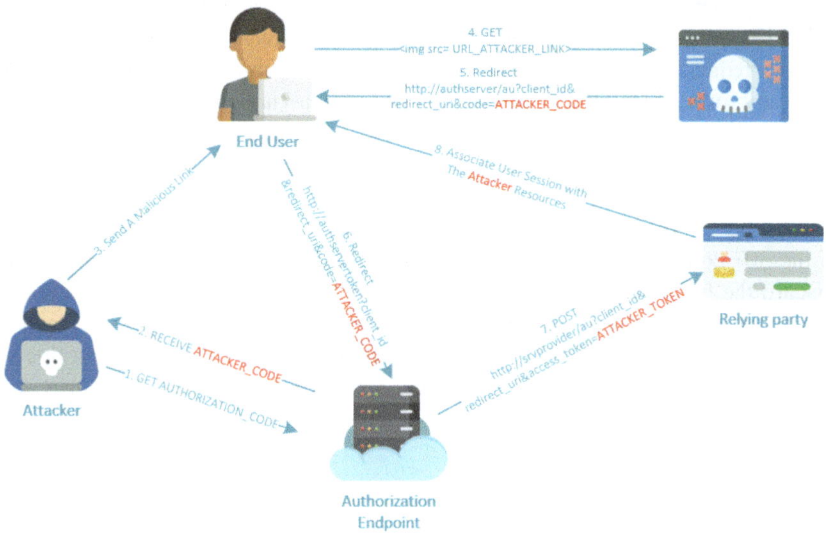

Fig. 12. OAuth 2.0 CSRF attack.

5.1.3 IdP Mix-Up

This particular attack, first identified by Fett et al. in 2016, targets the authentication process in a scenario where a Resource Provider (RP) interacts with multiple Identity Providers (IdPs), one of which is maliciously controlled by an attacker. The core objective of the attack is to deceive the RP into believing that an access token, issued by the attacker's IdP (AIdP), actually originates from a legitimate, Honest IdP (HIdP). This deception enables the attacker to illicitly obtain sensitive user information. The attack becomes feasible under certain conditions [23]:

- The attacker has the capability to proxy HTTP requests and responses between the end-user and the RP.

- The RP engages with multiple IdPs, including at least one compromised by the attacker.
- The RP utilizes a common redirect_uri for these multiple IdPs.

Attack Scenario:

The end-user initiates the authentication process by selecting an IdP on the RP's login page.

- The attacker intercepts this HTTP request, alters the IdP selection from HIdP to AIdP, and redirects it to the RP.
- The RP, unsuspecting of the change, sends an authorization request to the AIdP, including parameters such as client_id, redirect_uri, and state.
- The attacker then modifies the client_id to correspond with that of the HIdP and forwards this manipulated request, retaining the redirect_uri and state, to the HIdP.
- In response, the HIdP issues an authorization response with an access_token or code, directed back to the RP.
- The RP, misinterpreting the state parameter, erroneously attributes this response to the AIdP.
- The RP then attempts to retrieve user information from the AIdP using the received access_token or code. Meanwhile, the attacker leverages this code and access_token to extract sensitive data from the HIdP.

This attack highlights a significant vulnerability in scenarios where RPs engage with multiple IdPs, especially when one of these IdPs is under the control of an adversary (Fig. 13).

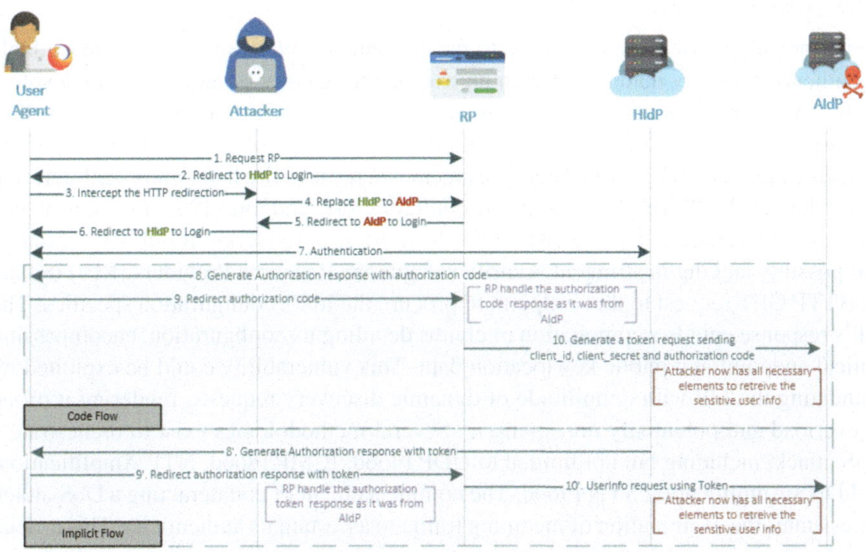

Fig. 13. OAuth 2.0 – IdP Mix-Up attack

To effectively prevent an IdP Mix-Up attack, several robust security measures can be implemented:

- **Unique Redirect URIs for Each IdP**: Assigning distinct redirect_uris to each Identity Provider (IdP) ensures accurate and secure routing of responses, eliminating confusion in attributing responses to the correct source.
- **Adoption of HTTPS Protocol**: Replacing HTTP with HTTPS, which encrypts data during transmission, can significantly reduce the risk of interception, particularly in the critical phase of the attack.
- **Verification of Issuer and Client ID in Responses**: It's essential to include and verify both the issuer and client_id in the authorization response. This should be cross-checked with the information registered with the Honest IdP (HIdP), to confirm the legitimacy of the communication.
- **Whitelisted Authorization Servers**: Maintaining a controlled list of trusted authorization servers narrows down interaction to known, credible entities, thereby mitigating risks associated with compromised or malicious IdPs.

5.2 OpenID Connect

OpenID Connect (OIDC), an extension of the OAuth 2.0 framework for authentication, inherits similar vulnerabilities as detailed in Sect. 5.1. However, OIDC also faces unique security risks, specifically Denial of Service (DOS) attacks and Cross-Site Scripting (XSS). This section will delve into these additional challenges, offering an in-depth exploration of the specific vulnerabilities associated with OIDC.

5.2.1 Denial of Service

The principal aim of these assaults is to incapacitate a service, rendering it unavailable to authentic users by deluging it with an exorbitant quantity of requests or malevolent traffic. An inspection of the process for acquiring configuration data from an OIDC identity provider unveils potential susceptibilities to a denial-of-service (DoS) attack. As delineated in Sect. 3.3.3, OIDC endorses metadata retrieval by hosting its configuration particulars at the "/.well-known/openid-configuration" endpoint [24]. *Frequently, this endpoint is accessible to all clients and relying parties,* rendering it publicly available and possibly lacking in stringent security safeguards. Resource Providers (RPs) initiate an HTTP GET request to this endpoint to procure the IdP's configuration specifics. The IdP's response entails a compilation of claims detailing its configuration, encompassing critical endpoint and public key location data. This vulnerability could be exploited by inundating the IdP with a multitude of dynamic discovery requests, rendering it prone to overload and potentially unresponsive. Several methodologies exist to orchestrate a DoS attack, including but not limited to UDP Flood, ICMP Flood, NTP Amplification, and DNS Amplification SYN Flood. The commencement of counteracting a DoS attack necessitates the examination of incoming traffic to ascertain its authenticity. This crucial initial step aids in determining the appropriate mitigation strategy. Several methods can be employed to mitigate such attacks, each tailored to specific types of threats:

- **Selective IP Allowance and Denial**: This method involves authorizing access for verified, legitimate IP addresses while concurrently blocking those identified as sources of attacks.
- **Rate Limiting**: Implementing a cap on the number of requests from users or specific IP addresses within a set time frame can effectively moderate the influx of potentially harmful traffic.
- **Web Application Firewall (WAF) Implementation**: Deploying a WAF as a reverse proxy offers a shield for the original server. Customizable rules within the WAF can be tailored to address specific vulnerabilities and threat patterns.
- **Upstream Filtering and DDoS Mitigation**: This technique involves the examination of incoming packets, comparing their IP addresses against databases of known attackers and botnets. The aim is to filter out malicious traffic while allowing legitimate data packets to pass through.

5.2.2 Cross-Site Scripting

Cross-Site Scripting (XSS) attacks are a specific type of injection attack where dangerous scripts are embedded into websites generally considered secure. In these attacks, attackers dispatch harmful code, often as client-side scripts, to users who are unaware of the threat. XSS attacks target web applications that improperly handle user input, failing to validate or safely encode it before including it in their output [25]. *This issue is critical because browsers, which cannot differentiate between safe and malicious scripts, end up executing the harmful scripts. This vulnerability is significant, as it allows attackers to access sensitive user data, such as cookies, session tokens, or other* private information stored in the user's browser for the affected website. Studies, including one analyzing Google's implementation of OpenID Connect, have identified vulnerabilities to XSS attacks. There are three main categories of XSS attacks: persistent, non-persistent, and DOM-based. In addressing Cross-Site Scripting (XSS) attacks, a set of basic but effective countermeasures is suggested:

- **Input Validation and Sanitization**: This involves a detailed check and cleansing of user inputs to ensure they do not contain harmful scripts.
- **Cookie Security**: Applying the HttpOnly attribute to cookies to block access through client-side scripts, thus increasing security.
- **Application of Specific Response Headers**: The use of 'Content-Type' and 'X-Content-Type-Options' headers helps specify the content type and reduce MIME-type confusion risk.
- Content Security Policy: Implementing the 'Content-Security-Policy' HTTP response header, crucial for controlling content sources and minimizing XSS attack risks.

5.3 SAML

The XML Signature standard is notably complex, especially in terms of the variability in the placement of the signature and the content being signed within the document. Consequently, this leads to a multitude of variations of the same XML document. This complexity introduces two primary vulnerabilities in the SAML protocol [26].

5.3.1 XML Signature Wrapping

XML Signature Wrapping (XSW) attacks were first described by McIntosh M et al. in 2005. In an XSW attack, the attacker can alter the content of a SAML message without affecting the validity of its signature [27]. *To exploit this vulnerability, the attacker needs an existing, validly signed SAML message. The purpose of XSW is to* relocate the signed content to a part of the document that is not displayed by the application upon opening but is still recognized and processed by the verification logic. Consequently, while the signature remains valid, the application ends up displaying content manipulated by the attacker. The placement of this malicious content varies based on where the original signed data is supposed to be included, which assertion the application processes, and which element is used to sign the SAML message (Fig. 14).

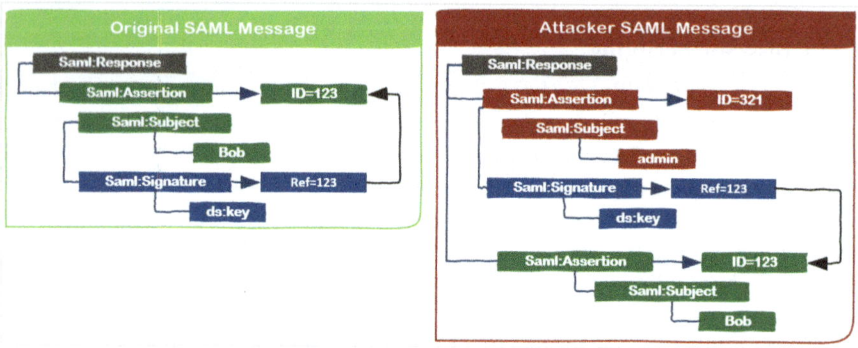

Fig. 14. XSW attack – content change.

To mitigate XML Signature Wrapping (XSW) attacks, several strategies can be implemented, although they might impact the flexibility that SAML protocols typically offer:

- **Structural Fixation**: This method can counteract XSW attacks but may interfere with the flexibility that SAML protocols are designed to provide.
- **Implementing Checks**: on the Receiver Side for SAML Messages: Acting as a filtering mechanism, these checks enhance security by ensuring the integrity and authenticity of the messages. The checks that could be applied include:
  - Element Presence: Verifying if certain expected elements are present in the message.
  - Element Position: Checking the correct positioning of elements within the message.
  - Quantity of Elements: Ensuring the number of specific elements aligns with expected counts.
  - Check Timestamps: Verifying the timestamps to ensure they are within acceptable ranges.
  - Check IDs: Confirming the uniqueness and validity of IDs used in the message.

5.3.2 XML External Entity

SAML's ability to include external sources in XML requests or responses is a key feature. It lets developers use XML data from outside sources in their applications without needing to store it on the same system. While this is useful in some cases, it also introduces a vulnerability called XXE (External Entity Attack). This issue doesn't directly impact SAML but shows the risks involved in using SAML with XML. Attackers can exploit this by using a wrongly set up XML parser. They do this by making a harmful XML document that refers to an external entity. When processed, this can lead to security issues like revealing sensitive information or allowing remote code execution. However, this risk mainly exists when attackers can interact with any type of XML requests. Mitigation measures for XXE include:

- **Disable DTDs (Document Type Definitions):** The most secure method to protect against XXE is to completely disable external entities within DTDs. This step effectively eliminates the primary avenue through which XXE attacks are executed.
- **Consult OWASP Guidelines:** If it's not possible to disable DTDs entirely, referring to the XXE Prevention Cheat Sheet by OWASP is recommended for alternative strategies and best practices. This resource offers detailed guidance on safeguarding against XXE vulnerabilities. The cheat sheet can be accessed at OWASP's GitHub repository[1] (Fig. 15).

```
Request:
<?xml version="1.0" encoding="UTF-8"?>
<!DOCTYPE productcheck [ <!ENTITY refname SYSTEM "file:///etc/passwd"> ]>
<stockCheck><productId>&refname;</productId></stockCheck>
Response:
Invalid product ID: root:x:0:0:root:/root:/bin/bash
daemon:x:1:1:daemon:/usr/sbin:/usr/sbin/nologin
bin:x:2:2:bin:/bin:/usr/sbin/nologin
```

Fig. 15. XXE attack example.

To summary (Table 2):

Table 2. Summary of attacks possible on SSO Protocols.

Attack	OAuth 2.0	OIDC	SAML
DOS	*	*	*
CSRF	*	*	*

(continued)

[1] OWASP's GitHub repository https://github.com/OWASP/CheatSheetSeries/blob/master/cheatsheets/XML_External_Entity_Prevention_Cheat_Sheet.md

Table 2. *(continued)*

Attack	OAuth 2.0	OIDC	SAML
307 Redirect	*	*	-
Idp Mix-UP	*	*	-
XSS	*	*	*
XSW	-	-	*
XEE	-	-	*
Phishing	*	*	*
Replay Attacks	*	*	*
Man-in-the-Middle	*	*	*

The table displays which security vulnerabilities are known to affect OAuth 2.0, OIDC, and SAML protocols. All three protocols are vulnerable to DOS, CSRF, Phishing, Replay Attacks, and Man-in-the-Middle attacks. OAuth 2.0 and OIDC are both susceptible to 307 Redirect and IdP Mix-Up attacks, while SAML is not. XSS attacks are noted as a risk for all three protocols. SAML is exclusively vulnerable to XSW and XEE attacks, which do not affect OAuth 2.0 and OIDC.

6 Conclusion

In summary, this discussion has covered various digital methods and standards for managing identities in cloud environments, such as LDAP, SSH Key, OAuth 2.0, OpenID Connect, and SAML. Each approach has its unique benefits for identity management but also presents specific security challenges. Over time, researchers have developed numerous solutions to overcome these security issues. Looking ahead, future studies will focus on addressing concerns related to privacy [28] **and single points of failure by investigating how Zero-Knowledge Proofs, Blockchain technology, and Machine Learning can improve security in these systems.**

References

1. Maidine, K., El-Yahyaoui, A.: Cloud identity management mechanisms and issues. In: 2023 IEEE 6th International Conference on Cloud Computing and Artificial Intelligence: Technologies and Applications (CloudTech), pp. 1–9. IEEE, November 2023. https://doi.org/10.1109/CloudTech58737.2023.10366178
2. Indu, I., Anand, P.M.R., Bhaskar, V.: Identity and access management in cloud environment: mechanisms and challenges. Eng. Sci. Technol. Int. J. **21**(4), 574–588 (2018). https://doi.org/10.1016/j.jestch.2018.05.010
3. El-Yahyaoui, A., Omary, F.: An improved framework for biometric database's privacy. Int. J. Commun. Netw. Inf. Secur. (IJCNIS) **13**(3), (2022). https://doi.org/10.17762/ijcnis.v13i3.5143
4. Tase, M.G.: Cloud Security: Identity Management Issues

5. El-Yahyaoui, A., Ech-Cherif El Kettani, M.D.: A verifiable fully homomorphic encryption scheme for cloud computing security. Technol. (Basel) **7**(1), 21 (2019). https://doi.org/10.3390/technologies7010021
6. Wilson, Y., Hingnikar, A.: Solving Identity Management in Modern Applications. Apress, Berkeley (2019). https://doi.org/10.1007/978-1-4842-5095-2
7. Xu, R., Yang, S., Zhang, F., Fang, Z.: MISO: legacy-compatible privacy-preserving single sign-on using trusted execution environments. In: 2023 IEEE 8th European Symposium on Security and Privacy (EuroS&P), pp. 352–372. IEEE, July 2023. https://doi.org/10.1109/EuroSP57164.2023.00029
8. Jones, M., Hardt, D.: The OAuth 2.0 Authorization Framework: Bearer Token Usage. https://tools.ietf.org/html/rfc6750
9. Flaws_in_Oauth_2_0_Can_Oauth_be_used_as
10. Hardt, D., Parecki, A., Lodderstedt, T.: The OAuth 2.1 Authorization Framework. https://datatracker.ietf.org/doc/html/draft-ietf-oauth-v2-1-08
11. Yousra, B., Yassine, S., Yassine, M., Said, S., Lo'ai, T., Salah, K.: A novel secure and privacy-preserving model for OpenID connect based on blockchain. IEEE Access **11**, 67660–67678 (2023). https://doi.org/10.1109/ACCESS.2023.3292143
12. Sakimura, N., Bradley, J., Jones, M., De Medeiros, B., Mortimore, C.: OpenID Connect Core 1.0 incorporating errata set 1. https://openid.net/specs/openid-connect-core-1_0.html
13. Navas, J., Beltrán, M.: Understanding and mitigating OpenID Connect threats. Comput. Secur. **84**, 1–16 (2019). https://doi.org/10.1016/j.cose.2019.03.003
14. Lewis, J.E., Lewis, K.D.: Web single sign-on authentication using SAML (2009). https://www.researchgate.net/publication/45872317
15. Somorovsky, J., et al.: On Breaking SAML: Be Whoever You Want to Be. www.wso2.com
16. Cabarcos, P.A., Mendoza, F.A., Marín-López, A., Díaz-Sánchez, D.: Enabling SAML for Dynamic Identity Federation Management
17. Talwana, J.C., Huang, J., Mulumba, C.M., Mijom, J.: Shibboleth based on SAML SSO Authentication 1 Shibboleth based on SAML SSO Authentication
18. Radha, V., Reddy, D.H.: A survey on single sign-on techniques. Procedia Technol. **4**, 134–139 (2012). https://doi.org/10.1016/j.protcy.2012.05.019
19. Naik, N., Jenkins, P.: Securing Digital Identities in the Cloud by Selecting an Apposite Federated Identity Management from SAML, OAuth and OpenID Connect
20. Li, W., Mitchell, C.J.: Security issues in OAuth 2.0 SSO implementations
21. Li, X., Xu, J., Zhang, Z., Lan, X., Wang, Y.: Modular security analysis of OAuth 2.0 in the three-party setting. In: Proceedings - 5th IEEE European Symposium on Security and Privacy, Euro S and P 2020, pp. 276–293. Institute of Electrical and Electronics Engineers Inc., September 2020. https://doi.org/10.1109/EuroSP48549.2020.00025
22. Li, W., Mitchell, C.J., Chen, T.: Mitigating CSRF attacks on OAuth 2.0 and OpenID Connect, January 2018. http://arxiv.org/abs/1801.07983
23. Li, W., Mitchell, C.J.: Does the IdP Mix-Up attack really work? (2016). http://self-issued.info/?p=1524
24. Fett, D., Hosseyni, P., Kuesters, R.: An Extensive Formal Security Analysis of the OpenID Financial-grade API, January 2019. http://arxiv.org/abs/1901.11520
25. Gupta, S., Gupta, B.B.: Cross-Site Scripting (XSS) attacks and defense mechanisms: classification and state-of-the-art. Int. J. Syst. Assur. Eng. Manag. **8**, 512–530 (2017). https://doi.org/10.1007/s13198-015-0376-0

26. Kaur, K., Bansal, D.: Security Vulnerabilities in SAML based Single Sign-On Authentication in Cloud (2013)
27. Jurreit, J., Fehrenbach, P., Kaspar, F.: Analysis of security vulnerabilities in Microsoft Office 365 in regard to SAML
28. Alenizi, B.A., Humayun, M., Jhanjhi, N.Z.: Security and privacy issues in cloud computing. J. Phys. Conf. Ser. (2021). https://doi.org/10.1088/1742-6596/1979/1/012038

Power Consumption in HPC-AI Systems

Roblex T. Nana[1(✉)], Claude Tadonki[1], Petr Dokladal[2], and Youssef Mesri[3]

[1] Mines Paris - PSL, Centre de Recherche en Informatique (CRI), Fontainebleau, France
{roblex.nana_tchakoute,claude.tadonki}@minesparis.psl.eu

[2] Mines Paris - PSL, Centre de Morphologie Mathématique (CMM), Fontainebleau, France
petr.dokladal@minesparis.psl.eu

[3] Mines Paris - PSL, Centre de Mise en Forme de Matériaux (CEMEF), Sophia Antipolis, France
youssef.mesri@minesparis.psl.eu

Abstract. Beside the need for speed that has led to HPC, energy has become a crucial concern that is addressed in HPC hardware and software solutions. There are several aspects when it comes to energy in computing systems: *cost, source, heat, carbon, lifetime*, and more. From the standpoint of *embedded systems*, devices are battery-powered, thus the available amount of energy to proceed with is limited. In critical cases like these, it is vital to optimize all sources of substantial power consumption. Regarding more standard computing systems, including supercomputers, the question of energy saving mainly translates into *electricity cost* and *carbon emission*. Indeed, the overall energy required to run an HPC infrastructure including cooling systems represents an important part of the maintenance budget. In addition, considering *heat dissipation*, lifetime of the hardware is reduced as well as MTBF (mean time between failures). One the major topic from the application standpoint that HPC has to consider carefully is *artificial intelligence* (AI). Indeed, this processing paradigm has tremendously grew up with more and more ambitious perspectives. The pervasiveness of AI solutions and the noticeable computing time required for at least the training phases exacerbate the concern of power consumption in this context. The purpose of this chapter is to explore and illustrate the issue of power consumption in HPC-AI systems so as to make the issue more clear to the reader and highlight the main solutions and perspectives.

Keywords: HPC · AI · energy · power consumption · carbon footprint

1 Introduction

In the current era of noteworthy technological advances, the rapid proliferation of High-performance Computing (HPC) devices and Artificial Intelligence (AI) applications has raised in transformative capabilities across industries. These

M. Zbakh et al. (Eds.): CloudTech 2023, LNNS 1220, pp. 89–116, 2024.
https://doi.org/10.1007/978-3-031-78698-3_6

systems, characterized by their immense computational power and data processing efficiency, have become indispensable tools driving innovation in fields including healthcare, finance, scientific research and entertainment. The HPC market is poised for significant growth, projected to reach USD 49.9 billion by 2027 with a compound annual growth rate (CAGR) of 6.7%, and USD 78.254 Billion by 2032 with 7.1% of CAGR [27].

However, great power comes with responsibility, and the unprecedented growth of HPC-AI systems has brought to the forefront pressing concerns regarding their energy consumption and environmental impact. For example, in 2020, the information and communication technologies sector accounted for approximately 1.8% to 2.8% of greenhouse gas (GHG) emission, a figure that surpassed even the emissions from the aviation sector [13]. As these systems evolve to meet the ever-increasing demands for computational resources, they exhibit a corresponding escalation in energy requirements, rising significant challenges in sustainability and resource management. The International Energy Agency (IEA) predicts that by 2026, data centers could consume twice as much electricity, mainly due to the rise of cryptocurrencies and artificial intelligence (AI) [20]. Data centers are crucial for storing vast amounts of digital content, and their proliferation for activities like Bitcoin mining and AI training has raised concerns about their environmental impact. In 2022, data centers, cryptocurrencies, and AI jointly accounted for 2% (using 460TWh) of the global electricity demand, with crypto mining alone responsible for nearly the quarter of that consumption. The IEA estimates that by 2026, data centers could use up to 1,050TWh of electricity, which is equivalent to adding the electricity demand of a country like Sweden or Germany [20]. Figure 1 show another global energy demand for ICT sector for 2030.

Over the years, the fastest HPC systems, as ranked by the Top500 list[12], have not only increased in speed but have also become more energy-intensive. For instance, in November 2007, the top system consumed 2.3 MW, rising to 8.2 MW in November 2012. By November 2017, the consumption surged to 15.4 MW for the leading supercomputer system. The trend continued with the first Exascale system deployed in 2022, consuming over 29 MW. While future technology advancements may lower this figure, there's a clear pattern of an increasing power consumption in HPC systems. The significant power consumption of HPC systems of top500 ranking in 2021 is estimated at 600 MW (or 5.2 terawatt-hours per year, with the top ten systems using around 100 MW). Their carbon footprint is about two million metric tons of carbon per year, equivalent to around 285,000 average households in USA [31]. As most electrical energy is generated from fossil fuels, this trend significantly contributes to carbon emissions. Operating these systems with maximum energy efficiency is crucial to minimize their environmental impact. Considering a conservative estimate, in the next five years, the power consumption of top HPC systems could still notably surpass the current levels, keeping the trend of increasing energy demand and carbon emissions.

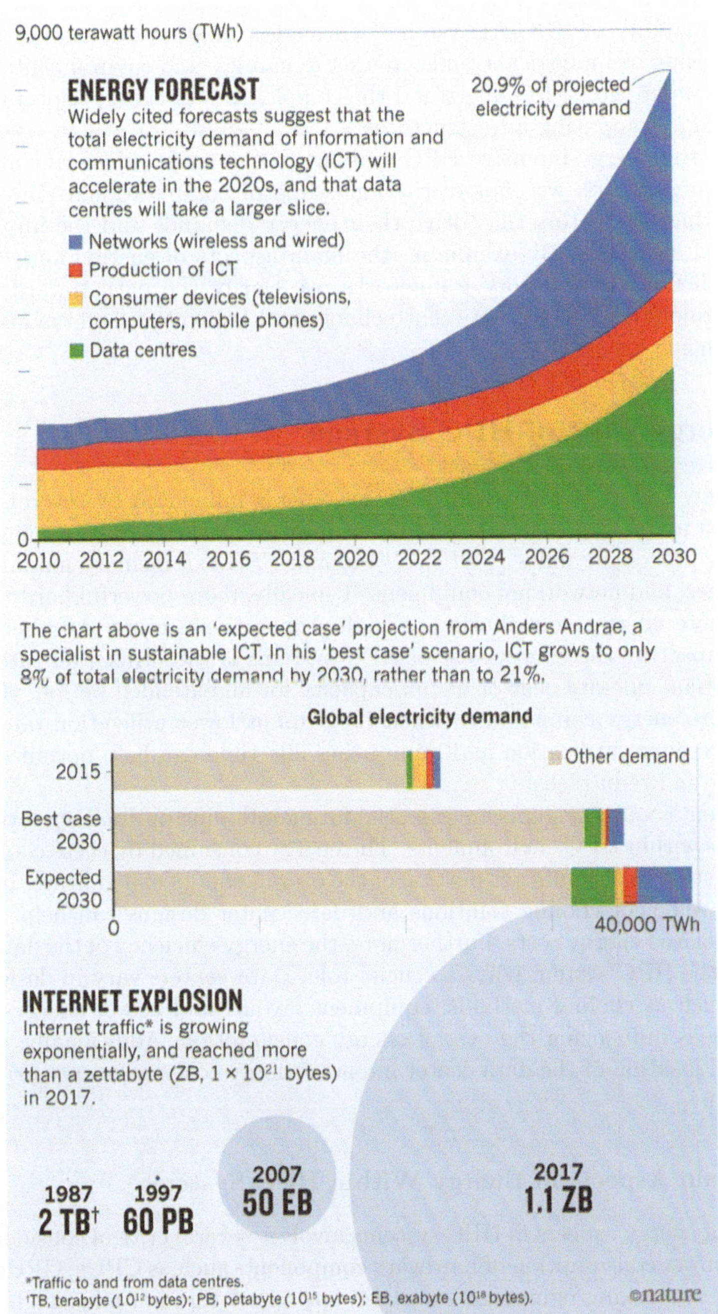

Fig. 1. Forecasting the demand for energy in the global electricity market, considering the impact of the internet boom. Source: [21]

This paper explores power consumption, energy efficiency, and environmental sustainability of HPC-AI systems. Through a multidimensional exploration encompassing technological trends, market dynamics, and environmental considerations, we are going to understand the complexities of energy aspects in these cutting-edge computing infrastructures.

From the energy-intensive HPC systems to the latest advancements in AI-driven data centers, we explore the landscape of computational powerhouses. Highlighting the factors that drive their energy demands and the implications on carbon emissions. By examining the main aspects of energy, financial cost, case studies, and regulatory frameworks, we clarify the path towards achieving a harmonious balance between technological innovation and environmental management.

2 Energy Cost of HPC Systems

The energy cost associated with HPC systems is influenced by several key factors. First and foremost is the power consumption of the hardware components, including processing units *(CPUs, GPUs, and FPGAs)*, memory modules, storage devices, and networking equipment. Typically, more powerful hardware consumes more energy, contributing to higher operational costs. Another critical factor impacting the energy cost of HPC systems is the utilization rate. When HPC systems operate near or at full capacity for an extended period, they consume more energy compared to when they run at lower utilization rates. Maximizing resource utilization and minimizing idle time can help optimize energy efficiency and reduce costs.

Efficient cooling systems are essential for maintaining optimal operating temperatures within HPC environments. The energy consumed by these cooling systems constitutes a significant portion of the overall energy expenditure. Investing in energy-efficient cooling solutions and data center designs can help mitigate cooling-related energy costs. Furthermore, the energy efficiency of the data center housing the HPC system plays a crucial role. Data centers vary in design, with factors such as cooling methods, equipment layout, and use of energy-efficient technologies influencing the overall energy consumption. Additionally, the geographical location of the data center impacts energy costs, as energy prices vary by region.

2.1 Main Aspects of Energy Within HPC Systems

Managing energy aspects in HPC systems involves several critical considerations. Firstly, power consumption of hardware components such as CPUs, GPUs, memory, and networking equipment must be optimized through efficient hardware selection and optimization strategies. Secondly, efficient cooling infrastructure is crucial to cool down the significant heat generated by HPC systems, requiring advanced cooling solutions. Efficient heat dissipation and thermal management techniques, including temperature monitoring and control, ensure the reliability and longevity of HPC hardware. Implementing energy-efficient cooling solutions and optimizing data center design contribute to the overall energy efficiency while maintaining adequate cooling performance. Integration of renewable energy sources and efficient power distribution systems further enhance sustainability and reduce environmental impact. Let us see some details through different components of the system.

1. **GPU (Graphics Processing Unit):** The energy consumption of GPUs in HPC systems depends on various factors. These include the GPU architecture and manufacturing process, with newer architectures often offering improved energy efficiency due to advancements in design and fabrication technology. The workload running on the GPU also affects energy consumption, as different applications have varying computational requirements. Additionally, factors such as *GPU utilization rates, clock speeds*, and *memory usage influence power usage.* Optimizing GPU utilization, employing power management features, and utilizing efficient algorithms and software can help reduce the energy consumption of GPUs in modern HPC systems. Furthermore, advancements in GPU technology, such as low-power modes and hardware acceleration for specific tasks, contribute to improving energy efficiency in HPC applications.
2. **CPU (Central Processing Unit):** The energy consumption of CPUs is influenced by the CPU architecture and manufacturing process as similar to GPUs components. Newer generations typically offer improved energy efficiency through advancements in design and fabrication technology, such as *smaller transistors* and *architectural optimizations* [37]. Additionally, the *workload* running on the CPU affects energy consumption, as different types of computation and processing tasks require varying kinds of resource and levels. Factors such as *CPU utilization rates, clock frequencies*, and *instruction-level parallelism* also impact energy usage. Techniques such as *Dynamic Voltage and Frequency Scaling (DVFS)*, which adjust CPU voltage and frequency based on workload demands, as well as power management features like *sleep states* and *idle power management*, can help optimize energy consumption in modern HPC systems. Furthermore, advancements in heterogeneous computing, where CPUs are combined with accelerators like GPUs or FPGAs, offer opportunities for energy-efficient parallel processing and improved performance-per-watt in HPC workloads.
3. **DRAM (Dynamic Random Access Memory):** The energy consumption of RAM in modern HPC systems is influenced by several factors. The type of RAM technology used, such as DDR4 or DDR5, can impact energy efficiency, with newer generations showing better improved power efficiency. RAM capacity also plays a role, as larger memory configurations typically consume more energy. The power consumption of DRAM exhibits a non-linear growth pattern, transitioning from sub-linear at low capacity to super-linear at high capacity. For instance, while a 16× increase in capacity from 8 GB to 128 GB results in a modest 3× rise in power consumption, the subsequent

4× increase to 1 TB leads to a significant 10× energy rise [5]. This increase is attributed to the necessity of using high-density DRAM DIMMs for achieving high capacities, coupled with the considerably higher power consumption of these high-density DIMMs compared to their low-density. Furthermore, memory access patterns and utilization rates affect energy consumption, with higher utilization leading to increased power usage. RAM consumes power continuously to maintain data stored in memory cells. While the power consumption of individual RAM modules remains relatively constant regardless of the workloads, the total power consumption can increase with the amount of RAM installed. Optimizing memory usage, employing power-saving features such as low-power modes, and utilizing advanced memory management techniques can help reduce the energy consumption of RAM in modern HPC systems.

4. **Storage:** Energy consumption in storage devices depends on the type of storage technology used. Traditional Hard Disk Drives (HDDs) consume significant power while spinning their platters and moving read/write heads, regardless of data access activity. Solid State Drives (SSDs), such as flash drives, consume less power compared to HDDs since they have no moving parts and only consume power when reading from or writing to memory cells. The capacity and speed of the storage system also impact energy usage, as larger and faster storage solutions tend to consume more power. Additionally, factors such as *data access patterns, utilization rates*, and *efficiency of storage protocols and software* can affect energy consumption. Optimizing storage architectures, employing energy efficient hardware, implementing data management strategies, and leveraging power saving features can all contribute to reducing the energy footprint of storage in HPC systems. Hybrid storage solutions *(combining HDDs and SSDs)* offer a balance between time performance and energy efficiency.
5. **Network Interface:** Network components, such as Ethernet controllers or Wi-Fi adapters, consume power during data transmission and reception. The energy consumption of network interfaces in modern HPC systems is influenced by factors such as the *type of interface used, data transfer rates, link utilization, networking technologies, protocols, software efficiency, network topology, system scaling*, and *power management features.* Varying data transfer rates and levels of link utilization affect power usage. Emerging technologies like *RDMA* and *efficient networking protocols* can mitigate energy consumption. Advanced power management features, such as *low-power modes* and *wake-on-LAN*, help reduce energy consumption during idle periods or low network traffic. Additionally, *network topology optimization, parallelism*, and *power management features* play crucial roles in minimizing energy in HPC systems. Efficient management of these factors is essential for improving the overall energy efficiency of HPC networks.
6. **Cooling System:** The escalating power consumption of modern CPUs, reaching levels of 300 W and projected to reach 500–750 watts [28], presents a significant challenge for data center infrastructure. 750 W GPUs packages are already being designed by AMD for Instinct MI300X APU series [4]. Air

cooling, traditionally used to dissipate heat, is reaching its limits in effectively managing these high power densities. Liquid cooling emerges as a dominant solution to address thermal challenges efficiently. Implementing an effective end-to-end liquid cooling solution becomes increasingly crucial in this context. Liquid cooling offers superior thermal conductivity compared to air, allowing for more efficient heat dissipation. By directly cooling components like CPUs and GPUs with liquid, data centers can better manage the escalating power demands of modern hardware while maintaining an optimal operating temperature. Furthermore, liquid cooling enables more precise temperature control, which is essential for maximizing performance and extending the lifespan of hardware components. However, deploying liquid cooling systems requires careful planning and investment in infrastructure to accommodate the additional complexity. The maintenance requirements associated with such systems is another factor to care about. As data center technology continues to advance, the adoption of liquid cooling solutions will likely become more widespread to meet the demands of increasingly power-hungry hardware, while ensuring optimal performance and reliability.

Unlike individual HPC system, which have a relatively short lifespan of around 5 years, the infrastructure of a data center typically lasts much longer, ranging from 20 to 30 years with only minor changes. Understanding the infrastructure provides insight about resources and key performance indicators (KPIs) that may not be visible at the system or application level. One notable aspect to mention is the allocation of resources based on maximum system power draw. Infrastructure components are designed to endure constant power draw from all components at their maximum load, even though this maximum load is rarely reached during regular operation. This approach ensures stability and readiness for peak performance scenarios, such as on benchmarking sessions like with the High Performance Linpack (HPL).

Aligning infrastructure with the needs of the HPC systems is crucial for optimizing performance and resource utilization. This also underscores the importance of efficient interfaces between infrastructure and the systems they support, all the way up to the application level. By understanding and optimizing these interfaces, HPC centers can better manage their resources and improve overall performance. Efforts to optimize algorithms and software for energy efficiency can also yield substantial savings.

2.2 Financial Cost Related to the Energy of HPC Systems

In addition to the environmental concerns associated with the rise of power consumption, there is a financial impact. A large portion of the budget allocated for HPC systems is allocated to electricity bills rather than hardware, impacting computational performance. HPC systems entail several costs related to *energy, equipment, maintenance*, and more. Here's a breakdown:

1. **Electricity Costs:** The primary cost for HPC systems is related to electricity used to power the *servers, cooling systems*, and other infrastructure components. This cost can be significant, especially for large-scale installations. Cooling systems are essential to prevent overheating of HPC hardware. They consume additional electricity and contribute to energy costs. Efficient power distribution systems are necessary to ensure reliable electricity supply for HPC systems. This includes *transformers, distribution panels*, and *cabling*, all of which contribute to energy consumption. The latest trends in operational costs reveal contrasting shifts: while power costs per kilowatt-hour remain on an upward trajectory for most operators, there has been a significant decrease in the number experiencing per-unit cost hikes for IT hardware compared to 2022 as show on Fig. 2. However, this favorable trend is expected to reverse soon. Specifically, the prices of certain chips, such as GPUs essential for AI training models, are anticipated to rise due to supply constraints. This impending rise in chip costs may indirectly impact the IT hardware market for data centers, as component vendors endeavor to integrate AI capabilities into their offerings.
2. **Equipment Costs:** The initial investment in HPC hardware, including servers, storage systems, networking equipment, and specialized components such as GPUs or accelerators, represents a substantial portion of the total cost. Beyond computing hardware, infrastructure components like *racks, cabinets, power supplies*, and *cooling units* contribute to equipment costs. As example, Sam Altman's, the OpenAI's CEO, as reported by The Wall Street Journal, seeks to mobilize 7 trillion USD from investors to tackle the global shortage of semiconductor chips [2]. This ambitious initiative aims to dramatically enhance chip-building capacity worldwide, with a particular focus on addressing the scarcity of AI chips such as Nvidia H100 [3]. Notably, Altman's strategy resonates with his ongoing investments in clean energy, given the massive energy consumption associated with AI facilities. However, important challenges lie ahead, including navigating the geopolitical complexities and persuading investors to establish chip factories in the U.S., despite hurdles like a shortage of skilled and high labor costs. Despite being labeled as wildly ambitious and optimistic, if successful, Altman's unprecedented initiative could fundamentally reshape the dynamics of the semiconductor market.
3. **Maintenance Costs:** Maintenance routine is required to ensure optimal performance and reliability of HPC systems. This includes *cleaning, inspection*, and *replacement of components* as needed. Inevitably, hardware failures or malfunctions occur, necessitating repair or replacement of faulty components. The associated labor and replacement costs contribute to the overall maintenance expenses. Periodical upgrades of hardware or software may be necessary to keep system functional with evolving technology and user demands. These upgrades incur additional costs but are essential for maintaining system efficiency and competitiveness.
4. **Personnel Costs:** Personnel costs include salaries for system administrators, network engineers, and other technical staff responsible for managing and maintaining HPC systems. Ongoing training and professional development

are essential to keep the staff up-to-date with the latest technologies and best practices, adding to staff costs.

5. **Facility Costs:** The physical space required to house HPC systems, including data centers or server rooms, may incur costs associated with *rental, construction*, or *renovation.* Beside cooling, facilities must maintain appropriate environmental conditions, including humidity and air quality, to ensure the reliable operation of HPC hardware.

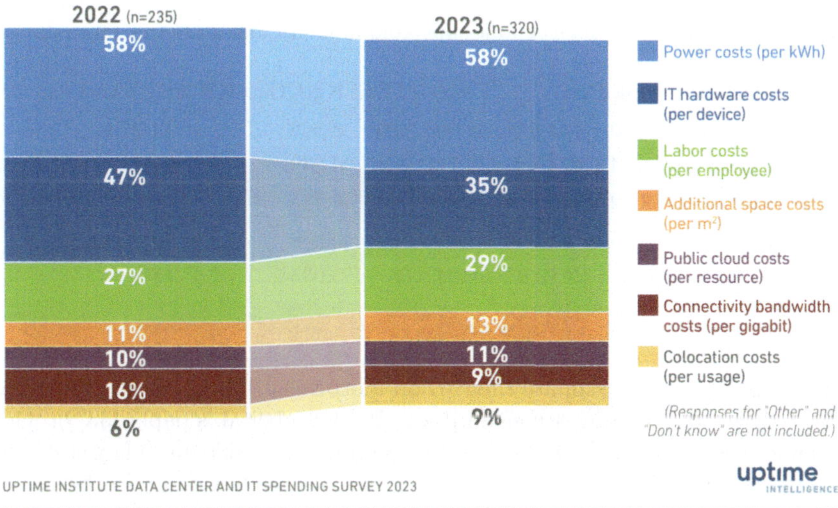

Fig. 2. Cost increase in data center by UPTIME INSTITUTE. [42] **"n"** is the number of respondent of the survey.

2.3 Carbon Footprint

The conventional metric for energy efficiency, measured in FLOPS/Watt, falls short in offering a thorough evaluation of the environmental footprint of HPC systems. This is because it neglects crucial factors like the embodied carbon and the carbon intensity of the energy utilized. Simply evaluating energy efficiency does not fully reflect the operational carbon footprint of an HPC system, as this footprint also depends on the carbon intensity of the energy used. For instance, a system with higher energy efficiency may still have a higher operational carbon footprint if it relies on energy sources with a higher carbon footprint. Therefore, the environmental impact of HPC systems cannot be solely determined by their energy efficiency metrics.

While acknowledging the meaning of cost considerations in the hardware upgrades, it is essential to note that low-cost systems can potentially offer a lower carbon footprint depending on factors like *system load* and *carbon intensity.* However, incorporating hardware costs into studies presents challenges due to a range of factors such as *geographical location*, *system scale*, and *market fluctuations*. This complexity underscores the need for careful consideration and the use of conservative assumptions when integrating these costs. Although the cost can be a valuable aspect to consider in sustainability analyses, its variability across different contexts underscores the need for careful and nuanced examination to ensure accurate and meaningful conclusions. Table 1 provides an illustration with the impact of carbon intensity of a region on total equivalent emissions estimate.

Table 1. CO_2 per hour for the first supercomputers of Jun 2024 top500 ranking.

Machine	Peak Perf.	Power (TDP)	Kg(CO_2)/KWh	Kg(CO_2)
AURORA	1.980 EFLOPS	38.698 MW	0.369	14 279
FRONTIER	1.714 EFLOPS	22.786 MW	0.369	8 408
FUGAKU	537.2 PFLOPS	29.899 MW	0.485	14 501
LUMI	531.51 PFLOPS	7.107 MW	0.079	561
ALPS	353.75 PFLOPS	5.194 MW	0.035	181

The typical way to estimate the carbon footprint of IT infrastructure activities is to derive it from power consumption. Patterson et al.'s paper [33] provides a valuable study of the carbon footprint of computing workloads. They state that CO_2 equivalent emissions (CO_2e) account for carbon dioxide and other greenhouse gases such as methane and nitrous oxide. This equivalent emission can be calculated from the electric power by multiplying it with the carbon intensity of the energy supply, as expressed through Eq. 1:

$$CO_2e = W_h \times (CO_2e_{per\ W_h}) \tag{1}$$

Carbon intensity (CO_2e per Wh) represents the amount of carbon dioxide (CO_2e) released to produce one watt-hour of electricity. The average data-center carbon emissions in 2020 were 0.429 ton per Megawatt hour (MWh), but the gross CO_2e per MWh can be 5× lower in some specific data centers [33].

Detailed modeling of the embodied carbon footprint, examination of the regional carbon intensity effects, and a comprehensive end-to-end characterizations of the carbon footprint throughout the system life cycle, can provides valuable insights into environmental impacts. A methodology and the findings presented by Li et al. [24] aim to encourage HPC researchers to prioritize sustainability efforts, promoting more research towards achieving carbon neutrality in the HPC community and contributing to the environmental initiatives.

3 Overview of the HPC as a Support for AI

HPC infrastructures for AI typically consist of specialized hardware and software configurations designed to efficiently handle the computational demands of AI workloads. These infrastructures are essential for training and deploying complex AI models, such as deep neural networks, across various domains like *computer vision, natural language processing*, and *reinforcement learning*. Here is an overview of the key components and characteristics of HPC infrastructures for AI:

1. **Hardware Accelerators**: HPC infrastructures often incorporate hardware accelerators like Graphics Processing Units (GPUs), Tensor Processing Units (TPUs), or Field-Programmable Gate Arrays (FPGAs). These accelerators are optimized for highly parallel tasks, making them ideal for accelerating the training and inference of AI models.
2. **High-performance Computing Clusters**: HPC infrastructures typically consist of clusters with thousands of interconnected compute nodes. Each equipped with multi-core CPUs and one or more accelerators. Interconnect technologies like InfiniBand or Ethernet provide high-speed communication between nodes, enabling efficient distributed computing.
3. **Storage Systems**: HPC infrastructures require robust and scalable storage systems to handle large volumes of data generated during AI training and inference. These storage systems may include high-speed parallel file systems, distributed storage solutions, and object storage systems that can accommodate petabytes of data.
4. **Software Stack**: HPC infrastructures for AI rely on a comprehensive software stack that includes frameworks, libraries, and tools optimized for AI tasks. The major frameworks include TensorFlow, PyTorch, and MXNet, which provide high-level abstractions for building and training AI models. Additionally, specialized libraries for distributed computing, such as MPI (Message Passing Interface), are often used to optimize performance across HPC clusters.
5. **Containerization and Orchestration**: Containerization technologies like Docker and orchestration platforms like Kubernetes are increasingly used in HPC infrastructures for AI. Containers provide a lightweight and portable way to package AI applications and their dependencies, facilitating reproducibility and scalability across diverse HPC environments.
6. **Job Scheduling and Resource Management**: HPC infrastructures employ sophisticated job scheduling and resource management systems to efficiently allocate compute resources and prioritize AI workloads. These systems ensure that compute nodes are utilized optimally and that AI jobs are executed in a timely manner (Fig. 3).

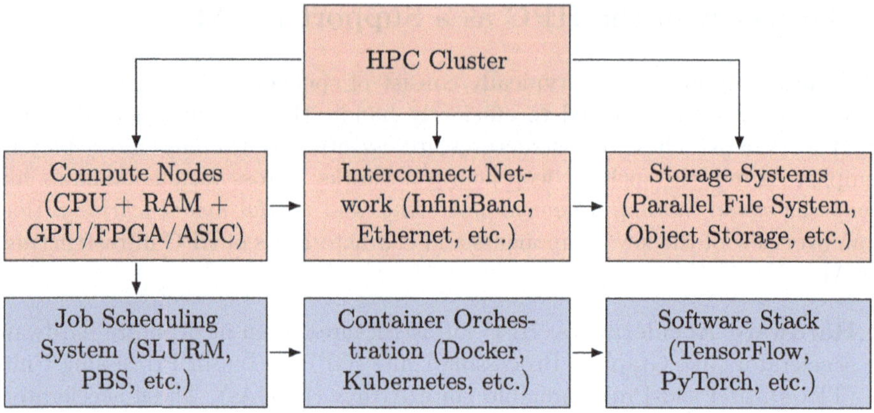

Fig. 3. Illustration of HPC-AI architecture from hardware (red) and software (blue) standpoints (Color figure online)

4 Energy of HPC-AI Solutions

AI system development cycle typically consists of four stages: data collection and preprocessing, model training, model deployment, and model updating. Training and Inferences are the most computational and power hungry part for this process, as depicted in Fig. 4. The intensity of model usage, particularly in terms of how frequently the compressed model is utilized, significantly impacts the overall energy consumption. The concept of model usage intensity (MUI) refers to the number of inferences conducted in one life-cycle. If the model requires frequent updates and only a few inferences can be made in each life cycle, a large portion of the total energy cost will arise from the training phase. Conversely, if the number of inferences per cycle is high, the energy cost of the inference will dominate the overall energy consumption. The assertion that inference can surpass the cost of training in pervasive systems, with inference accounting for up to 90% of machine learning costs for deployed AI systems [10], is consistent with observations in the field. In many real-world scenarios, especially in pervasive systems where models are deployed at scale, the cost of inference can indeed outweigh the cost of training. This is primarily due to the continuous and often high-volume nature of inference tasks in deployed systems, leading to significant computational and energy expenses.

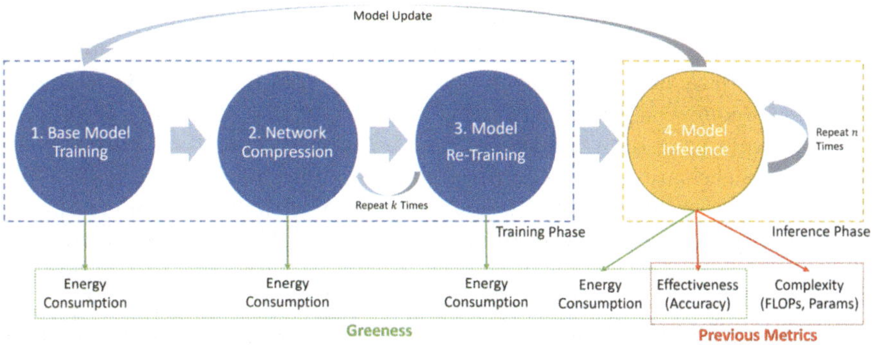

Fig. 4. Full Life cycle of efficient deep learning model [25]

Many existing approaches focus solely on obtaining energy-efficient models for inference, neglecting the energy costs associated with the earlier stages of the life cycle. To facilitate a fair and comprehensive comparison of the energy efficiency of different deep learning methods, it's crucial to consider the energy costs across the entire life cycle and under varying levels of MUI. This holistic approach ensures a more accurate assessment of the energy efficiency of AI systems.

HPC solutions for AI have unique characteristics that impact energy consumption and power supply. Let's delve into these particularities and provide illustrations:

- **Workload Characteristics:** Tasks associated with AI, especially deep learning training and inference, frequently involve processes like matrix multiplications and nonlinear operations. These operations align well with parallel processing capabilities offered by GPUs. These workloads are characterized by irregular computation patterns and large datasets [22,32]. Traditional HPC w.r.t Scientific computing workloads typically involve simulations, modeling, and numerical computations, often based algorithms that are highly parallelizable [39]. These workloads may vary in terms of computational intensity but typically involve regular computation patterns and well-defined algorithms.
- **Hardware Accelerators:** AI workloads heavily rely on specialized hardware accelerators such as GPUs, TPUs (Tensor Processing Units), and FPGAs (Field Programmable Gate Arrays) for efficient parallel processing of neural networks. Scientific HPC often utilizes general-purpose processors (CPUs) and sometimes accelerators like GPUs for specific tasks. However, the reliance on specialized accelerators is typically less pronounced compared to AI workloads. For example, modern data center GPUs can consume 2 times power than CPUs per SoC unit as illustrated in Fig. 6. Note about this comparison that a modern HPC-AI cluster node generally use 8 GPUs SoC with 1 or 2 CPUs SoC.

- **Data Movement and Communication:** AI workloads involve frequent data movement between processing units, such as GPUs and memory modules. This data movement incurs energy costs. Traditional HPC workloads may also require data movement, but the patterns are often more predictable and the volume of data transferred may be lower compared to AI workloads. The differences are indeed significant when it comes to data handling. When it comes to AI-related tasks, it is common to deal with a large number of small files that are gathered and examined in real-time. This contrasts with HPC, where data processing typically occurs in batches with pre-existing datasets. The nature of AI workloads demands a file system that is capable of managing these small files efficiently. Each file generates metadata, and when their volume is above a certain threshold, performance can be severely impacted. Therefore, having a scalable metadata management engine becomes crucial for storing AI workload data effectively. Moreover, AI operates in real-time, analyzing data as they are being generated from sensors or devices. This requires not only high performance to handle incoming data quickly but also excellent output performance to provide timely results. Consequently, the file system supporting AI applications must excel in both aspects. Bandwidth is another critical factor for both HPC and AI, but the difference lies in the size and real-time nature of data in AI workloads. Streaming numerous small files at high speed can strain the metadata performance of the file system. Therefore, addressing this challenge is essential to ensure efficient processing of AI workloads. In summary, the unique characteristics of AI workloads, such as real-time processing and handling of millions of small files, rises challenges to file systems compared to traditional HPC applications. Addressing these challenges is vital for optimizing performance and responsiveness in AI environments.
- **Software Stack and Algorithm Complexity:** AI workloads use complex deep learning algorithms and frameworks like TensorFlow, PyTorch, or CUDA for GPU programming. These frameworks are optimized for parallel processing and often require specialized optimization techniques. Scientific computing typically involves using established numerical libraries and algorithms tailored to specific domains. While some of these workloads may be computationally intensive, the algorithms are often well understood and optimized for traditional HPC architectures.
- **Energy Efficiency Challenges:** In the landscape of HPC for AI, the energy efficiency challenge focus to the optimization of specialized hardware accelerators such as GPUs, alongside the effective management of energy costs associated with data movement and communication overhead. Conversely, energy efficiency challenges in traditional HPC systems center on enhancing CPU utilization, minimizing idle power consumption, and implementing efficient cooling solutions to dissipate the heat generated by high-performance computing nodes. This juxtaposition underscore distinct focuses between the two types of systems: one emphasizing hardware accelerator optimization and energy-aware data handling, while the other prioritizes CPU-centric efficiency and thermal management.

- **Power Supply Considerations:** Power supply considerations in HPC for AI involve ensuring adequate power delivery to GPU clusters and specialized accelerators, which often have high power requirements. As example, the Nvidia DGX H100 server node, powered by eight H100 GPUs, projects a maximum power consumption of approximately 10.2 kW [30]. Figure 5 show's a breakdown of the physical server's node power consumption for modern HPC-AI datacenter. More than 55% of power is used for the 8 × 700 Watt GPUs. In addition, managing power distribution and cooling infrastructure to support these accelerators is crucial. Power supply considerations in traditional HPC systems focus on delivering stable power to CPU clusters and ensuring efficient distribution and cooling to maintain optimal operating conditions.

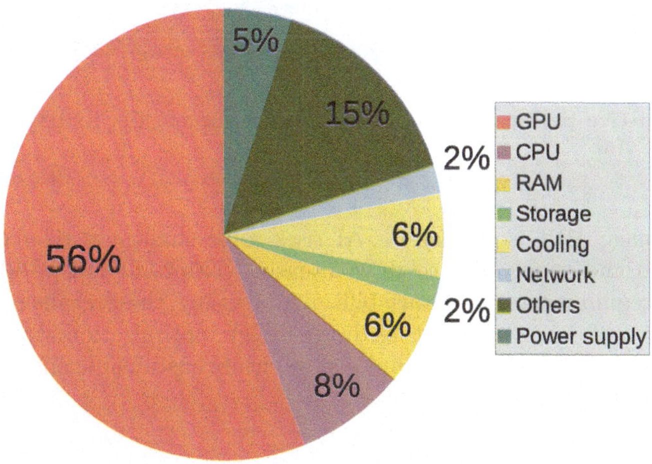

Fig. 5. Typical power breakdown inside a modern data center server node.

Illustrations
In an AI-focused HPC environment, training a large deep learning model on a GPU cluster may require significant power consumption due to the large processing demands and data movement overhead. In contrast, a traditional HPC system running scientific application may have a more predictable power consumption profile, primarily driven by CPU utilization. Optimizing energy efficiency in both cases involves different strategies tailored to specific workload characteristics and hardware configurations.

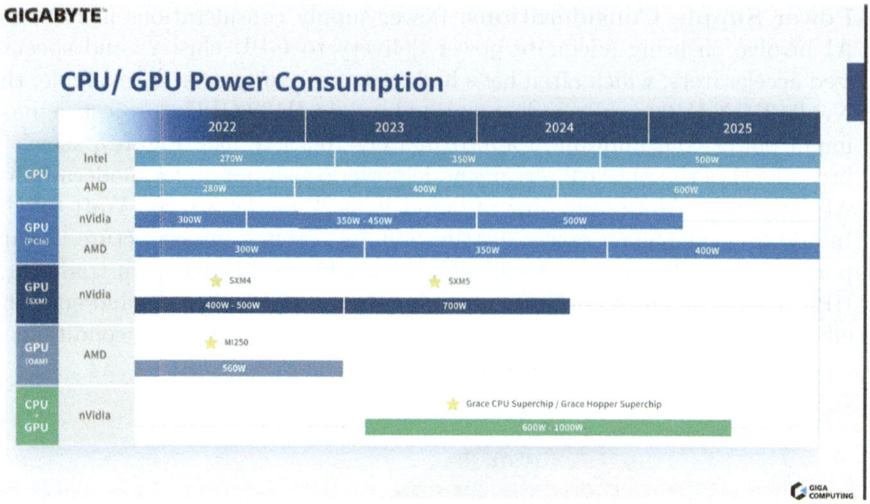

Fig. 6. Next-Gen Server CPUs & GPUs Could Consume Up To 1000W Power By 2025. Source: [16]

In summary, HPC solutions for AI rises some challenges related to power consumption and supply due to their intensive computational workloads, data movement requirements, and variability. Addressing these challenges requires a combination of energy-efficient hardware design, optimization algorithms, and intelligent power management techniques to ensure sustainable and environmental friendly operation.

5 Selected HPC-AI Cases w.r.t. to Energy

Deep learning models, especially those used in natural language processing (NLP) or computer vision, require massive amounts of computational power for training. This process often involves using large clusters of GPUs or TPUs to handle the immense computational workload. Training these deep learning models can consume a vast amount of energy. Running multiple GPUs or TPUs continuously for days or weeks during the training process can lead to substantial energy consumption. Additionally, as models grow and data sets increase in size, the energy requirements for training these models also increases significantly. Let us examine two pervasive model in the field of AI: *Convolutional Neural Network (CNN)* and *Large Language Model (LLM)*.

5.1 Case 1: Convolutional Neural Network (CNN)

CNNs are a class of deep neural networks primarily used in the field of computer vision. They are designed to automatically and adaptively learn spatial hierarchies of features from input data. CNNs are particularly powerful for tasks like image recognition, object detection, and image segmentation due to their ability to capture spatial dependencies between data.

In the realm of HPC and AI, CNNs benefit significantly from parallel processing capabilities offered by advanced GPUs and other accelerators like TPUs. These high-performance computing devices allow for efficient training and inference of CNN models, reducing the running time required and enabling the handling of massive datasets. However, the increasing reliance on HPC-AI tasks also raises concerns about energy consumption and environmental impact. CNN models, especially large architectures like ResNet (Residual networks, proposed by He et al. [17], utilize skip connections to address the vanishing gradient problem and enable the training of extremely deep networks), VGG [38] (The VGG family of architectures, characterized by their deep stacks of convolutional layers with small filter sizes, are known for their simplicity and effectiveness), EfficientNet (A family of CNN architectures introduced by Tan and Le [40], which achieve state-of-the-art performance by scaling model depth, width, and resolution in a principled manner), and Unet [35], demand substantial computational resources during both training and inference phases. This leads to high energy consumption and contributes to the carbon footprint of AI systems.

As example of case study, Castro et al. [6] provides a detailed analysis of the performance, energy efficiency, and accuracy of various CNN models on Nvidia GPUs, focusing on the Maxwell and Pascal architectures. Despite Nvidia's claims, the study found that peak power efficiency could not surpass 55% of the advertised values, reaching 20 GFLOPS/watt on Pascal and 12 GFLOPS/watt on Maxwell in the worst-case scenario with 64-bit data types. This highlights a significant performance per watt gap between the two GPU generations, which could reach up to 60% depending on the CNN model and batch size. Pascal GPUs demonstrated notable differences in performance and energy efficiency, ranging from 30% to 40% depending on the batch size and CNN model, aligning with Nvidia's estimations. Both forward and backward steps in CNN processing showed similar behavior across scenarios, indicating potential extensions of performance and power gains with larger batches. Moreover, the study found that while small batch sizes were preferred on small datasets for accuracy, larger batches sometimes maintained stability without compromising the results. On larger datasets, larger batch sizes were shown to minimize the trade-off between performance, energy consumption, and accuracy. The role of datasets emerged as crucial in influencing CNN model performance, with complexities leading to thermal stress on GPU hardware. In multi-GPU environments, batch size played a critical role due to reduced arithmetical intensity, necessitating larger sizes to mitigate communication and synchronization overhead. Larger models were required for effective communication overlap, and heterogeneity in GPU hardware introduced additional challenges. Looking ahead, the paper envisions GPUs playing an increasingly significant role in high-performance and low-power CNN and deep learning applications, especially with advancements such as 3D memory and the Volta generation. Nvidia Volta's architecture enhancements, including increased core count and the introduction of half precision and Tensor cores, are anticipated a positive impact on performance-energy efficiency and accuracy, setting the stage for promising scalability.

Efforts are being made to mitigate this concern [10,45]. Researchers are developing more energy-efficient algorithms and architectures to reduce the computational cost of CNNs without sacrificing neither performance, nor precision. Techniques like model pruning [46], quantization [14], and knowledge distillation [15] aim to streamline CNNs, making them more suitable for deployment on resource-constrained devices and reducing their environmental impact. Moreover, advancements in hardware technology, such as low-power processors and specialized AI accelerators, are also helping to address energy concerns [7,41]. Additionally, the use of renewable energy sources to power data centers and HPC facilities can further reduce the carbon footprint associated with CNN training and inference.

Overall, while CNNs play a crucial role in advancing computer vision applications, it is essential to consider the environmental impact of their deployment. By leveraging HPC resources efficiently, optimizing algorithms, and adopting sustainable practices, energy consumption and carbon footprint of CNN-based AI systems can be minimized.

5.2 Case 2: Large Language Model (LLM)

LLMs have gained immense popularity in recent years for their ability to generate human-like text and perform a variety of natural language processing tasks. Models like GPT (Generative Pre-trained Transformer) [34], BERT (Bidirectional Encoder Representations from Transformers) [11], and others are based on the Transformer architecture, which relies heavily on attention mechanisms [43] for processing sequential data.

In the context of HPC and AI, LLMs pose similar challenges and opportunities as CNNs. They require significant computational power for training and inference, especially due to the complexity of the Transformer architecture and attention mechanisms [43]. HPC systems equipped with advanced GPUs and TPUs play a crucial role in accelerating the training of large LLMs. These systems leverage parallel processing capabilities to handle massive amounts of data and optimize the computation of attention mechanisms, which are inherently parallelizable. However, like CNNs, the energy consumption and the environmental impact of training and deploying LLMs are significant concerns. Training large LLMs such as GPT-3 or BERT on extensive datasets can consume vast amounts of energy over extended periods [1]. The inference phase, although less computationally intensive than training, still requires substantial resources, especially for real-time or high-throughput applications.

One of the main studies of energy and carbon footprint was done by Luccioni et al. [26] for the BigScience Large Open-science Open-access Multilingual Language Model (BLOOM). BLOOM is a massive language model, consisting of 176 billion parameters. It was trained on 1.6 terabytes of data in 46 natural languages and 13 programming languages over a period of 118 days, utilizing a significant amount of computing resources. The training process consumed a substantial amount of energy, with a total energy usage of 433,196 kWh and 1,082,990 GPU hours. The GPUs used for training were Nvidia A100 80GB models. The carbon intensity of the energy grid used for training was 57 gCO_2e/kWh. Beyond the direct energy consumption during training, there are also embodied emissions associated with the manufacturing of the computing equipment used. The estimated embodied emissions for the servers and GPUs used in training BLOOM add up to approximately 11.2 t of CO_2e. The dynamic power consumption, which refers to the energy used for powering the GPUs during training, accounted for approximately 24.69 t of CO_2e. Idle power consumption, representing the energy needed to maintain the computing infrastructure when not actively training, was estimated to contribute around 14.6 t of CO_2e. The total emissions from different energy sources of the BLOOM model life cycle (contributions from embodied emissions, dynamic consumption, and idle consumption) were estimated to 50 t of CO_2e. We can see the comparison with most popular transformer's model from Table 2 to remember the impact of carbon intensity in the process of carbon estimation.

Table 2. carbon footprint for training SOTA NLP Transformer models. Adapted from [8,26,33]

model	parameters	carbon intensity	Energy	emissions
GPT-3	175B	429 gCO_2e/kWh	1,287 MWh	552 t
BLOOM	176B	57 gCO_2e/kWh	433 MWh	30 t
GLAM	1162B	88 gCO_2e/kWh	456 MWh	40 t
PALM	540B	79 gCO_2e/kWh	3,435 MWh	271 t

The carbon emissions associated with deploying BLOOM for inference were also estimated. Over an 18-day period, deploying BLOOM on a Google Cloud Platform instance resulted in approximately 340 kg of CO_2e emissions (914 kWh of electricity) for 230,768 requests in total. A detailed result shows that for this amount, 22.7% was consumed by the RAM (207.2 kWh), 2% by the CPU (18.5 kWh) and 75.3% by the GPU (688.38 kWh) Understanding the environmental impact of large language models like BLOOM is crucial for responsible development and deployment.

Efforts to mitigate power consumption, such as optimizing hardware usage and exploring renewable energy sources, can help to minimize the carbon footprint. Efforts are underway to address these challenges at software level too. Researchers are exploring techniques to optimize the efficiency of LLMs, including model distillation [36], knowledge pruning, and neural network architecture modifications. These approaches aim to reduce the computational complexity and memory footprint of LLMs without compromising their performance. Furthermore, advancements in hardware technology, such as low-power processors and specialized accelerators for natural language processing tasks, contribute to mitigating energy concerns associated with LLMs.

In summary, while LLMs offer remarkable capabilities in natural language processing, their widespread adoption raises important considerations regarding energy consumption and environmental impact. By *leveraging HPC resources efficiently, optimizing algorithms*, and *embracing sustainable practices*, we can work towards reducing the environmental footprint of large LLMs while keeping the ongoing advances on the topic.

6 Perspective of HPC-AI w.r.t. Energy

6.1 Energy Efficiency in HPC

Efficiency in HPC systems requires a skillful approach, addressing both hardware and software aspects. The "four Pillar Framework for energy-efficient HPC data centers" [44] outlines four critical areas for optimization: *applications, system software, system hardware*, and *infrastructures.* Hussain et al. [19] extended it to seven pillars to achieve energy efficiency in HPC Data Centers by adding considerations about *network energy, management policies* and *load balancing.* Let us break down the key points:

- **Applications Optimizations:** Optimizing applications to the specific hardware on which they are running. Optimization techniques can be employed to improve energy efficiency, for instance choosing algorithms that yield the same results with less energy. Using techniques like checkpointing and restart to avoid the need to re-run entire calculations if there is a failure.
- **System Software:** System software encompasses all software components required to execute applications on a particular hardware system. It includes the operating system, runtime libraries, and workload schedulers. The goal of optimization in system software is to reduce energy consumption without impacting application performance. System software facilitates the use of power-saving features in hardware and enforces energy-saving policies.
- **HPC System Hardware:** System hardware choice significantly impacts the overall energy consumption of the system. It includes computational hardware, networks, and auxiliary components such as storage systems. Hardware selection needs to be done wisely as later optimizations might be difficult to achieve. Computer architecture should allow for efficient execution of the variety application that will run on the system.
- **Building Infrastructures:** Building infrastructures include non-IT infrastructures required to operate an HPC system, such as power delivery and cooling systems. Optimizations in this pillar cover power delivery, cooling, and reuse of waste heat.
- **Holistic Approach:** It is generally more efficient to cover multiple pillars simultaneously in a holistic approach. For example, energy efficiency features of system hardware can be properly exploited with optimized system software. Operational parameters of building infrastructure could be fed to the workload scheduler to execute energy-hungry applications only when there is sufficient cooling capacity (Fig. 7).

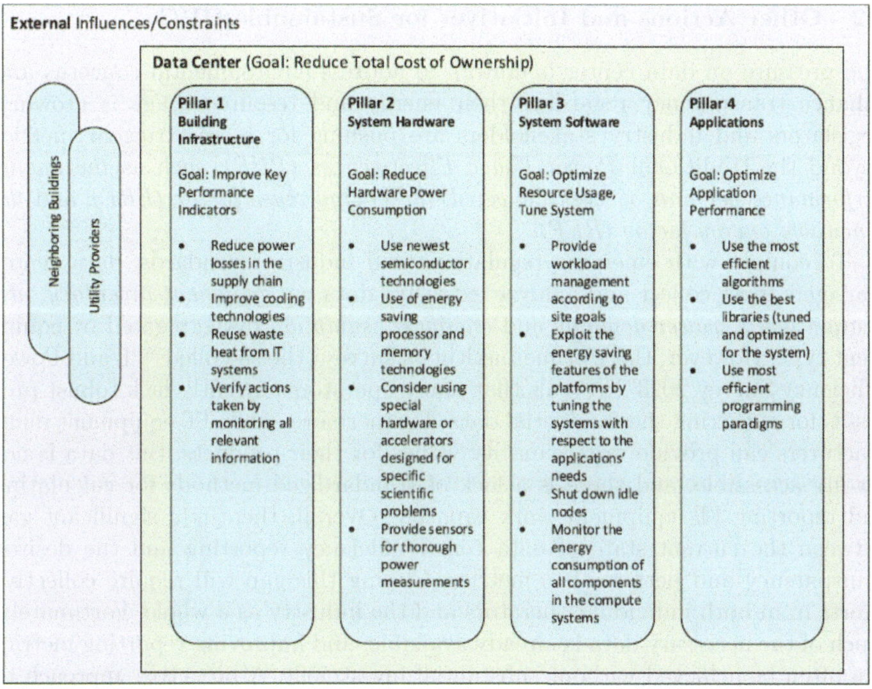

Fig. 7. 4 Pillar framework for energy saving by T. Wilde et al. [44]

In a pure HPC data center, where there is a limited control over the running applications, addressing the applications pillar of energy efficiency can indeed be challenging. However, there are still ways to encourage energy-saving practices among users by providing access to energy consumption metrics (offer users visibility into the energy consumption of their jobs). This can raise awareness and encourage them to consider energy efficiency when designing or selecting algorithms and workflows. Most batch scheduling systems already allow for reporting consumed energy as part of the standard job report, which can be leveraged to provide users with this information. Another way to encourage energy-saving practice is to provide fine-grained energy measurements. Offer users access to more detailed energy measurements, allowing them to identify specific regions of their code that consume more energy. With this information, users can optimize their code by focusing on energy-intensive sections, potentially reducing overall energy consumption without sacrificing performance. A survey on energy management in HPC was done by Nana et al. [29], with emphasis on a taxonomy of energy management from measurements to optimization approaches for sustainable computing.

6.2 Other Actions and Initiatives for Sustainable HPC

The pressure on data center operators to address environmental concerns and enhance transparency regarding their energy and resource usage is growing. Legislators and industry stakeholders are pushing for more stringent metrics beyond the traditional *Power Usage Effectiveness (PUE)*, such as *the energy performance per unit of work delivered, the energy reuse factor (ERF)*, and *the renewable energy factor (REF)*.

To comply with emerging regulations and industry standards, data center managers must collect and analyze extensive data on *equipment inventory, utilization levels, power demand*, and *energy consumption* disaggregated by equipment type. However, the Uptime Institute's survey (the so-colled "IT and Power Efficiency Survey 2023") reveals that many operators currently lack robust processes for gathering these essential data. Furthermore, while IT equipment manufacturers can provide work capacity values for their products, this data is not readily accessible, and there is a lack of standardized methods for calculating and reporting IT equipment work capacity. Overall, there is a significant gap between the current state of data center efficiency reporting and the desired transparency and performance metrics. Closing this gap will require collective efforts from both individual operators and the industry as a whole. Fortunately, much of the necessary data is already available, and improving reporting metrics can often be achieved without substantial investment. A proactive approach to not only ensure compliance with regulatory requirements, but also contributes to long-term sustainability goals was provided as data map framework by Uptime Institute in this survey report (See Fig. 8).

The Climate Neutral Data Centre Pact [9] initiative aimed at making data centers in Europe climate neutral by 2030. It was launched in November 2020 by the European Commission, along with major cloud and data center providers, industry associations, and NGOs. The pact commits signatories to achieving three main goals: **(a)** *Transitioning to climate-neutral data centers*, **(b)** *Ensuring all new data centers are designed to be climate neutral* and **(c)** *Renovating existing data centers to be climate neutral where possible by 2030.* This pact is crucial in addressing the significant energy consumption and carbon emissions associated with data centers, which have been growing rapidly due to the increasing demand for digital services and cloud computing. By making data centers climate neutral, the pact aims to reduce the environmental impact of digital infrastructure while supporting the EU's broader climate goals, such as achieving carbon neutrality by 2050.

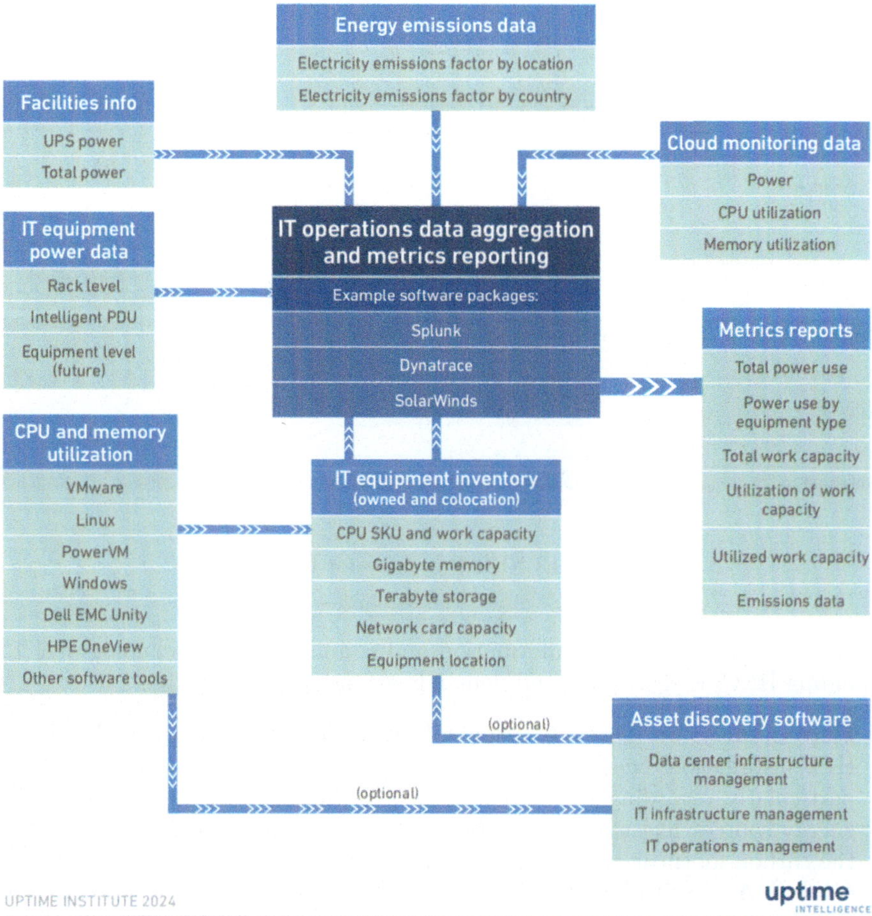

Fig. 8. IT operations and sustainability data map

Scientific computing has a notable carbon footprint, but there is promise for environmentally sustainable computational science (ESCS) to thrive. Achieving this requires a coordinated effort to tackle current challenges by increasing awareness, transparency, and estimation accuracy of environmental impacts. Lannelongue et al. [23] provides an overview of ESCS's current status and introduce the GREENER principles (see Fig. 9), offering guidance on adopting best practices for sustainable growth in computer science.

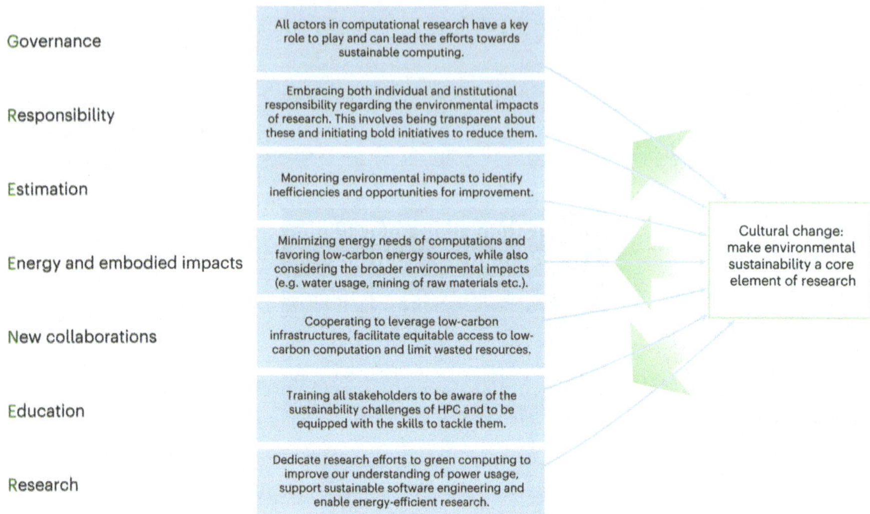

Fig. 9. GREENER Principles for ESCS [23]

The carbon footprint of information and communication technology (ICT), including HPC, is significant and demands attention. Here is why:

- **Dependency on Data Center Choices:** The environmental impact of HPC installations heavily relies on the decisions made at the data center level, particularly regarding the source of energy they use. Opting for renewable energy sources like *wind, solar, hydro,* or *nuclear* power can greatly reduce the carbon footprint of HPC operations.
- **Global Accessibility:** HPC systems are not bound by physical location; they can be located anywhere while serving users worldwide. This flexibility allows for strategic placement in areas where renewable energy sources are abundant, further enhancing their green credentials.
- **Carbon-neutral Operations:** Achieving carbon-neutral or even negative carbon footprint in HPC operations is indeed feasible today. By exclusively using 100% carbon-neutral energy sources, HPC facilities can significantly mitigate their environmental impact.
- **Heat Reuse:** The excess heat generated by HPC operations represents a substantial amount of energy. By capturing and re-purposing this heat for heating purposes or other industrial processes, HPC facilities can maximize energy efficiency and minimize waste.
- **Re-purposing Existing Infrastructure:** Instead of constructing new data centers from scratch, re-purposing existing buildings and utilizing brownfield solutions can significantly reduce the environmental impact associated with construction, land use, and infrastructure development.

Adopting a holistic approach that encompasses renewable energy utilization, heat reuse strategies, and sustainable infrastructure practices is essential for mitigating the carbon footprint of HPC operations and advancing the green agenda in the ICT sector.

The development and adoption of tools for tracking carbon emissions associated with computational tasks are invaluable for researchers aiming to minimize their environmental footprint. These tools can provide valuable insights into energy-intensive processes and guide decisions toward more eco-friendly alternatives. Moreover, initiatives like the Green Software Foundation [18] underscore the importance of prioritizing sustainability in software development, fostering a culture of environmental responsibility within the computational community.

Addressing the environmental challenges of computational research requires a multifaceted approach, encompassing not only carbon footprint estimation but also considerations of governance, responsibility, and cultural change. It is essential to tackle issues related to hardware production costs and e-waste management, recognizing the broader environmental implications of technological advancement.

Encouraging researchers to report the carbon footprint of their work is a positive step toward fostering accountability and driving innovation in sustainable computational practices. By raising awareness and promoting energy-efficient design principles, we can collectively work towards a more sustainable future. Every small step counts, and collective efforts within the scientific community can make a significant impact in mitigating climate change.

7 Conclusion

As highlighted throughout this Chapter, the exponential growth in computational demands has led to a corresponding increase in energy requirements, rising serious challenges in sustainability and resource management. From the early stage of energy-intensive HPC systems to the latest advancements in AI-driven data centers, we have traversed a landscape with complexities and implications for environmental considerations.

The pressure to address these challenges has spurred a multifaceted approach encompassing technological innovation, market dynamics, policy interventions, and financial considerations. Advances in *hardware design, optimization algorithms*, and *cooling systems* offer promising avenues for enhancing energy efficiency in HPC-AI systems. Moreover, initiatives such as energy *efficiency standards, renewable energy initiatives*, and *carbon pricing mechanisms* play a pivotal role in driving industry-wide efforts towards sustainability.

While significant steps have been made, the journey towards achieving a harmonious balance between technological innovation and environmental stewardship is far from over. Continued collaboration and dialogue among stakeholders (from researchers and industry leaders to policymakers and environmental advocates) are essential to navigate within the complexities of energy utilization in HPC-AI systems.

Looking ahead, the future trajectory of HPC-AI systems has both challenges and opportunities. Emerging technologies such as quantum computing and neuromorphic computing offer tantalizing prospects for further enhancing computational efficiency and reducing environmental impact. However, realizing these benefits will require concerted efforts and a steadfast commitment to sustainability principles. Finally, as we stand at the precipice of a new era of technological advancement, let us remain regular in our commitment to harnessing the power of HPC and AI for the betterment of humanity, while simultaneously safeguarding the health of our planet for futures generations.

Acknowledgements. This research was supported by The Transition Institute 1.5 driven by École des Mines de Paris - PSL.

References

1. Alberto, R.: Meet m6 - 10 trillion parameters at 1% GPT-3's energy cost (2021). https://towardsdatascience.com/meet-m6-10-trillion-parameters-at-1-gpt-3s-energy-cost-997092cbe5e8
2. Azhar, A.: https://www.hpcwire.com/2024/02/12/sam-altman-seeks-7-trillion-to-supercharge-chip-production/
3. Azhar, A.: https://www.hpcwire.com/2024/02/06/nvidias-dominance-in-ai-chips-challenged-by-big-tech-companies/
4. AMD. https://www.amd.com/en/products/accelerators/instinct/mi300.html
5. Appuswamy, R., Olma, M., Ailamaki, A.: Scaling the memory power wall with dram-aware data management. In: Proceedings of the 11th International Workshop on Data Management on New Hardware (2015)
6. Castro, F.M., Mata, N.G., Marín-Jiménez, M.J., Serrano, J.P., Ujaldón, M.: Energy-based tuning of convolutional neural networks on multi-GPUs. Concurr. Comput.: Pract. Experience **31** (2018)
7. Chen, Y.H., Krishna, T., Emer, J.S., Sze, V.: Eyeriss: an energy-efficient reconfigurable accelerator for deep convolutional neural networks. IEEE J. Solid-State Circuits **52**(1), 127–138 (2017). https://doi.org/10.1109/JSSC.2016.2616357
8. Chowdhery, A., et al.: Palm: scaling language modeling with pathways. J. Mach. Learn. Res. **24**, 240:1–240:113 (2022)
9. Climate Neutral Data Centre. https://www.climateneutraldatacentre.net/
10. Desislavov, R., Martínez-Plumed, F., Hernández-Orallo, J.: Trends in AI inference energy consumption: beyond the performance-vs-parameter laws of deep learning. Sustain. Comput. Inform. Syst. **38**, 100857 (2023)
11. Devlin, J., Chang, M.W., Lee, K., Toutanova, K.: BERT: pre-training of deep bidirectional transformers for language understanding. In: North American Chapter of the Association for Computational Linguistics (2019)
12. Erich, S., Jack, D., Horst, S., Martin, M.: https://www.top500.org/
13. Freitag, C., Berners-Lee, M., Widdicks, K., Knowles, B., Blair, G.S., Friday, A.: The real climate and transformative impact of ICT: a critique of estimates, trends, and regulations. Patterns **2**(9) (2021)
14. Gholami, A., Kim, S., Dong, Z., Yao, Z., Mahoney, M.W., Keutzer, K.: A survey of quantization methods for efficient neural network inference. arXiv abs/2103.13630 (2021)
15. Gou, J., Yu, B., Maybank, S.J., Tao, D.: Knowledge distillation: a survey. Int. J. Comput. Vis. **129**(6), 1789-1819 (2021). https://doi.org/10.1007/s11263-021-01453-z

16. Mujtaba, H.: https://wccftech.com/gigabyte-server-power-consumption-roadmap-points-600w-cpus-700w-gpus-by-2025/
17. He, K., Zhang, X., Ren, S., Sun, J.: Deep residual learning for image recognition. In: 2016 IEEE Conference on Computer Vision and Pattern Recognition (CVPR), pp. 770–778 (2015)
18. Hussain, A.: https://greensoftware.foundation/articles/what-is-green-software
19. Hussain, S.M., et al.: Seven Pillars to Achieve Energy Efficiency in High-Performance Computing Data Centers, pp. 93–105. Springer, Cham (2019)
20. International Energy Agency: IEA, Paris. https://www.iea.org/reports/electricity-2024
21. Jones, N.: How to stop data centres from gobbling up the world's electricity. Nature **561**, 163–166 (2018). https://doi.org/10.1038/d41586-018-06610-y
22. Kwon, S., Bahn, H.: Classification and characterization of memory reference behavior in machine learning workloads. In: 2022 IEEE/ACIS 23rd International Conference on Software Engineering, Artificial Intelligence, Networking and Parallel/Distributed Computing (SNPD), pp. 103–108 (2022). https://doi.org/10.1109/SNPD54884.2022.10051800
23. Lannelongue, L., et al.: Greener principles for environmentally sustainable computational science. Nat. Comput. Sci. **3**(6), 514–521 (2023)
24. Li, B., Basu Roy, R., Wang, D., Samsi, S., Gadepally, V., Tiwari, D.: Toward sustainable HPC: carbon footprint estimation and environmental implications of HPC systems. In: Proceedings of the International Conference for High Performance Computing, Networking, Storage and Analysis, SC 2023. ACM (2023). https://doi.org/10.1145/3581784.3607035
25. Li, B., et al.: Full-cycle energy consumption benchmark for low-carbon computer vision. arXiv abs/2108.13465 (2021)
26. Luccioni, A.S., Viguier, S., Ligozat, A.L.: Estimating the carbon footprint of bloom, a 176b parameter language model. arXiv abs/2211.02001 (2022)
27. Market Research Future. https://www.marketresearchfuture.com/reports/high-performance-computing-market-2698
28. Motivair. https://www.motivaircorp.com/products/liquid-cooling-systems/
29. Nana, R., Tadonki, C., Dokladal, P., Mesri, Y.: Energy concerns with HPC systems and applications. arXiv abs/2309.08615 (2023)
30. Nvidia. https://docs.nvidia.com/dgx/dgxh100-user-guide/introduction-to-dgxh100.html
31. Peckham, O.: https://www.hpcwire.com/2021/12/09/with-a-carbon-footprint-like-hpcs-it-matters-when-and-where-you-step/
32. Park, S., Bahn, H.: Memory access characteristics of neural network workloads and their implications. In: 2022 IEEE Asia-Pacific Conference on Computer Science and Data Engineering (CSDE), pp. 1–6. IEEE Computer Society, Los Alamitos, CA, USA (2022)
33. Patterson, D., et al.: The carbon footprint of machine learning training will plateau, then shrink. Computer **55**(7), 18–28 (2022). https://doi.org/10.1109/MC.2022.3148714
34. Radford, A., Narasimhan, K.: Improving language understanding by generative pre-training. arXiv (2018)
35. Ronneberger, O., Fischer, P., Brox, T.: U-net: convolutional networks for biomedical image segmentation. In: Medical Image Computing and Computer-Assisted Intervention (MICCAI). LNCS, vol. 9351, pp. 234–241. Springer (2015)
36. Sanh, V., Debut, L., Chaumond, J., Wolf, T.: DistilBERT, a distilled version of BERT: smaller, faster, cheaper and lighter. arXiv:1910.01108 (2019)

37. Schöne, R., Ilsche, T., Bielert, M., Gocht, A., Hackenberg, D.: Energy efficiency features of the intel skylake-sp processor and their impact on performance. In: 2019 International Conference on High Performance Computing & Simulation (HPCS), pp. 399–406 (2019).https://doi.org/10.1109/HPCS48598.2019.9188239
38. Simonyan, K., Zisserman, A.: Very deep convolutional networks for large-scale image recognition. In: Bengio, Y., LeCun, Y. (eds.) 3rd International Conference on Learning Representations, ICLR 2015, San Diego, CA, USA, 7–9 May 2015, Conference Track Proceedings (2015). http://arxiv.org/abs/1409.1556
39. Tadonki, C.: What Do HPC Applications Look Like?, pp. 27–51. Springer, Cham (2023). https://doi.org/10.1007/978-3-031-29769-4_3
40. Tan, M., Le, Q.V.: Efficientnet: rethinking model scaling for convolutional neural networks. arXiv abs/1905.11946 (2019)
41. Ting, Y.S., Teng, Y.F., Chiueh, T.D.: Batch normalization processor design for convolution neural network training and inference. In: 2021 IEEE International Symposium on Circuits and Systems (ISCAS), pp. 1–4 (2021). https://doi.org/10.1109/ISCAS51556.2021.9401434
42. Uptime Institute. https://datacenter.uptimeinstitute.com/rs/711-RIA-145/images/2023.SpendingSurvey.Report.pdf?
43. Vaswani, A., et al.: Attention is all you need. In: Guyon, I., et al. (eds.) Advances in Neural Information Processing Systems, vol. 30. Curran Associates, Inc. (2017)
44. Wilde, T., Auweter, A., Shoukourian, H.: The 4 pillar framework for energy efficient HPC data centers. Computer Science - Research and Development **29** (2013). https://doi.org/10.1007/s00450-013-0244-6
45. Xu, Y., Martinez-Fernandez, S., Martinez, M., Franch, X.: Energy efficiency of training neural network architectures: an empirical study. arXiv abs/2302.00967 (2023)
46. Yang, T.J., Chen, Y., Sze, V.: Designing energy-efficient convolutional neural networks using energy-aware pruning. In: 2017 IEEE Conference on Computer Vision and Pattern Recognition (CVPR), pp. 6071–6079 (2016)

Integrated Architecture for Cloud and IoT with Logical Sensors and Actuators - Logical IoT Cloud

Takayuki Kushida[1(✉)] and Yuka Kato[2]

[1] School of Computer Science, Tokyo University of Technology, Tokyo, Japan
kushida@acm.org

[2] Division of Mathematical Sciences, Tokyo Woman's Christian University, Tokyo, Japan

Abstract. Cloud has become a common infrastructure for IT systems and has been adopted for many mission-critical systems. IoT applications usually provide IoT services with data collected from physical sensors and control commands to physical actuators such as Cyber-Physical Systems. When IoT applications run on Cloud, the integrated systems management for Cloud and IoT is required to manage physical sensors and actuators with ordinary cloud resources such as CPU, memory, storage, and network. It can provide the same level of services as ordinary cloud resources. This paper proposes an innovative cloud architecture called Logical IoT Cloud (LIC) that provides integrated systems management for Cloud and IoT resources. LIC also enables Logical Sensors (LSs) with multiple virtual sensors and Logical Actuators (LAs) with numerous virtual actuators. Those LSs and LAs can perform the same functions as physical sensors and actuators running on cloud-native environments. They can support mission-critical IoT applications on LIC. As a result, IoT applications for mission-critical usages on LIC can collect sensor data via LSs and send physical actuators a set of execution commands via LAs.

1 Introduction

Cloud was introduced as a service model with computing resources and has been rapidly deployed with a large number of installations in enterprises by Cloud Service Providers (CSPs). Cloud has three basic types of services such as Infrastructure as a Service (IaaS), Platform as a Service (PaaS) and Software as a Service (SaaS) [1]. Those services are provided to get access via the Internet. The Cloud has been a common infrastructure in the IT industry. Different research studies have been conducted on the Cloud [2]. Their architectural designs are generally decided with characteristics, deployment models, service models, and others-as-a-service for the Cloud. Open technical challenges for the Cloud was studied and reported [3]. They are handling uncertainties, handling dynamic variations in workload, optimization of virtual network topologies, efficient VM consolidation, scientific workflow scheduling, public auditing and data

M. Zbakh et al. (Eds.): CloudTech 2023, LNNS 1220, pp. 117–136, 2024.
https://doi.org/10.1007/978-3-031-78698-3_7

availability. As container technology has been introduced to reduce footprints of deployed images, cloud-based container architectures are provided in the Cloud [4]. In addition, the container orchestration supports managing multiple different containers as a single management entity. Those multiple containers can be deployed on distributed cloud environments.

Internet of Things (IoT) was defined as an integrated deployment of physical and virtual things seamlessly integrated into the network [5]. Those identifiable things can be communicated with each other. IoT applications get access to physical sensors and actuators to interact with physical space. Physical environmental data are captured by physical sensors, and digitalized sensing data are transferred to the central server on the Cloud. The server analyzes the sensing data in the IoT application. After they are analyzed in an IoT application on the Cloud, the actionable commands are generated by the IoT application and sent to the physical actuators via the network. The physical actuators, by actionable commands, act for the physical environments. This closed loop of software among physical sensors, IoT applications on the Cloud, and physical actuators is essential for IoT usages and activities [6].

Many IoT Cloud platforms support the transfer of sensing data on IoT devices and analyze them in the Cloud. They need to have quick interactions between the Cloud and IoT. One of the research studies is the integration that uses an agent-oriented and cloud-assisted approach [7]. The approach has adopted the layered architecture. The survey report for integrating Cloud and IoT covers twenty-six Cloud IoT platforms in ten different domains [8]. The study of the integration for Cloud and IoT is called the CloudIoT paradigm [9]. The main drivers for the integration can be categorized as communication, storage, and computation. The purpose of CloudIoT is to ease its use and reduce costs. In addition, typical IoT applications are deployed on the Cloud as a vertical solution. They can isolate each other even when standard underlying components are installed.

The study for Sensor Cloud (SC) proposed a Virtual Sensor (VS) that enables a single physical sensor to be virtualized [10]. Multiple VSs could be shared with multiple IoT applications as a cloud resource. As VS in SC is implemented as a software object, multiple VSs for a single physical sensor can be quickly created on SC. The survey summarized to report many research papers for SC-related studies after SC was originally proposed [11]. A single IoT application exclusively uses a physical sensor with a legacy architecture and design; other IoT applications cannot access the physical sensor. VSs for the physical sensor were created to support multiple IoT applications at the same time. As a result, SC could support multiple IoT applications that would like to access a single physical sensor simultaneously. The architecture and its implementation for SC with experimental evaluations were studied and reported [12–14].

Issues and Challenges

Integration of the Cloud resources and IoT devices is required to support systems management for physical sensors and actuators. The systems management needs to support the same level of infrastructure service as ordinary cloud resources such as CPU, memory, storage, and network.

When IoT applications for mission-critical usage are deployed and operated on SC, SC has several fundamental issues in supporting them since it only supports virtualization for physical sensors. The following items summarize those issues when IoT applications for mission-critical usage are deployed and operated on SC.

1. **Actuator support:** SC can create multiple VSs for a single physical sensor on the Cloud, but it does not support actuators with the virtualization in the SC architecture. Typical IoT applications require both physical sensors and actuators for their use-case scenarios. Physical sensors make digitized data for environmental measurements in physical space. On the other hand, physical actuators execute specific actions executed by a set of commands, and IoT applications create and send them physical actuators. When the commands are sent to the physical actuator, the specific action with the physical actuator is executed in physical space.
   For example, the physical actuator, such as the servo motor is deployed and connected to the IoT application. IoT application sends the physical actuator multiple command data to control the servo motor. When an IoT application that needs a physical actuator runs on the Cloud, the control of the physical actuator is required for IoT applications as a part of cloud resources. When multiple IoT applications for a single physical actuator are deployed on the Cloud at the same time, Cloud needs to support a virtualized resource for the single physical actuator.
2. **Characteristics, properties, and statuses of hardware:** As both physical sensors and actuators are IoT hardware devices, they have management data such as characteristics, properties, and statuses in hardware devices. Since physical sensors and actuators have different characteristics, properties, and statuses, their data should be dynamically retrieved from those hardware devices when IoT applications require them for their usage.
   When IoT applications collect management data from those hardware devices, human manipulations are required to retrieve them in a legacy operation. In addition, management data are usually described in the device manual in human-readable format and need to be in a more well-defined format as a software input. As a result, there is no shared access method or format for retrieving management data in physical sensors and actuators.
3. **Lifecycle management for IoT software objects:** Ordinary cloud resources such as CPU, memory, storage, and network are managed with the lifecycle. Lifecycle management is defined as the status of creation, execution, stop/resume, and deletion. It is usually done in a unified manner on the management console. When SCs with VC were deployed on the Cloud, the lifecycle management was also required to support multiple VSs with ordi-

nary cloud resources. In addition, there is no common access method with a user interface provided for the lifecycle management of VSs.

In summary, those issues and challenges for supporting IoT hardware device seamlessly as a part of Cloud resources need to be resolved with a new cloud architecture.

2 Related Studies

This section describes related studies in the research area of integrated systems management for Cloud and IoT.

As the number of IoT deployments has increased recently, the detailed analysis of IoT platform architectures was studied and reported [15,16]. Different approaches, standards, and use-cases need help selecting the right IoT platform with customer's requirements and environmental restrictions. The study also describes underlying concepts, similarities, and differences among IoT platforms. It proposed the reference architecture that consists of a Sensor, Actuator, Device, Gateway, IoT Integration Middleware and Application. Four open-source platforms and four proprietary IoT solutions were compared as a reference architecture for the IoT platform.

Many IoT platforms have different capabilities and functions. The method to select the suitable and adequate IoT platform from available IoT platforms was proposed with its demands [17]. It describes the essential blocks in IoT platforms as components: the sensing component, communication, and identification component, computation and Cloud component, and services and applications component. It proposed the reference IoT architecture that consists of those components.

The role of IoT platforms was reported as enabling the development and deployment for IoT applications and services [18]. It also describes three critical impacts of IoT applications on several industries. The impacts are automation, integration, and servitization, and the servitization means that organizations can move from a primarily product-centered business model to a service-oriented business model. The report describes how IoT platforms can remove the complexity in the development and deployment phase for the lifecycle management of IoT applications and solutions. IoT platforms also support a new service provider in adding value-added services that provide end-to-end IoT applications and solutions for new end users.

IoT virtualization with the container technology is proposed to support the scenarios of IoT Cloud [19]. The study focuses on combining Cloud and IoT to create a new opportunity for service delivery and systems management for IoT operation. It also proposes server virtualization for IoT devices using Linux Container Virtualization (LCV). The container-based virtualization of smart IoT objects can support use-case scenarios on IoT Cloud.

IoT services need two mandatory features for dependability, and they are reliable and available. The study proposes a model-based approach to support two critical features on the application layer in the IoT system [20]. It has adopted virtualized IoT services that can support a service abstraction model for dependability requirements. The virtualization for IoT devices enables sharing sensor services among multiple applications. In addition, this study could add the integration of dependability to the virtualized IoT services as a distinguished approach from other studies.

The IoT architecture is proposed with four layers: network management, perception layer, transmission layer, and application layer [21]. The perception layer is the device, sensor, or recognition layer. The transmission layer is also called the transport layer or network layer. The application layer is focused on managing IoT applications that provide personalized services for end users.

Distributed Internet-like Architecture for Things (DIAT) is proposed and can support the deployment and operation for large-scale IoT systems [22]. It adopts a policy model for usage control to support security and privacy in distributed environments. The architecture has six layers: the Physical and Communication layer, Virtual Object (VO) layer, Composite Virtual Object (CVO) layer, Service layer, and Application layer. It also provides advanced functions such as Dynamic Service Creation, Dynamic Service Modeling and Security Management.

IoT applications usually need access to physical sensors and actuators in the Cloud. Although many new architectures for IoT and Cloud platforms have been proposed in related studies, the dominant architecture still needs to be created. The IoT and Cloud platform architecture requires an integrated approach without boundaries among cloud resources. This paper is focused on proposing a new architecture for the integration of IoT and Cloud platform with systems management.

3 Architecture of Logical IoT Cloud (LIC)

The proposed architecture to support IoT devices for IoT applications in the Cloud are called Logical IoT Cloud (LIC) in Fig. 1. It can support logical IoT devices as software objects and logical IoT devices have the same functions as physical sensors and actuators. They can be executed for IoT applications with ordinary cloud resources such as CPU, memory, storage, and network. The logical IoT device for a physical sensor is defined as a Logical Sensor (LS), and one for a physical actuator is defined as a Logical Actuator (LA). LSs, LAs, and ordinary cloud resources work together to provide IoT services on LIC.

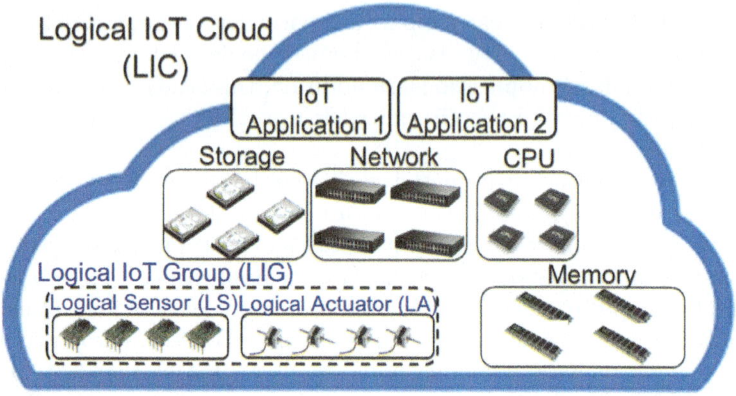

Fig. 1. Logical IoT Cloud (LIC) including ordinary cloud resources

Figure 1 shows the basic concept of LIC. Ordinary cloud resources such as CPU, memory, storage, and network are also illustrated in Fig. 1. LSs and LAs can be managed as a part of cloud resources on LIC. Logical IoT Group (LIG) is a group function of LS and LA. LIG supports making a logical group with multiple LSs and LAs on LIC since individual LS and LA are tiny devices compared with ordinary cloud resources. LIG is required as a basic unit for integrated systems management. LS receives digitized data from a physical sensor, and LA sends the commands to a physical actuator. After LIG that includes LSs and LAs is deployed with ordinary cloud resources on LIC, IoT applications can be accessed to them on LIC.

3.1 LIC Components

Figure 2 shows the architecture of LIC. It consists of LS, LA, Connection Management (CM), Integrated Systems Management (ISM), Cloud Resources, and IoT Applications. Sensor Nodes (SNs) and Actuator Nodes (ANs) are connected to LIC via the Internet in Fig. 2. Ordinary cloud Resources are CPU, memory, storage, and network. IoT Applications as shown in Fig. 2 get access to LSs and LAs to interact with SNs and ANs regarding their usages. Those cloud resources are provided to execute IoT applications on LIC. LIG is a group of multiple LSs, ANs, and LIC's provisioning and management unit. LIGs, including LSs and LAs, are managed by Integrated Systems Management (ISM). IoT applications get access to LSs and LAs in LIG and provide IoT services for end-users.

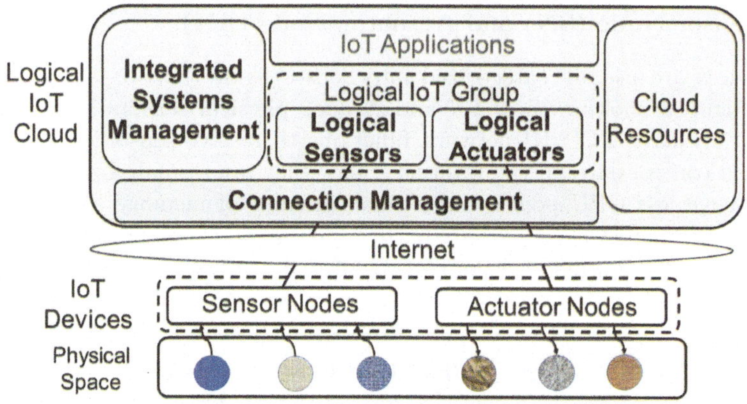

Fig. 2. Architecture of Logical IoT Cloud

The proposed architecture provides a software object such as LS, LA, and LIG for IoT applications. When IoT applications need hardware management data such as characteristics, properties, and status of physical sensors and actuators, IoT applications can manage the executions for physical sensors and actuators.

For example, a physical sensor called BME280[1], which provides temperature, humidity and pressure sensors generate digitized sensed data for IoT applications. When its data are analyzed in the IoT application on the Cloud, the IoT application does not know the hardware device, its precisions, and its status. It only receives digitized sensed data for the analysis.

The IoT application has to get predefined IoT device values from the configuration file since they are needed for usage in the IoT application. On the other hand, LIC, including LIG, LS, and LA, dynamically provides hardware management data for IoT applications when the IoT applications request to receive hardware management data from LS and LA.

When an IoT application is operated as a part of mission-critical system, it needs to have high availability for a non-functional requirement to maintain continuous IoT service for end-users. The IoT application is required to configure high availability. The data path between the IoT application and the physical space must support continuous data transfer. Sensed data from the physical sensor and control command data to the physical actuator are required without any interruptions when it is a mission-critical system. When LS and LA for the application would have high availability on LIC, it can be integrated into mission-critical systems and operations.

[1] https://www.bosch-sensortec.com/media/boschsensortec/downloads/datasheets/bst-bme280-ds002.pdf.

3.2 Sensor Node (SN) and Actuator Node (AN)

IoT devices are usually categorized into Sensor Node (SN) or Actuator Node (AN). Figure 3 shows typical SN and AN for physical sensors and actuators. SN and AN have TCP/IP network functions to receive sensor data and send command control data on the Internet. Since they are a small form factor, they usually have less CPU power and a small amount of memory and storage. In addition, as some SNs and ANs are battery-powered devices, power consumption control also needs to manage the lifetime of IoT devices.

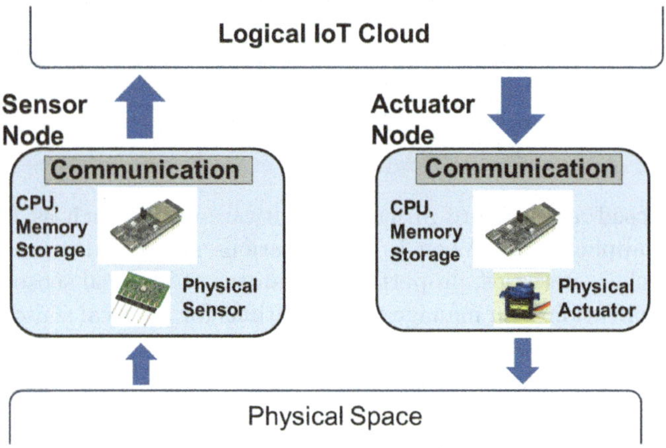

Fig. 3. Sensor Nodes (SNs) and Actuator Nodes (ANs)

Both SNs and ANs are usually called physical IoT devices. Although several physical IoT devices have SN and AN functions in the same hardware, the proposed architecture describes either SN or AN. The devices in the architecture are simplified to describe SN and AN functions illustrated in Fig. 3.

Configuration of SN
Figure 3 shows SN, which can capture environmental parameters such as temperature, humidity, and air pressure in physical space. It consists of sensors and a tiny control unit with CPU and memory. SN also has the network function of "Communication." to LIC via the Internet. The software stack in SN digitizes environmental values in physical space with the physical sensor, which becomes sensor data. It also sends sensor data to LIC after the sensor data are ready to use in SN hardware.

Configuration of AN
Figure 3 shows AN that can receive a set of command control data from LIC and execute specific actions for the physical actuator in the physical space. AN also has a network function called "Communication" that connects to LIC via the Internet. After the software in AN receives those command control data for the physical actuator from LIC, command control data are applied to the physical actuator. After AN gets the execution results and status data from the physical actuator, they are also sent to LIC via the Internet in Fig. 3.

3.3 Logical Sensor (LS) and Logical Actuator (LA)

LSs and LAs support the provision of management data in physical sensors and actuators. As IoT applications require management data, LSs, and LAs provide management data to execute further actions on IoT applications. IoT application usually changes their workflow processes with management data. For example, IoT applications can receive the hardware specifications of the physical sensors to generate sensor data, analyze sensor data according to the hardware specifications, and get better results. When IoT application also receives hardware limitations or constraints with management data in the physical actuator, command control data can be sent to the physical actuator with the limitations and constraints.

Since LS and LA are logical software objects, they can quickly be replicated to multiple objects with user requests. They can provide the same function as physical sensors and actuators for IoT applications. Figure 4 shows an overview of LSs and LAs on LIC.

LS and LA are software-based; they can be replicated and connected with a single physical sensor and actuator. As a result, LSs and LAs are simultaneously provided for multiple IoT applications on LIC. LIC can create multiple LSs and LAs without restriction since they consume only a tiny amount of memory and less CPU power when a logical service with LS and LA provides IoT applications. Multiple IoT applications can get access to them. LIC also provides the service by separating LSs and LAs among multiple IoT applications. Each IoT application would get access to LSs and LAs without concern for other usages.

LS Configuration
LS is a software object that consists of two VSs, a High Availability (HA) Switch, and a Communication component in Fig. 4. HA Switch has the switch function of two VSs to support high availability and provide continuous service of LS for IoT applications. The communication component in LS can interact with CM that are connected to SN. LS supports REST API to provide sensor and management data for IoT applications, as shown in Fig. 4.

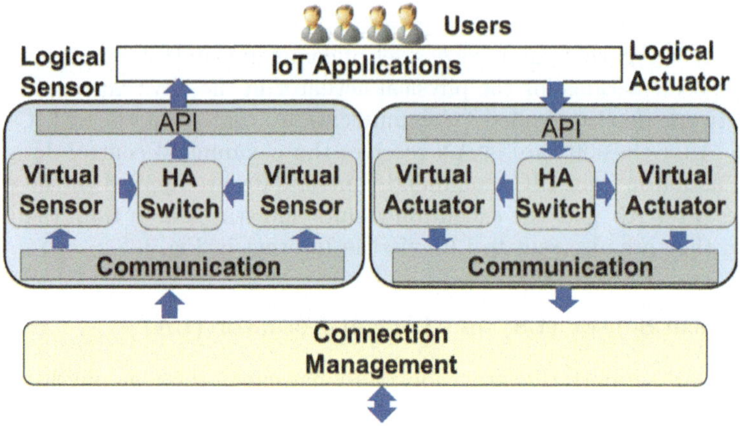

Fig. 4. Logical Sensor (LS) and Logical Actuator (LA)

Management data in LS are characteristics, properties, and the statuses of the physical sensor. For example, BME280, which is a physical sensor, has the sensor functions to monitor temperature, humidity, and pressure. It has management data for the sensing capability, the hardware specification, and the status of hardware registers. LSs provide management data when IoT applications request them for their usage. IoT applications can obtain both sensor and management data via REST API.

In summary, LS has the following features:

- LS provides the same sensor and management data as the physical sensor does via REST API. IoT applications can get access to management data in physical sensors.
- LS provides a continuous sensor service for IoT applications since the high availability is provided with HA Switch and two VSs.

LA Configuration

LA is a software object that consists of two Virtual Actuators (VAs), HA Switch and Communication in Fig. 4. LA has the same function as a physical actuator, and provides the logical actuator's service to IoT applications. It is highly available with HA Switch and two VAs. When one VA is down and unavailable, another VA can be up and running to accept command data from IoT applications via REST API and control the physical actuator via CM.

LA needs to support the physical actuator control that is different from LS. Although multiple LAs can be replicated for a single physical actuator, the actual operation for LA would be an exclusive use of the physical actuator since it needs to be only controlled by a single software entity at a certain period. Therefore,

LA provides the schedule function for IoT applications and allows the single IoT application for exclusive operation. As a result, the specific LA can control the physical actuator with the set of commands.

LA has the following features:

- LA provides high availability with HA Switch and two VAs, and supports a continuous logical actuator service for the IoT application.
- LA provides the same command data and operation as the physical actuator. IoT applications can control and manage those physical actuators with LSs via REST API.
- LS provides the exclusive use for the specific physical actuator with the schedule function. The physical actuator needs to receive the command data from a single IoT application at a certain period.

3.4 Logical IoT Group (LIG)

LIG is a group function to support multiple LSs and LAs, and a logical management unit by ISM. LIG usually is a group by the same or similar attributes of LSs and LAs. For example, LSs and LA have the exact attributes of the same city, building, company, or department. They can get together as LIG since they have the exact attributes. LSs and LAs also have the same type of hardware, such as temperature/humidity sensors and servo motors. They can also be LIG with the same type of hardware.

LIG consists of multiple LSs and LAs as independent management units with ordinary cloud resources. The user can request to create LIG on the cloud management console. As the list of LIGs is shown in the cloud catalog, users can select to create the specific LIG with ordinary cloud resources on the cloud management console.

Figure 5 shows LIG that consists of three LSs and LAs. As a software object supporting multiple LSs and LAs on LIC, it can be created without limitations for user requests.

When users request to create multiple LSs and LAs directly, a large number of LSs and LAs are deployed on LIC. As a result, ISM has to manage those LSs and LAs independently, and it is hard to manage those software objects directly. As LIG supports multiple LSs and LAs as a single group on LIC, ISM can manage LIGs with ordinary cloud resources. As LIG can reduce the processing workload in ISM, it contributes to the scalability of ISM. As LIG is based on a software object, it has the flexibility to support a wide variety of LSs and LAs.

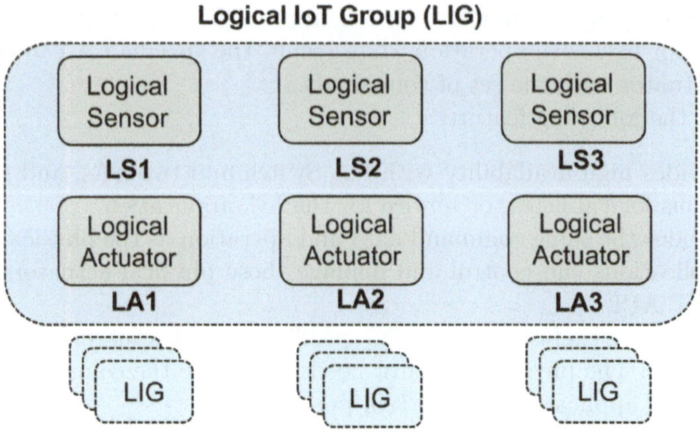

Fig. 5. Logical IoT Group

For example, fifty physical temperature sensors are deployed in Hachioji City, Tokyo, JAPAN. Fifty LSs are created for fifty physical temperature sensors. LIG is also created for those LSs and can manage them as a single management entity. When a user requests the single LIG for fifty physical temperature sensors in Hachiouji-city, LIC provides both LIG and ordinary cloud resources for users. As a result, IoT applications can access LIG for fifty LSs that can monitor temperature and control management data for the physical temperature sensor.

LIG is summarized as following features.

- **Group of multiple LSs and LAs:** LIG is a group of multiple LSs and LAs and a minimum unit for systems management on LIC. LIC can create, delete, and update LIG instances at the user's request since LIG has become a part of cloud-managed resources. When a new catalog entry of LIG is created on LIC, it can be added to the catalog list on LIC. Users can request the LIG with ordinary cloud resources.
- **Group with the same type of characteristics:** LIG can usually be configured with the same kind of multiple LSs and LAs. For example, LIG is created as a group of multiple LSs with temperature sensors and multiple LAs with servo motors. Another example of LIG is that those LSs and LAs are located in the same area or location. They can be a single LIG for the management unit on LIC. For example, multiple LSs and LAs are located in Hachioji City, Tokyo JAPAN. They are in the exact location and can be a single LIG on LIC. Both LSs and LAs have the same purpose of usage. Multiple LSs monitor weather conditions such as temperature, humidity, and atmospheric pressure, and LIG can be provided to enable multiple LSs for weather conditions.
- **Duplicated LSs and LAs for LIGs:** As LSs and LAs are software objects, many LIGs with LSs and LAs can be created on LIC. For example, LSs are created to monitor the room temperatures on LIC. LIG can be provided for those temperature LSs. Another LIG can be provided to monitor room

temperature, humidity, and pressure with LSs. The same LSs for room temperature can be categorized into different LIGs.

3.5 Connection Management (CM)

Figure 6 shows CM in LIC. It creates logical communication links among LSs, LAs, SNs, and ANs with request commands from ISM since ISM manages those logical links to transfer data among them in CM.

CM receives management data such as characteristics, properties, and statuses of SNs and ANs. It also transfers control command and configuration data to LSs and ANs since those logical communication links are already established to specified SNs and ANs in CM.

For example, when the Switch Controller in CM receives the connection request between the specific SN and LS from ISM, the controller establishes the logical communication link between them in Switch Fabric. After the logical connection is established, Switch Fabric in CM exchanges the sensor and its management data between SN and LS in Fig. 6. The same procedure steps can be executed in logical communication the link between LA and AN.

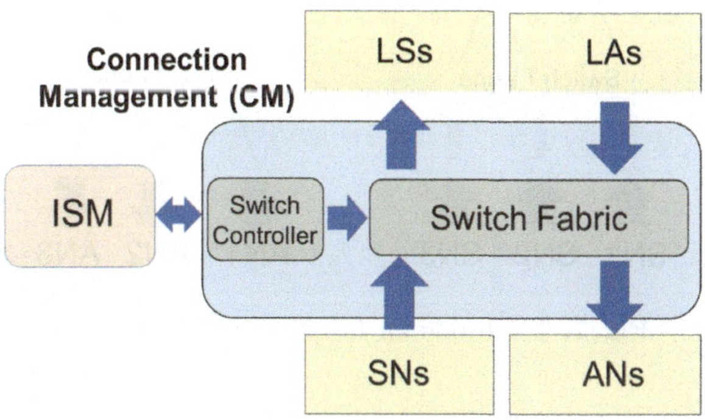

Fig. 6. Connection Management

SN has the physical sensor device that is attached to the CPU device. As LS is a software object that can be replicated on LIC, many LSs can be created for a single SN, that is, the hardware device, and the single SN is connected to CM. CM transfers sensor and management data from the single SN to multiple LSs simultaneously when multiple logical connections for those LSs are established in CM.

For example, CM receives the connection request among SN and LSs, it creates logical connections. SN can monitor temperature data in the physical sensor, and those LSs are ready to provide temperature data for IoT applications. IoT applications get access to LS via API when sensor data are required for their application usage. Figure 7 shows the inside of CM after multiple connections are created among SNs, ANs, LSs, and LAs.

When logical connections are created among SN{1,2,3} and LS{1,2}, sensor and management data can be seamlessly transferred among them. SN1 is connected to LS1, and SN3 is connected to LS2 in Fig. 7. Switch Fabric starts transferring sensor and management data among them. When SN2 is connected to LS1 and LS2 in Switch Fabric, sensor, and management data are duplicated in Switch Fabric in Fig. 7.

When the connection is established among LA{1,2} and AN{1,2,3} in Fig. 7, actuator and management data are transferred between them. As LA1 is connected to AN1 and AN2 in Fig. 7, Switch Fabric starts transferring data among them. After LA1 connects to AN2, LA2 also requests to connect to AN2 and AN3. AN2 has two logical connections in Switch Fabric in Fig. 7. LA2 needs to wait for the operation of AN2 when LA1 finishes the logical connection to AN2. Requesters need to use the physical actuator exclusively. After AN2 is available for the operation of LA2, LA2 starts connecting to AN2 for getting access to the physical actuator.

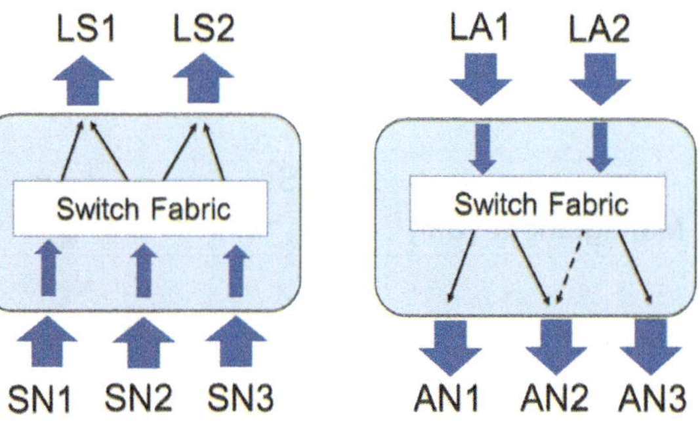

Fig. 7. CM inside for LSs/SNs and LAs/ANs

Switch Fabric in CM creates logical connections and transfers data from SNs and LSs. It also creates them and transfers data from LAs to ANs. When IoT application needs faster data transfer for IoT devices as a requirement, the transfer delay needs to be minimized on Switch Fabric in CM.

As the number of active logical connections among those components, such as SN, LS, LA, and AN, is increased, the scalability for Switch Fabric is required to support many logical connections in CM. As multiple Switch Fabrics are deployed on LIC, they can transfer data on several logical connections.

CM is summarized as following features.

- Switch Controller establishes logical connections in Switch Fabric.
- Switch Fabric transfers sensor and management data from SNs to LSs.
- Switch Fabric transfers actuator and management data from LAs to ANs.

3.6 Integrated Systems Management (ISM)

ISM is an LIC component that manages Cloud and IoT resources with a unified access method. Figure 8 shows the configuration of ISM. When a new LIG is created on LIC, ISM sends the commands for the logical connection to CM and LIG and CM creates a logical connection among SN, AN and LIG for LSs and LAs. LIG starts exchanging data from SN and AN, which can be integrated with IoT applications. ISM has functions of systems management that are provisioning, logging, monitoring, accounting, registration, Graphical User Interface (GUI) and others in Fig. 8.

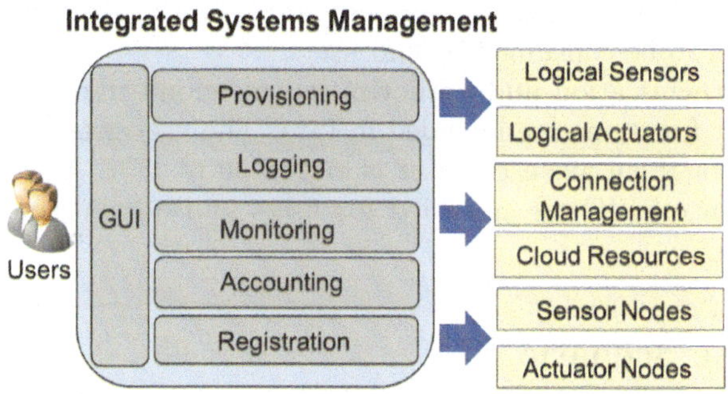

Fig. 8. Integrated Systems Management (ISM)

ISM is summarized as following features.

- **Provisioning:** Users can request those resources for deployment on the cloud management console with GUI on ISM. When a user requests to create a new LIG on LIC, ISM creates LIG that consists of LSs and LAs and registers LIG. ISM provides LIG's lifecycle management by provisioning and de-provisioning LIGs and ordinary cloud resources.
- **Logging:** After LSs and LAs for a new LIG are up and running, log messages in LSs and LAs are generated for their activities and stored in a log file. Those log messages are forwarded to a central log server in ISM. ISM provides the logging service to administrators.
- **Monitoring:** When LSs and LAs are up and running on LIC, metric data such as usage and performance are generated in LSs and LAs and forwarded to a central monitoring server in ISM. ISM provides the monitoring service to administrators.
- **Accounting:** As users access LIG for LSs and LAs, usages of users' activities are collected for accounting purposes. The central accounting server in ISM collects those usages from LSs and LAs that belong to LIG on LIC. ISM provides the accounting service with usages to administrators.
- **Registration:** When a new SN with the physical sensor and AN with the physical actuator are attached to LIC, the registration process is executed.

This process registers SN with the physical sensor and AN with the physical actuator, and they have the logical link to LS and LA in LIG. In addition to the registration of SN and AN, LIG is also registered for multiple LSs and LAs and is deployed on LIC with the user's request.

- **GUI:** Web Graphical User Interface (GUI) is provided to users and administrators for the cloud management console. User can manage their cloud resources with Web GUI, and administrator works for systems management with Web GUI. Web GUI is the integrated interaction point with users and administrators who want to access LIC and manage cloud resources.
- **Service catalog:** ISM supports the service catalog for LIGs instead of many LSs and LAs directly. When a user's request creates a new LIG, a new cloud resource for LIG is added to the service catalog on LIC. It consists of multiple LSs and LAs to be consumed by end users.
- **others:** Authentication and authorization for cloud resources and systems management functions are included in ISM. The user gets the proper access permission for them when the user is logged on to ISM. Authentication and authorization on LIC are provided for users and administrators who want access to LIC.

4 Use-Case Scenario

Figure 9 shows a typical use-case scenario for LIC, and it is a greenhouse with smart agriculture [23,24]. Temperature sensors to monitor the inside of the greenhouse temperature and a motor drive to control windows and awnings in the greenhouse are installed and deployed [25]. Physical temperature sensors are attached to SNs that are logically linked to LSs on LIC. Motors for windows and awnings in the greenhouse are logically linked to ANs directly, and they are also linked to LAs. A single LS includes two VSs, and a single LA includes two VAs in Fig. 9.

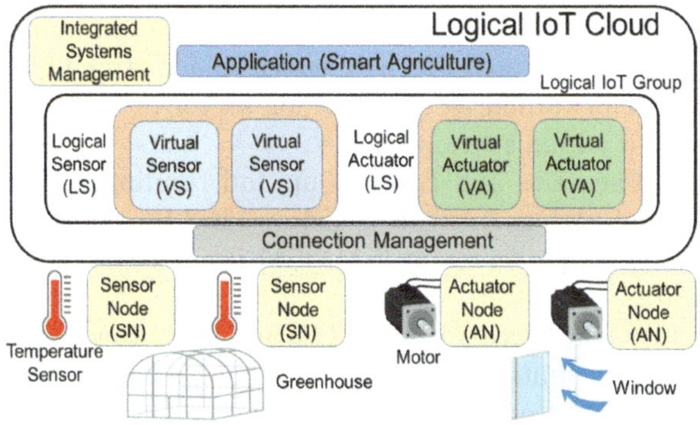

Fig. 9. Use Case Scenario on LIC.

LIG is a logical unit that has those LSs and LAs. When a user requests to create LIG for a greenhouse with ordinary cloud resources, LIG, and ordinary cloud resources are deployed on LIC. IoT application is also deployed on the same LIC.

LIG is deployed on the same process as ordinary cloud resources do. When IoT application to control the greenhouse is deployed, it gets access to the greenhouse LIG to monitor temperatures and send control commands to servo motors for windows and awnings. When IoT application monitors greenhouse temperature via LS in LIG, it can control the open and closed windows and awnings via LA.

As LSs for SNs receive sensor and management data from CM, IoT applications can be obtained via API in LSs. LAs also receive them from CM, and IoT applications can send control commands to the physical actuators attached to ANs. In addition to sensor and actuator data, management data are exchanged among LSs, LAs, and ISM.

5 Discussion

Delay for Sensor and Command Data

IoT application needs to collect sensor data from the physical sensor and send the control command to the physical actuator in real-time when it has a real-time requirement in physical space. When IoT applications can get sensor data on LS, the physical sensor and management data are collected on SN and transferred to LS via CM. It takes a processing time in SN, CM, and LS. In addition, it takes a network transfer time among those components. Both the processing and network transfer times cause delays for IoT applications. Therefore, the delay should be minimized when The IoT application receives sensor and management data from LIC.

When IoT application controls LA for the physical actuator, a similar delay exists on LIC. It takes a processing time on LA, CM, and AN. It also takes a network transfer time among those components. The LA to AN delay should also be minimized when the IoT application controls the actuator.

One of the methods to minimize the transfer delay is to prioritize data when they are processed and transferred among those components. The prioritized data are handled for prioritized processing and network transfer, and other data have less priority for processing and network transfer on those components.

Actuator for Exclusive Use

The physical actuator is attached to the AN and can execute specific actions, such as the servo motor in physical space. Therefore, usage for AN needs the exclusive use for the hardware device of the physical actuator. Cloud resources on LIC are shared among users when they request cloud resources, including LSs and LAs for their applications. LA, one cloud resource, must accept the request for access to the physical actuator.

Switch On and Off for Actuators

When the usage of the physical actuator is rather casual, there is no conflict in getting access to the physical actuator. For example, the bright light switch on the corridor has a switch that is attached to AN. The smart switch is changed either "on" and "off" with the actuator when the control command receives on AN. Multiple IoT applications can get access to the same switch since the usage of the light switch is not critical for anyone. When the smart light switch is"off," it can be changed to "on" or stay "off." The actuator in the light switch does not have to have the exclusive use. Multiple LAs can access the same AN to control the physical actuator for the smart light.

Multiple LAs Get Access to the Single AN at the Same Time

On the other hand, physical actuators which do not allow us multiple access from LAs. For example, there are physical actuators that are attached to ANs in robots and specific LAs in robots need to have exclusive access to those physical actuators for ANs. In general, robot control is done by a single application. When those actuators are attached to ANs linked to LA on LIC, they are required to support the schedule function for ANs since multiple LAs would like to get access to the specific AN simultaneously. CM can manage AN using the schedule function for access requests among multiple LAs.

Scalability of LIC

When a large number of physical sensors and actuators are configured on LIC, scalability is required for LIC. CM has the message bus to transfer data among LSs, LAs, SNs, and ANs. When the number of LSs and LAs that are created on LIC is increased, scalable software components are required to enable multiple CM deployments on servers. When LIC has multiple CM deployments, inter-CM communication is also required to manage many connections among LSs, LAs, SNs, and ANs. In addition, multiple CM deployments need to keep data transfer time among them to a minimum. The interconnections among message buses in multiple CMs need to be configured to share sensor and actuator data among multiple CM deployments.

6 Conclusion

This paper proposes an innovative LIC architecture that seamlessly supports Cloud and IoT resources. It has a novel architecture for integrated system management of Cloud and IoT. LIC enables a logical software object called Logical Sensors (LSs) and Logical Actuators (LAs). LSs are linked to physical sensors, and LA is to physical actuators. LSs and LAs support high availability with two VSs and VAs. In addition to enabling LSs and LAs, LIC supports a group function with IoT Group (LIG) that allows multiple LSs and LAs to be made as a single management unit on LIC. LIG is a minimum management resource for Integrated Systems Management. LIC provides lifecycle management for LIG and ordinary cloud resources such as CPU, memory, storage, and network.

Multiple IoT applications can access LSs and LAs that are logically linked to physical sensors and actuators. LSs and LAs also provide management data such as the characteristics, properties, and statuses of physical sensors and actuators since IoT applications need to get the latest IoT environmental data and need to manage their executions instantly.

LIC can support IoT applications that have mission-critical usages and public usage activities.

Acknowledgment. This work was supported by JSPS KAKENHI Grant Numbers JP23K11073 and JP23K11087.

References

1. Mell, P., Grance, T.: The nist definition of cloud computing. NIST Spec. Publ. **800**(145), 7 (2011). http://csrc.nist.gov/publications/nistpubs/800-145/SP800-145.pdf
2. Odun-Ayo, I., Ananya, M., Agono, F., Goddy-Worlu, R.: Cloud computing architecture: a critical analysis. In: 2018 18th International Conference on Computational Science and Applications (ICCSA), 2018, pp. 1–7 (2018)
3. Puthal, D., Sahoo, B.P., Mishra, S., Swain, S.: Cloud computing features, issues, and challenges: a big picture. In: 2015 International Conference on Computational Intelligence and Networks. IEEE, 2015, pp. 116–123 (2015)
4. Pahl, C., Brogi, A., Soldani, J., Jamshidi, P.: Cloud container technologies: a state-of-the-art review. IEEE Trans. Cloud Comput. **7**(3), 677–692 (2019)
5. Vermesan, O., et al.: Internet of things strategic research roadmap. In: Internet of things-global technological and societal trends from smart environments and spaces to Green ICT. River Publishers, pp. 9–52, 2022
6. Pallavi Sethi, S.R.S.: Internet of things: architectures, protocols, and applications. J. Electr. Comput. Eng. **2017** (2017)
7. Babu, S.M., Lakshmi, A.J., Rao, B.T.: A study on cloud based internet of things: cloudiot. In: 2015 Global Conference on Communication Technologies (GCCT), 2015, pp. 60–65 (2015)
8. Ray, P.P.: A survey of iot cloud platforms. Futur. Comput. Inform. J. **1**(1), 35–46 (2016)
9. Botta, A., De Donato, W., Persico, V., Pescapé, A.: Integration of cloud computing and internet of things: a survey. Futur. Gener. Comput. Syst. **56**, 684–700 (2016)
10. Yuriyama, M., Kushida, T.: Sensor-cloud infrastructure-physical sensor management with virtualized sensors on cloud computing. In: 2010 13th International Conference on Network-Based Information Systems. IEEE, 2010, pp. 1–8 (2010)
11. Alamri, A., Ansari, W.S., Hassan, M.M., Hossain, M.S., Alelaiwi, A., Hossain, M.A.: A survey on sensor-cloud: architecture, applications, and approaches. Int. J. Distrib. Sens. Netw. **9**(2), 917923 (2013)
12. Yuriyama, M., Kushida, T.: Integrated cloud computing environment with it resources and sensor devices. Int. J. Space-Based Situated Comput. **1**(2–3), 163–173 (2011)

13. Yuriyama, M., Kushida, T., Itakura, M.: A new model of accelerating service innovation with sensor-cloud infrastructure. In: 2011 Annual SRII Global Conference, pp. 308–314. IEEE, 2011
14. Takahashi, H., Kushida, T.: Innovative IAAS management system for sensor devices and it resources. Int. J. Distrib. Sens. Netw. **10**(6), 931968 (2014)
15. Guth, J., Breitenbücher, U., Falkenthal, M., Leymann, F., Reinfurt, L.: Comparison of iot platform architectures: a field study based on a reference architecture. In: 2016 Cloudification of the Internet of Things (CIoT), pp. 1–6, 2016
16. Guth, J., et al.: A Detailed Analysis of IoT Platform Architectures: Concepts, Similarities, and Differences. Springer, pp. 81–101, 2018
17. Hejazi, H., Rajab, H., Cinkler, T., Lengyel, L.: Survey of platforms for massive iot. In: 2018 IEEE International Conference on Future IoT Technologies (future IoT). IEEE, 2018, pp. 1–8 (2018)
18. Lucero, S., et al.: IoT platforms: enabling the internet of things, White paper, 2016
19. Celesti, A., Mulfari, D., Fazio, M., Villari, M., Puliafito, A.: Exploring container virtualization in iot clouds. In: 2016 IEEE International Conference on Smart Computing (SMARTCOMP), 2016, pp. 1–6 (2016)
20. Dar, K.S., Taherkordi, A., Eliassen, :F. Enhancing dependability of cloud-based iot services through virtualization. In: 2016 IEEE First International Conference on Internet-of-Things Design and Implementation (IoTDI), pp. 106–116, 2016
21. Gupta, B., Quamara, M.: An overview of internet of things (iot): architectural aspects, challenges, and protocols. Concurr. Comput. Pract. Exp. **32**(21), e4946 (2020). e4946 CPE-18-0159.R1. https://onlinelibrary.wiley.com/doi/abs/10.1002/cpe.4946
22. Sarkar, C., SN, A.U.N., Prasad, R.V., Rahim, A., Neisse, R., Baldini, G.: DIAT: a scalable distributed architecture for iot. IEEE Internet Things J. **2**(3), 230–239 (2015)
23. Maraveas, C., Piromalis, D., Arvanitis, K.G., Bartzanas, T., Loukatos, D.: Applications of iot for optimized greenhouse environment and resources management. Comput. Electron. Agric. **198**, 106993 (2022). https://www.sciencedirect.com/science/article/pii/S0168169922003106
24. Subahi, A.F., Bouazza, K.E.: An intelligent iot-based system design for controlling and monitoring greenhouse temperature. IEEE Access **8**, 125 488–125 500 (2020)
25. Riskiawan, H.Y., et al.: Artificial intelligence enabled smart monitoring and controlling of iot-green house. Arab. J. Sci. Eng. **49**, 3043–3061 (2024)

The Need for HPC in AI Solutions

Imane Ettifouri[1(✉)], Mostapha Zbakh[1], and Claude Tadonki[2]

[1] National School of Computer Science and Systems Analysis, Mohammed V University, in Rabat, Morocco
imane_ettifouri@um5.ac.ma, m.zbakh@um5r.ac.ma
[2] Mines ParisTech - PSL University, Centre de Recherche en Informatique (CRI), 35, rue Saint-Honoré, 77305 Fontainebleau Cedex, France
claude.tadonki@mines-paristech.fr

Abstract. High-Performance Computing (HPC) aims at providing the most efficient computing support for problems solving. It is basically a conjunction of three components: *methods*, *programming*, and *machines*. When considered as a support to a given paradigm, the main focus is on the last two components. Artificial Intelligence (AI) is becoming a major topic in the computer science landscape, with a broad range of applications that is continuously increasing. Cutting-edge AI is very ambitious and the required computing power to reach the expected level of efficiency is very high, thus the vital need for HPC support. The aim for this chapter is to illustrate the importance of HPC in AI investigations and solutions through a number of selected applications. For each application or AI-oriented study, we illustrate the critical aspect of the execution time, considering *training* and *inference*.

Keywords: HPC · AI · Efficiency · Parallelism

1 Introduction

In the realm of Artificial Intelligence (AI), the last decade has witnessed an unparalleled surge in the development of algorithms and models that fundamentally transform society's fabric. From enabling advanced predictive analytics in healthcare to powering complex decision-making processes in autonomous vehicles, AI technology is leading the fourth industrial revolution [1]. However, the escalating complexity and scale of AI models, exemplified by the latest generational leaps in machine learning and deep learning frameworks, pose substantial computational challenges. These challenges necessitate the deployment of High Performance Computing (HPC) infrastructures to facilitate the training, inference, and application of AI systems at a larger scale [2].

The emergence of AI as a dominant force in technology and innovation traces back to early explorations into machine learning and algorithmics. Initially focused on rule-based systems and basic pattern recognition, the field of AI has evolved through various phases, marked by significant milestones such as the development of the perceptron in the 1950s and the subsequent advancements in neural networks and machine learning algorithms [3]. This evolutionary

M. Zbakh et al. (Eds.): CloudTech 2023, LNNS 1220, pp. 137–159, 2024.
https://doi.org/10.1007/978-3-031-78698-3_8

journey has been characterised by an increasing sophistication in model architecture, from simple linear models to more complex, multi-layered structures of today's deep learning networks. Each leap forward in AI capabilities has been matched by a corresponding increase in computational demands, highlighting the growing importance of HPC in supporting these technological advancements [4].

The critical juncture in AI's evolution was reached with the advent of deep learning, which revolutionised the field by enabling models to learn from data in ways that mimic human cognition. Deep learning's potential, however, is fully unleashed only through the power of HPC, which provides the computational horsepower required to process and analyse the enormous datasets essential for deep learning models. The synergy between AI and HPC is thus not merely a technical requirement but a vital couple that drives the continuous innovation and application of AI across diverse domains [5]. This introduction sets the stage for a deeper exploration into how HPC facilitates cutting-edge advancements in AI, underscoring the symbiotic relationship between these two fields.

The rest of the chapter is structured in the following manner: Sect. 2 discusses the major HPC systems and their suitability for AI, including shared memory machines, distributed memory machines, clusters, and GPUs. Next, Sect. 3 offers specific examples of HPC-AI applications, such as climate modeling, genomics, drug discovery, and large-scale language models. Following that, Sect. 4 examines the computational aspects of AI, exploring the intricacies of AI models and the impact of their extensive use. Section 5 focuses on AI-specific HPC configurations, detailing specialized setups to enhance AI research and operations. Furthermore, Sect. 6 explores the computational performance of HPC AI applications, providing an overview of results as well as specific benchmark outcomes. Finally, Sect. 7 concludes by discussing the future implications of HPC in AI, including possible technical advancements such as incorporating quantum computing.

2 Major HPC Systems and Their Suitability for AI

High Performance Computing systems play a critical role in supporting the complex computations required by Artificial Intelligence projects. Each system's architecture significantly influences its ability to handle specific AI tasks. It's important to recognize the strengths and limitations of various HPC configurations-including shared memory systems, distributed memory machines, clusters, and GPUs-to effectively adapt AI workflows for enhanced performance and efficiency.

2.1 Shared Memory Machines

Shared memory machines, also known as Symmetric Multiprocessors (SMPs), feature a uniform memory access model that allows each processor equal access time to all memory nodes. This architecture excels in supporting AI applications that require intricate data interdependencies and substantial memory bandwidth, such as certain neural network training scenarios where frequent data exchange across processors is crucial. Nevertheless, the scalability of SMPs is constrained by their memory architecture, which can restrict performance when managing exceptionally large AI models.

2.2 Distributed Memory Machines

Distributed memory machines, often realized through Massively Parallel Processor Arrays (MPPs), consist of multiple processing units each with its local memory, connected through a network. This configuration excels in running large-scale AI models by distributing the computation across multiple nodes, thereby enhancing scalability and processing power. The challenge lies in the increased complexity of data communication and synchronization, which requires efficient parallel algorithms and network topologies to minimize latency.

2.3 Clusters

Clusters represent a flexible HPC configuration, where multiple computers (nodes) are connected through a high-speed network to operate as one cohesive system. Such systems can be customized with nodes that are specialized for particular AI tasks, including those equipped with high-performance GPUs for deep learning. The modularity of clusters not only facilitates tailored solutions to meet the specific computational needs of AI applications but also enhances system robustness through fault tolerance. Moreover, the ability to expand resources easily – known as horizontal scalability – is crucial for handling increasingly complex AI tasks, and this is viable thanks to being able to distribute loads efficiently across resources, resulting in less processing time. Research highlighted by Becchi et al. [6], Song and Dongarra [7] illustrates the effectiveness of GPU-equipped clusters in speeding up AI computations, showing notable performance gains in deep learning workloads.

2.4 GPUs

Graphics Processing Units (GPUs) play a crucial role in the advancement of AI. They offer parallel processing capabilities that far surpass those of traditional CPUs. GPUs are specifically designed for efficient matrix and vector operations, which are essential in deep learning algorithms. This allows for multiple calculations to be carried out simultaneously.

In addition to their speed, GPUs also provide the accuracy required for complex AI computations. As a result, they are an integral part of high-performance computing (HPC) systems dedicated to AI. The Table 1 highlights the significant technological advancements in GPU design over the years, illustrating the rapid evolution of hardware that supports increasingly complex AI computations. Each generation of GPUs has brought substantial improvements in processing power, memory capacity, and efficiency, making them indispensable in fields requiring high computational resources, such as AI and deep learning.

Figure 1 shows a simplified view of the GPU architecture, emphasizing its suitability for deep learning and other intensive AI tasks.

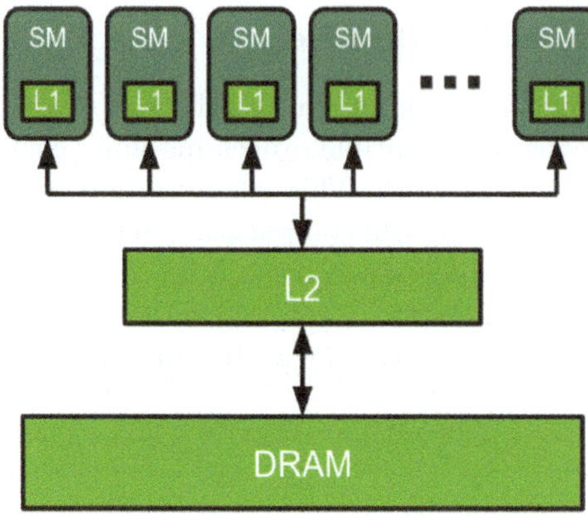

Fig. 1. Simplified view of the GPU architecture.

Table 1. Evolution of GPU Technology from the First GPU to 2024.

Year	Model	Clock Speed (MHz)	GFLOPS (GOps)	Memory	Cache
2024	Radeon RX 7900 GRE	2245	45,978	16 GB GDDR6	L0: 64 KB per WGP/L1: 256 KB per Array/L2: 6 MB/L3: 64 MB
2022	GeForce RTX 4090	2520	82,575	24 GB GDDR6X	L1: 128 KB (per SM)/L2: 72 MB
2020	GeForce RTX 3090	1695	35,581	24 GB GDDR6X	L1: 128 KB (per SM)/L2: 6 MB
2012	GeForce GTX 690	1019	6,261	2 GB GDDR5	L1: 16 KB (per SMX)/L2: 512 KB
2009	Radeon HD 5970	725	4,640	1024 MB GDDR5	L1: 8 KB (per CU)/L2: 512 KB
2006	GeForce 8800 GTS (640MB)	1188	228	640 MB GDDR3	L2: 80 KB
1999	GeForce 256 SDR*	120	0.5	32 MB SDR	N/A
1998	Rage 128 GL*	80	0.2	32 MB SDR	N/A

* Denotes performance measured in "GOps"-gigaoperations per second-as opposed to GFLOPS.

2.5 Selecting Optimal High-Performance Computing Systems for Advanced AI Applications

When choosing an HPC system for AI, it is crucial to take into account various factors such as for instance the type of task, amount of data, complexity of models, and the requirement for immediate processing. Shared memory machines may be suitable for applications requiring intensive data exchange and lower-scale parallelism. In contrast, distributed memory machines and clusters can tackle larger, more distributed workloads typical of big data analytics and large-scale neural network training. GPUs, given their computational prowess, are almost universally beneficial for AI, significantly reducing training times and enabling more complex model architectures.

Choosing a high-performance computing system greatly affects the success, expenses, and productivity of artificial intelligence projects. As artificial intelligence algorithms become more advanced and datasets increase in volume, there will be an increasing demand for high-performing and specialized computing resources. The ongoing advancements in HPC configurations, highlighted by works such as those by Kindratenko et al. [8], are pivotal in meeting these demands, driving forward the capabilities and applications of artificial intelligence.

3 Selected HPC-AI Cases

Integrating High-Performance Computing with Artificial Intelligence: A Perspective In their quest for scientific research and complex problem-solving, the union of High-Performance Computing with Artificial Intelligence has led to a fundamental discovery. From these concepts, various sectors have made significant transformational achievements. This section offers a detailed analysis of specific case studies where HPC-AI convergence has played a critical role. Areas of focus include Climate Modulating, Drug Recovery and Genomic, and Large-Scale Language Models. The three exhibits case specific powers rapidly interact, with AI and computational power. Cases difference in sizes and complexity address how combinations can tackle extensive and complex problems. By relying on large amounts of data and computational output, climate modelling accurately predicts and simulates environmental transformations. Finally, there is AI that requires HPC due to the amount of data and thus processing. AI is applied mainly in medical sciences for drug discovery and genomics, in news and report generation as well as in many smaller applications and, importantly, in language models. Clearly, HPC significantly accelerates the processes of developing innovations and directly influences the scientific progress of the recent years.

3.1 Climate Modelling

Artificial Intelligence (AI) is making a significant leap forward in unraveling the intricate web of climate systems around the globe and in the improvement of climate models. AI, specifically machine learning algorithms, provide a distinct ability to look at large sets of data and uncover patterns and trends that traditional analytical tools cannot detect. This ability has been invaluable in predicting very accurately the short term weather conditions. N. Vanitha and J. Haritha [9] showcase this through their research, where AI models trained on historical weather data from numerous stations predict future weather conditions with significantly reduced computational demands compared to traditional methods. This leap in efficiency and accuracy marks a transformative step in how climate scientists approach the prediction and analysis of weather and climate phenomena, setting a new standard for precision in short-term forecasts.

Parallel to AI's rise in climate modeling is the indispensable role of High Performance Computing (HPC) in powering complex climate simulations. The intricacies of climate models, which simulate interactions across atmospheric, oceanic, and terrestrial components, require the substantial computational resources that only HPC can provide. As detailed by ndré et al. [10], the reliance of European Earth System Models on HPC infrastructure underscores the computational challenge inherent in climate modeling. HPC systems enable the processing of large datasets and the execution of large scale simulations, enhancing model resolution and the accuracy of climate predictions. This computational power is critical not only for developing models but also for ensuring that these models can run efficiently, thus making HPC a cornerstone of contemporary climate science.

A significant transformation takes place when artificial intelligence and high-performance computing come together in climate modeling, merging problem solving abilities of AI with the processing capabilities of HPC. Thanks to this collaboration, we are able to effectively and accurately model and forecast climate systems. [11] demonstrate the integration of HPC systems to enhance the performance modeling of ML models used in weather as well as climate forecasting, showcasing the potential for substantial improvements. The combination allows for the most effective and efficient use of high-performance computing resources, making it possible to create, test, validate, and deploy machine learning models for climate prediction with minimal delay. The merging of HPC and AI enhances climate science by increasing the precision of weather and climate forecasts while also streamlining time and resources needed for these calculations. This union is initiating a transformative phase in climate science. This approach enables more detailed simulations of complex weather patterns, leading to significant improvements in the precision of weather forecasts and our ability to predict and mitigate the impacts of climate change. A comprehensive theoretical and also technical exploration, as illustrated in Tables 2 and 3, underscores the substantial impact of high performance computing (HPC) on the fields of artificial intelligence (AI) and climate science. This study spans numerous functions which together highlight how HPC capabilities are crucial in enhancing AI algorithms and models.

3.2 Drug Discovery and Genomics

Artificial Intelligence (AI) in drug discovery stands as a pivotal element in modern pharmaceutical research, providing robust computational capabilities that accelerate the identification and optimization of novel drug candidates. Through the employment of advanced algorithms and machine learning techniques, AI has the capacity to sift through extensive biochemical data, predicting molecular interactions and effectiveness with notable accuracy and efficiency. This development not only streamlines the initial stages of drug discovery but also brings about an unprecedented level of precision in disease mechanism targeting, significantly diminishing the timeframe and financial investment traditionally associated with drug development [15].

The incorporation of High-Performance Computing (HPC) into drug discovery processes further propels this paradigm shift. HPC delivers the essential computational strength required to manage the vast datasets and intricate simulations that characterise contemporary pharmaceutical research [16]. This combination facilitates the swift progression of drug-target interaction simulations, drug candidate optimisation, and the analysis of genomic and proteomic data on an unprecedented scale and velocity. The rapid processing and analysis capabilities of HPC amplify AI's predictive analytics, collectively fostering a more dynamic, efficient, and holistic approach to drug discovery.

The combination of AI, HPC, and drug discovery ushers in a major development within the pharmaceutical sciences domain, introducing a synergistic framework identified as HPC+. This integrated approach not just speeds up the drug development process but additionally refines the accuracy of therapeutic interventions, setting the stage for personalized medicine and also precise treatments [17]. In light of global health challenges, such as the COVID-19 pandemic, the capabilities of AI have played a crucial role in expediting the development of vaccines and therapeutic strategies, showcasing the transformative potential of these technologies in swiftly and effectively responding to emergent health crises, knowing that to design the vaccines it is important to be able to define the epitopes that could be potentially predicted by multiple neural network solutions like MARIA, NetMHCPan4 [18,19] and Long Short Term Memory (RNN) [20–22] shown some hopeful results, The NLP was also taken into consideration by applying different approaches of pre-trained transformers that were exploited to predict protein interactions [23]. Figure 2 shows how deep learning had been an integral part of the COVID-19 drug discovery pipeline [24].

Mirroring the insights from the prior subsection on climate science, a detailed theoretical and also technical examination provided in Tables 4 and 5 illustrates the crucial part of high performance computing (HPC) in drug discovery and genomics. This analysis, that synthesizes findings from numerous functions, highlights the utilization of HPC to improve AI algorithms and also computational models which are important for studying large scale genomics data and expediting drug development processes. As demonstrated in previous studies, the melding of HPC with AI not only boosts analytical capabilities in these fields but also promises substantial breakthroughs, similar to those observed in AI applications for climate studies.

Table 2. AI-HPC Climate Applications Part 1

Application	Focus Area	Data Size	Computing Infrastructure	Computing Load	Technical Metrics
Using ML at Scale in HPC Simulations: An Application to Ocean Climate Modeling [12]	Scaling machine learning applications in HPC for simulations	The project conducted a set of ocean simulations across 120 simulated years, utilizing a 12-member ensemble that processed 970 billion inferences, indicating a massive scale of data generation and analysis in climate research	Hardware: Cray XC50 system with database nodes featuring Nvidia P100 GPUs, 64 GB DDR4-2400 memory, 18-core 2.3 GHz Intel Broadwell CPUs. Client nodes with 48-core 2.1 GHz Intel SkyLake CPUs and 192 GB DDR4-2666 memory. Connectivity through the Aries interconnect Software: Uses RedisAI for ML runtimes, supporting a comprehensive performance tuning setup including I/O threads, GPU and CPU worker threads. The infrastructure is capable of connecting from 960 to 7,680 clients simultaneously to the database, with scalability demonstrated from 4 to 16 database nodes	High, due to managing data flows between HPC simulations and ML models for online inference at runtime. Supports scalability across various HPC environments	Throughput: 1.86 million online inferences per second. Efficiency: 6% computational overhead for ML integration. Stability: All 12 ensemble members completed without numerical instability
Performance Modelling for Weather Forecast ML [11]	Optimising weather and climate forecast ML models on HPC	On average, 10 TB of data will be gathered for each application involved in the MAELSTROM project, indicating the large volume of data for training, testing, and running the applications	HPC Systems Used: Aion Supercomputer: Constructed by AMD with 318 computing nodes, 40,704 compute cores, and 81,408 GB RAM, with a peak speed of around 1.70 PFLOP/s. Iris Supercomputer: An Intel supercomputer with 196 computing nodes, 5,824 compute cores, and 52,224 GB RAM, with a peak speed around 1.072 PFLOP/s, and 96 NVIDIA V100 GPU-AI accelerators	High, involves handling data flows: Postprocessing (5 TB), Radiation (2 TB), ENS10 (2.6 TB), Downscaling (300 MB), Power production (1 TB)	Throughput: N/A, Latency: N/A, Efficiency: N/A

Table 3. AI-HPC Climate Applications Part 2

Application	Focus Area	Data Size	Computing Infrastructure	Computing Load	Technical Metrics
FourCastNet: Accelerating Global High-Resolution Weather Forecasting using Adaptive Fourier Neural Operators [13]	Accelerating global high-resolution weather forecasting	For model training, the system uses data from the years 1979–2017, as for the remaining data from 2018 was held for the testing phase, a single year of data could easily exceed tens of petabytes	Utilized three major HPC systems: Perlmutter, JUWELS Booster Module, and Selene. Perlmutter features 1,536 nodes with AMD EPYC 7763 CPUs and NVIDIA A100-40 GB GPUs, connected via NVLink3 and Slingshot 10 interconnect. JUWELS Booster Module includes 936 nodes with AMD EPYC 7402 CPUs and NVIDIA A100-40 GB GPUs, leveraging NVLink3 and HDR200 InfiniBand. Selene comprises 560 DGX-A100 nodes with AMD EPYC 7742 CPUs and NVIDIA A100-80 GB GPUs, utilizing NVLink3 and HDR200 InfiniBand. Supported by an all-flash LUSTRE global file system and employs NVIDIA NGC PyTorch container for software environment across systems.	High, due to extensive model and data parallelism for training and inference at kilometer-scale resolution	Throughput: 160 TFLOP/s on a single GPU as an indicative measure. Latency: Average step time around 517 ms for model instance size 1. Efficiency: Approximately 51% of dense FP16 tensor core peak performance on a single GPU
Canonical Workflows for ML in Climate and Weather Prediction: DeepRain project [14]	Developing reproducible and FAIR ML workflows for climate and weather prediction	More than 1.3 PBytes of meteorological data	Jülich Supercomputing Centre (JSC) implemented Jupyter-JSC with access to CPU and GPU on JUWELS and other HPC systems Integrated with data storage and version control (git)	High, due to processing massive datasets and executing complex ML algorithms for precipitation forecast	Throughput: N/A, Latency: N/A, Efficiency: N/A (Not specified in detail, focus on ensuring FAIR principles and reproducibility of ML practices.)

Table 4. AI-HPC Drug Discovery Applications - Part 1

Application	Focus Area	Data Size and Computing Load	Computing Infrastructure	Technical Metrics
Hybrid ML and Physics-Based Simulations for Drug Discovery [25]	Accelerating COVID-19 drug discovery through integration of machine learning with computational simulations. Targeted crucial SARS-CoV-2 proteins: 3CLPro, PLPro, ADRP, and NSP15	The project managed large-scale datasets typical for high-throughput virtual screening and molecular dynamics simulations, essential for evaluating millions of compounds and generating extensive training data for surrogate models. Demonstrated significant computational demand, executing workloads at 95% usage on Summit and Theta, Managing 24,552 diverse executable tasks and 126×106 python function tasks. Utilized up to 392,000 cores and 24,582 GPUs for these tasks, with a record performance of 144M/hour in docking hits screening. This level of performance, managing both the vast datasets and the intensive computational requirements, underscores the project's substantial utilization of computing resources to process and analyze the data efficiently, achieving a twofold improvement in screening capabilities using over 8000 compute nodes	Utilized the ORNL Summit supercomputer among other HPC resources. Summit is equipped with IBM AC922 nodes, each containing two IBM Power9 CPUs and six NVIDIA Tesla V100 GPUs, interconnected with NVIDIA's high-speed NVLink. The computing environment is optimized for hybrid computational tasks that integrate ML models with physics-based simulations, reflecting a purpose-built infrastructure to support intensive computational chemistry and machine learning workflows	Achieved a twofold improvement in performance for docking hits screening, reaching 144M/hour and screening 1011 ligands using over 8000 compute nodes. This improvement doubled the capacity for screening large compound libraries and producing training data for surrogate models. RCT (RADICAL-Cybertools) was fully functional on Summit and Theta, executing workloads with 95% usage efficiency

Table 5. AI-HPC Drug Discovery Applications - Part 2

Application	Focus Area	Data Size and Computing Load	Computing Infrastructure	Technical Metrics
DeepDriveMD Framework [26]	Biomolecular simulations for drug design and understanding protein dynamics	The DeepDriveMD framework demonstrated significant performance advantages when leveraging concurrent (stream-based) over sequential (file-based) execution. Specifically, in use case 1 (UC1) involving the folding of the BBA protein, DeepDriveMD-S (stream-based) outperformed DeepDriveMD-F (file-based) by conducting 8.4X and 74.9X more training and inference iterations per hour, respectively, and generating 1.5 times more MD data per hour. This increased concurrency allowed DeepDriveMD-S to learn from a larger dataset (50 training epochs vs. 20; 1000–5000 vs. 500–700 outliers), showcasing its efficient use of computing resources and effective data processing capabilities. In this setup, DeepDriveMD-S also avoided writing ~50 GB to disk by streaming data, highlighting its capability to handle large-scale data more effectively without creating I/O bottlenecks	The DeepDriveMD framework's experiments were conducted on the LLNL Lassen HPC platform, utilizing 30 and 33 compute nodes for sequential (DeepDriveMD-F) and concurrent (DeepDriveMD-S) executions, respectively. Both implementations leveraged 120 simulation tasks running on GPUs, with DeepDriveMD-S achieving greater efficiency through continuous simulation execution and avoiding I/O bottlenecks by using ADIOS SST for streaming data communication among components. This setup underscores the framework's adaptable use of computing resources to enhance data processing efficiency. Furthermore, UC2 scaling studies expanded to using 120 to 6120 GPUs on both Summit and Lassen platforms, reflecting the framework's scalability across different HPC environments	DeepDriveMD-S significantly outperformed DeepDriveMD-F by: Producing 1.5X more MD data per hour, highlighting the efficiency of concurrent execution Increased training and inference iterations by 8.4 times and 74.9 times more per hour, respectively, demonstrating the rapid learning and decision-making capabilities enabled by HPC Reducing I/O times to 0.8% and optimizing read times in training and inference from over 1400 s to under 12 s, showcasing significant gains in computational efficiency

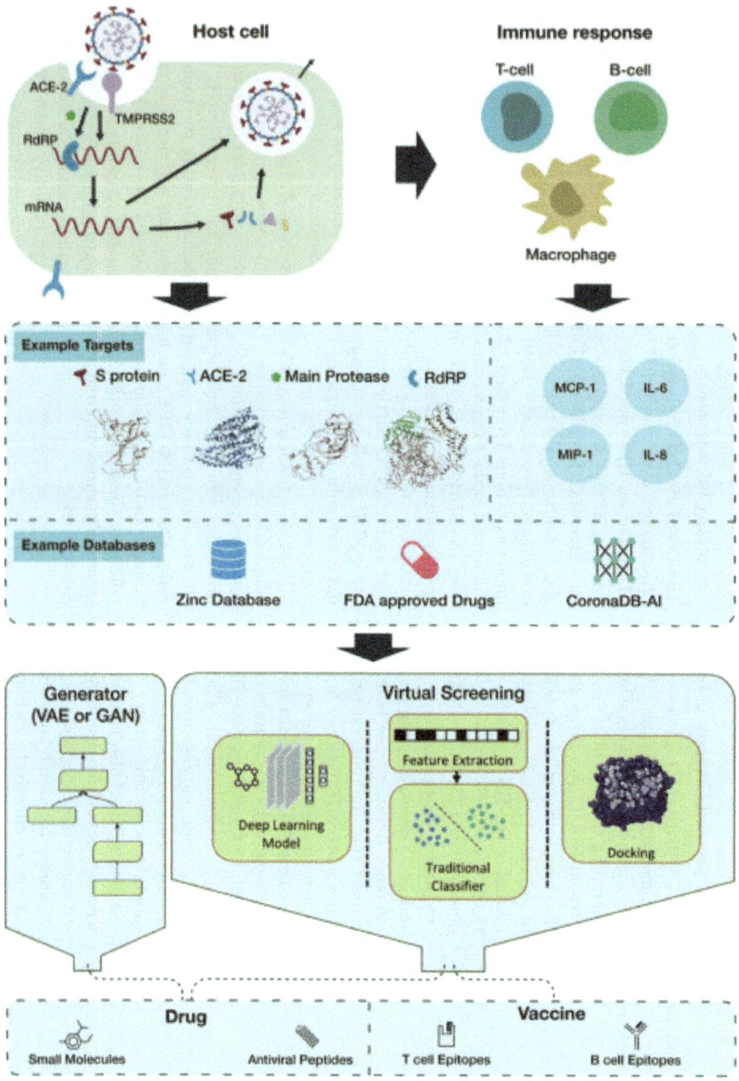

Fig. 2. Vaccine discovery AI-based pipeline

3.3 Large-Scale Language Models

The integration of Artificial Intelligence (AI), into Language Models (LLMs) like GPT and BERT has greatly improved the abilities of natural language processing and machine learning. These models utilize AI to process and learn from extensive datasets, enabling them to recognize complex patterns and generate human-like responses. This function is crucial in a variety of areas, including content creation, language translation services and interactive chatbots, which enhance the interaction between machines and human. The advancement of AI algorithms and model structures continue to boost the effectiveness and precision of these models, pushing their boundaries and broadening their range of uses [27].

LLMs benefit greatly from High Performance Computing (HPC) as it offers the necessary computational power to analyse extensive data sets and perform intricate calculations quickly. Parallel processing techniques are utilised in high-performance computing to improve the speed of calculations, particularly for models containing billions of parameters like LLMs. Making use of advanced computing resources such as GPUs, Tensor Processing Units (TPUs), and extensive distributed computing networks can reduce training times from several weeks to a few days. Such capabilities are crucial for developing more complex models that can capture and interpret the subtleties of human language more effectively, thanks to faster training processes and the ability to experiment with larger and more intricate neural network architectures [28,29].

Furthermore, the synergy between HPC and AI in training LLMs substantially boosts their performance. For instance, the Hanayo framework [30], which incorporates wave-like pipeline parallelism, has achieved a throughput improvement of up to 30.4% (as shown in Table 6) over conventional models like Chimera [30]. Additionally, advanced HPC techniques have improved the scalability and efficiency of models such as GPT and BERT, allowing them to process larger datasets more effectively [31]. This integration not only enhances computational speed and scalability but also optimises resource usage, which is essential for the efficient training of state-of-the-art LLMs. This section examines two pivotal advancements: the application of the Hanayo framework's wave-like pipeline parallelism and the implementation of sophisticated HPC techniques to enhance the capabilities of GPT and BERT models, highlighting the combined improvements brought by HPC and AI.

A Enhanced Efficiency and Scalability through Hanayo [30]

Hanayo Framework Enhancements:

- **Increased Throughput:** Hanayo makes use of wave - like pipeline parallelism to optimise computational flow across several GPU clusters and can boost throughput by as much as 30.4% compared to standard models like Chimera.
- **Memory Optimisation:** Efficiently distributing and utilising loads among devices helps decreasing memory usage, which is essential for handling large models with restricted GPU memory.
- **Scalability and Flexibility:** Shows improved scalability by efficiently increasing the number of pipeline stages (waves) without efficiency loss, enabling the use of bigger models and greatly reducing downtime.

B Advanced HPC Strategies for the GPT and BERT Models [32]

Strategic HPC Implementations

- **Dynamic Model Scaling:** Investigates adaptive scaling strategies that adjust the model's dimensions in response to available high-performance computing resources, guaranteeing optimal computational power utilisation while maintaining model performance.
- **Parallel Training Enhancements:** Introduces improved parallel training approaches that cut training durations while also improving resource utilisation, which is critical for extremely big models like GPT-3 and BERT large.
- **Optimised Data Handling:** It focuses on improving data throughput and reducing latency through advanced data batching and handling techniques, ensuring faster and more reliable model training processes; As the sophisticated batching is often achieved by multiple aspects such as consistency and balance through techniques like stratified sampling where batches represent the overall distribution of the data, making sure that batches are well balanced in terms of classes, which in its turn prevent the model from being biased and thus making the training process more reliable. The same goes for quality control as part of Advanced Data management, where performing it efficiently through handling missing values, outliers, and errors in data leads to ensure cleaner data, resulting into a more trustworthy model training.

C Comparative Performance Metrics. The Hanayo framework and enhanced HPC techniques for LLMs have led to notable enhancements in performance as well as resource efficiency, as outlined in the Table 6.

Table 6. Comparison Performance of different pipeline parallelism models.

Metric	GPipe	DAPPLE	Chimera	Hanayo (Improved)
Throughput Increase (%)	Base	Comparable to GPipe	+20%	+30.4%
Memory Usage (Maximum GB)	Elevated	Moderate	Low	Low (But More Effective)
Memory Balance (Variance)	High	High	Moderate	Low (Best Balanced)
Scalability (Across Devices)	Low	Moderate	High	Very High
Adaptability (Various Clusters)	Poor	Moderate	Good	Excellent
Efficiency Increases (Reduction in Bubble Size)	None	Low	Moderate	High

The Hanayo framework and advanced HPC techniques are examples that demonstrate how HPC significantly affects the development and effectiveness of LLMs. These advancements improve the efficiency and performance of models like GPT and BERT, leading to lower expenses and quicker training. By combining high-performance computing with artificial intelligence during the training of large language models, their abilities are greatly enhanced, enabling them to better adjust to different computing setups and handle more challenging assignments. This cooperation is crucial for advancing the next wave of AI technologies and applications.

4 Computational Aspects of AI

4.1 Growing Computational Complexity of AI Models

The development of AI models, like the GPT series and complex neural networks, has greatly increased the amount of computing power needed. These models have many layers and parameters, requiring a lot of computational resources for both training and inference. There are two new types of computing that could help with this: quantum computing and neuromorphic computing. Quantum computing can do many calculations at once, while neuromorphic computing is designed to work like the human brain. If these technologies continue to advance, they could make AI algorithms much faster and more efficient. This shows how important it is to find new ways to improve the speed and efficiency of AI models, so that they can handle complex data more effectively [33,34].

4.2 How Widespread AI Impact the Computational Demand

The widespread use of AI in various industries has led to a greater need for computing power. With AI being used in everything from instant data analysis to self-driving systems, it is crucial to have a computing infrastructure that is efficient, scalable, and easy to access. This is especially important as the number of users relying on AI applications continues to grow.

To meet this demand, we require robust computing solutions that can handle the heavy workloads generated by AI applications serving millions of users at the same time. One approach to tackling this challenge is through the integration of AI with edge computing. This strategy involves performing data processing tasks closer to where the data is generated, reducing delays and improving overall efficiency.

Furthermore, to support the operational demands of expansive AI initiatives, innovative computing architectures are essential. Techniques such as distributed computing and the use of high-performance servers are instrumental in accommodating the surge in processing requirements. These frameworks facilitate the rapid adjustment of computing resources to match fluctuating demands, ensuring the continuous availability and optimal performance of AI systems [34,35].

5 AI Specific HPC Setups

The demand for accelerated computing in AI applications necessitates the development of specialized High-Performance Computing (HPC) systems, tailored specifically towards supporting GPU-intensive and cloud-based infrastructures. GPUs are pivotal in HPC setups for AI because of their superior parallel processing capabilities. Unlike traditional CPUs, GPUs facilitate a dramatic reduction in the time required to train complex deep learning models, which is crucial for advancing both AI research and practical applications.

For instance, GPU-accelerated tensor operations are central to training neural networks, where each operation can be parallelized across thousands of small cores. The architecture of GPUs, thus, aligns well with the matrix and vector computations that are fundamental to most AI algorithms. This capability is not only beneficial for accelerating the computational speed but also for handling the vast amounts of data typical in deep learning scenarios.

Cloud-based HPC solutions further enhance AI research and deployment flexibility through scalable and on-demand computing resources. These environments allow AI practitioners to efficiently manage workloads with access to advanced hardware, such as GPUs for parallel processing tasks and TPUs for optimized tensor operations in neural network training. An excellent example of cloud HPC in action is the utilization of Google Cloud's TPU technology, which provides specialized hardware designed to accelerate the training phases of deep learning models without requiring significant local hardware investments [36].

Moreover, specific clusters designed for AI-related tasks exemplify the trend towards creating HPC environments that are fine-tuned for AI requirements. For instance, FlinkCL, a system detailed by Chen et al. [37], integrates OpenCL-based in-memory computing with heterogeneous CPU-GPU clusters, optimizing big data processing tasks common in AI applications. Similarly, Naikodi and Sanjay [38] discuss predictive modeling on CPU-GPU clusters using machine learning techniques to forecast the performance of HPC applications, demonstrating effective resource utilization and latency reduction in real-world AI tasks.

In conclusion, the evolution of high-performance computing systems, specifically designed for AI, underscores the critical role of adaptive and highly specialized computational resources in meeting the growing complexities of AI applications. Modern HPC systems not only integrate GPUs and cloud services to tackle these challenges but also evolve continuously to support the dynamic nature of AI research and its application needs. These advancements significantly contribute to the efficiency and scalability of AI technologies, paving the way for more sophisticated and widespread AI applications.

6 What About Computational Efficiency?

6.1 In-Depth Exploration of Computational Efficiency in HPC AI Applications

The computational efficiency of HPC AI applications is paramount, directly impacting the feasibility, scalability, and sustainability of scientific investigations. Evaluating this efficiency requires a nuanced understanding of benchmark metrics, which serve as the yardsticks for performance, accuracy, and resource utilisation.

6.2 Exascale Achievement with HPL-AI on Fugaku

The supercomputer Fugaku, developed by RIKEN and Fujitsu, represents a significant milestone in high-performance computing. Fugaku is based on ARM architecture and uses the A64FX processor, which is optimized for high performance and energy efficiency. It consists of 158,976 nodes and integrates high-bandwidth memory (HBM2), enabling exceptional performance for AI applications [39,40].

Fugaku's system configuration includes a total of 158,976 nodes, providing a memory capacity of 4.85 PiB and a total memory bandwidth of 163 PB/s. The interconnect system, Tofu Interconnect D, achieves low latency and high throughput through a 6D mesh/torus structure [40].

In terms of performance, Fugaku operates in two modes. In normal mode, with a CPU clock speed of 2 GHz, it achieves a theoretical peak performance of 488 PFlops for double precision, 977 PFlops for single precision, 1.95 EFlops for half precision (AI learning), and 3.90 EOps for integer operations (AI inference). In boost mode, with a CPU clock speed of 2.2 GHz, Fugaku achieves 537 PFlops for double precision, 1.07 EFlops for single precision, 2.15 EFlops for half precision, and 4.30 EOps for integer operations [40].

Fugaku's achievement in the HPL-AI benchmark, reaching 1.42 EFlop/s in 2019 with a speedup of 3.42, to reach 2.0 EFlop/s with a speedup of 4.5 in 2020 surpassing its record significantly [41], stands as a landmark in computational efficiency. This performance was highlighted in the work by Kudo et al. [42], which underpins the importance of numerical optimization and precision handling in large-scale computations.

Metrics and Methodologies: The key to Fugaku's performance lies in its approach to mixed-precision arithmetic, optimizing computational workflows to utilize FP16 (half precision) for matrix multiplication, FP32 (single precision), and FP64 (double precision) for ensuring numerical accuracy in iterative refinement processes. This strategy leverages the hardware capabilities efficiently, balancing computational speed and precision to achieve exascale performance.

The HPL-AI benchmark has been released in November 2019, which measures the speed to solve a large system of linear equations with the mixed-precision technique [42]. It refines a core concept from the LINPACK benchmark, but extends to allow low-precision floating point arithmetic like FP32 and FP16. The benchmark includes two significant components, first an LU factorization of a matrix, second an iterative refinement method.

The LU factorization must consist of $\frac{2}{3}n^3 + O(n^2)$ floating-point operations, where accuracy/precision at this stage are not critical. The iterative refinement (IR) solves the system of linear equations with 64-bit precision accuracy. The backward-error is defined as:

$$\frac{\|Ax - b\|_\infty}{\|A\|_\infty \|x\|_\infty + \|b\|_\infty} \times (n \times \epsilon)^{-1} < 16$$

where $Ax = b$ is the target equation, n is the size of the matrix and $\epsilon \approx 1.1 \times 10^{-16}$ is the machine-precision of FP64. The flops-rate is computed as $\left(\frac{2}{3}n^3 + \frac{3}{2}n^2\right)$ /total time-to-solution.

Outcomes: Achieving approximately 90% of theoretical peak performance, Fugaku demonstrated an unparalleled level of computational efficiency. The benchmark underscores the potential of thoughtful numerical analysis and hardware-optimized computing techniques in pushing the boundaries of what's computationally feasible. The effective performance of Fugaku on the HPL-AI benchmark, using 126,720 nodes (approximately five-sixths of the full system), demonstrated 1.421 EFlop/s, approximately 90% of the theoretical peak, marking the world's first achievement to exceed the exascale barrier in a floating-point arithmetic benchmark [42].

6.3 Comprehensive Evaluation with HPC AI500

The HPC AI500 benchmark suite, introduced by Jiang et al. [43], provides a holistic framework for evaluating HPC AI systems across diverse scientific domains. This suite assesses computational efficiency not only in terms of raw performance but also considering the accuracy, power consumption, and cost-effectiveness of AI applications in scientific computing.

Metrics and Methodologies: HPC AI500 uses a comprehensive method to benchmarking, including metrics like time-to-accuracy, which measures how quickly a system can reach a predefined accuracy threshold, and power-to-accuracy, assessing the energy efficiency of achieving the target results. These metrics, along with cost-to-accuracy, provide a complete understanding of computational efficiency, showing the trade-off between performance, resource usage, and operational expenses [44].

The benchmark suite is designed to address the scalability issues inherent in AI workloads. It incorporates the Bagging approach (Bootstrap Aggregation), a method that leverages the collective wisdom of an ensemble of base models to ensure both relevancy and scalability [45]. By utilizing this approach, HPC AI500 can adaptively scale with different sizes of HPC systems, making it suitable for evaluating next-generation scalable HPC AI systems. HPC AI500 V3.0, for example, includes advanced features such as bagging management and model parallelism management, allowing users to customize the number of base models and the degree of parallelism to optimize the trade-off between model quality and training speed [45].

Outcomes: Through its detailed benchmarking framework, HPC AI500 enables a nuanced comparison of HPC AI systems, facilitating insights into how different architectures and optimizations affect efficiency. By covering a wide range of scientific fields and utilizing real-world deep learning (DL) applications, this suite illustrates the practical challenges and achievements in current HPC AI research, guiding future developments toward more efficient computational models [44].

The HPC AI500 benchmark suite has shown near-linear scalability in both FLOPS (Floating point operations per second) and VFLOPS (Valid FLOPS), the latter emphasizing the importance of achieving state-of-the-art quality. The VFLOPS metric integrates system throughput and model quality, penalizing or rewarding the FLOPS based on the achieved quality compared to the target quality. This dual focus on performance and quality ensures that the benchmark provides a realistic measure of an HPC AI system's effectiveness [44,45].

Benchmarks and Applications: The benchmark suite includes workloads from both business and scientific computing. For instance, Image Classification using ResNet-50 and the ImageNet dataset represents a well-known benchmark in the commercial field. In scientific computing, the Extreme Weather Analytics workload uses the Faster-RCNN model and the Extreme Weather Dataset to detect various extreme weather patterns. These benchmarks are selected based on their representativeness, repeatability, and relevance to real-world applications, ensuring a comprehensive evaluation of HPC AI systems [44].

Overall, the HPC AI500 benchmark suite is a versatile and robust tool for assessing the performance and efficiency of HPC AI systems, providing valuable insights for both current and future developments in the field of high-performance computing and artificial intelligence [44].

6.4 Discussion

The exploration of computational efficiency through the lens of benchmarks like HPL-AI on Fugaku and HPC AI500 reveals a multi-dimensional landscape where efficiency transcends raw computational speed. In this context, efficiency encapsulates the adeptness of HPC AI systems in leveraging computational resources to achieve high levels of accuracy in scientific applications, all while minimizing power consumption and operational costs.

These benchmarks illustrate the importance of optimization both in terms of hardware utilization and algorithmic precision handling. They also highlight the evolving requirements of scientific computing, where the ability to conduct fast, accurate, and cost-effective computations is critical.

As artificial intelligence advances scientific research, the knowledge acquired from these standards will be crucial in shaping the evolution of future high-performance computing AI systems. Researchers and engineers are able to guarantee the sustainability and accessibility of advanced tools for a variety of applications by focusing on computational efficiency metrics.

7 Conclusion

Throughout this chapter, we have explored the significant role of High-Performance Computing (HPC) in advancing Artificial Intelligence (AI) across various domains and applications. The symbiotic relationship between AI and HPC is clear; cutting-edge HPC systems are essential for enhancing the efficiency and effectiveness of modern AI models, given their increasing complexity and computational demands. HPC architectures like Shared Memory Machines (SMPs), Distributed Memory Machines (MPPs), Clusters, and Graphics Processing Units (GPUs) provide the computational backbone for these extensive data and algorithm-driven tasks.

These systems are crucial across diverse fields such as climate modeling, drug discovery, genomics, and the development of large-scale language models, supporting tasks that require high throughput and low latency, and adapting to operational demands and performance metrics. The pursuit of exascale computing, particularly highlighted by Fugaku's supercomputer [39] achievements in the HPL-AI benchmark, showcases the peak capabilities of current HPC systems to execute AI applications at unprecedented scales, optimizing time-to-accuracy and cost-to-efficiency ratios.

Looking to the future, the integration of quantum computing with HPC and AI presents a promising frontier. Quantum computing offers the potential to perform complex computations at speeds unattainable by classical systems, potentially revolutionizing fields that rely heavily on AI, such as cryptography, material science, and complex system modeling. Additionally, advancements in neuromorphic computing, which seeks to emulate the neural structure and functioning of the human brain, could further enhance the efficiency of AI computations by reducing energy consumption and increasing processing speeds.

The convergence of green energy and AI within HPC data centers will play a pivotal role in shaping a sustainable future. By leveraging AI-driven algorithms to analyze and optimize complex workflows, HPC systems will achieve unprecedented levels of energy efficiency, reducing carbon emissions and minimizing energy waste. This synergy between green energy and AI will not only contribute to environmental sustainability but also drive cost savings and operational efficiency in HPC data centers.

Moreover, the democratization of AI and HPC through open-source software platforms and cross-platform accelerated programming models will empower a new generation of innovators and researchers. By breaking down barriers to entry and fostering a collaborative ecosystem, these advancements will accelerate the pace of discovery and innovation across various domains, from scientific research to industrial applications.

Advancements in hardware architectures, software frameworks, and algorithmic techniques will continue to evolve, enhancing the integration of AI and HPC. The emergence of specialized AI accelerators will further improve the performance and efficiency of AI workloads within HPC environments. Additionally, the development of hybrid CPU-GPU systems and the rise of edge computing will enable more distributed and localized AI-HPC deployments, bringing computational power closer to data sources and end-users.

Finally, as AI models become more integrated into everyday applications and critical systems, ensuring their security, reliability, and ethical use will be paramount. HPC will play a vital role in developing fast and robust AI systems that can operate securely and ethically in diverse environments.

In conclusion, the future of HPC and AI is intertwined, with advancements in one field driving progress in the other. By harnessing the synergies between these two powerful technologies, we will unlock new frontiers of discovery, innovation, and sustainability. As researchers, developers, and industry leaders, we must continue to collaborate, innovate, and push the boundaries of what is possible, ensuring that the advancements in HPC and AI benefit humanity as a whole.

Acknowledgment. This research was supported by the National Center for Scientific and Technical Research (CNRST) of Morocco under the "PhD-ASsociate Scholarship - PASS" program.

References

1. Deskera, AI revolution in production scheduling for manufacturing, 2023. https://www.deskera.com/blog/ai-revolution-production-scheduling-manufacturing/. Accessed 10 Apr 2024
2. Kasabov, N.K.: Neuroinformatics, neural networks and neurocomputers for brain-inspired computational intelligence. In: 2023 IEEE 17th International Symposium on Applied Computational Intelligence and Informatics (SACI), pp. 000 013–000 014, 2023
3. Buchanan, B.: A (very) brief history of artificial intelligence. AI Mag. **26**, 53–60 (2005)
4. Al-Sahaf, H., et al.: A survey on evolutionary machine learning. J. R. Soc. N. Z. **49**(2), 205–228 (2019)
5. Husin, J.: AI and HPC: The Dynamic Duo Reshaping the Future of Artificial Intelligence, August 2023. publication Title: Sesterce, https://medium.com/sesterce/ai-and-hpc-the-dynamic-duo-reshaping-the-future-of-artificial-intelligence-518bac031f15
6. Becchi, M., Sajjapongse, K., Graves, I., Procter, A., Ravi, V., Chakradhar, S.: A virtual memory based runtime to support multi-tenancy in clusters with GPUs. In: Proceedings of the 21st international symposium on High-Performance Parallel and Distributed Computing, ser. HPDC '12. New York, NY, USA: Association for Computing Machinery, pp. 97–108, June 2012. https://doi.org/10.1145/2287076.2287090
7. Song, F., Dongarra, J.: A scalable framework for heterogeneous GPU-based clusters. In: Proceedings of the Twenty-Fourth Annual ACM Symposium on Parallelism in Algorithms and Architectures, ser. SPAA '12. New York, NY, USA: Association for Computing Machinery, pp. 91–100, 2012. https://doi.org/10.1145/2312005.2312025
8. Kindratenko, V., et al.: GPU clusters for high-performance computing, pp. 1–8, January 2009
9. Vanitha, N., Haritha, J.: An investigation of weather forecasting using machine learning techniques. Int. J. Eng. Manag. Res. **11**(1), 72–78 (2021). number: 1. https://ijemr.vandanapublications.com/index.php/ijemr/article/view/138
10. André, J.-C., et al.: High-performance computing for climate modeling. Bull. Am. Meteorol. Soc. **95**(5), ES97–ES100 (2014). publisher: American Meteorological Society Section: Bulletin of the American Meteorological Society. [Online]. Available: https://journals.ametsoc.org/view/journals/bams/95/5/bams-d-13-00098.1.xml
11. Selvam, K.P., Brorsson, M.: Performance modeling of weather forecast machine learning for efficient HPC. In: 2022 IEEE 42nd International Conference on Distributed Computing Systems (ICDCS), pp. 1268–1269. Bologna, Italy: IEEE, July 2022. https://ieeexplore.ieee.org/document/9912160/
12. Partee, S., et al.: Using machine learning at scale in HPC simulations with SmartSim: an application to ocean climate modeling, April 2021. arXiv:2104.09355 [physics]. http://arxiv.org/abs/2104.09355

13. Kurth, T., et al.: FourCastNet: Accelerating Global High-Resolution Weather Forecasting using Adaptive Fourier Neural Operators, August 2022. arXiv:2208.05419 [physics]. http://arxiv.org/abs/2208.05419
14. Mozaffari, A., et al.: HPC-oriented canonical workflows for machine learning applications in climate and weather prediction. Data Intell. **4**(2), 271–285 (2022). https://direct.mit.edu/dint/article/4/2/271/109844/HPC-oriented-Canonical-Workflows-for-Machine
15. Majeed, A., Lee, S.: Applications of machine learning and high-performance computing in the era of COVID-19. Appl. Syst. Innov. **4**(3) (2021). https://www.mdpi.com/2571-5577/4/3/40
16. Oracle, "Bringing the power of high-performance computing to drug discovery and healthcare," 2021
17. Lewandowski, N., Koller, B.: Transforming medical sciences with high-performance computing, high-performance data analytics and AI. Technol. Health Care: Off. J. Eur. Soc. Eng. Med. **31**(4), 1505–1507 (2023)
18. Chen, B., et al.: Predicting hla class ii antigen presentation through integrated deep learning. Nat. Biotechnol. **37**(11), 1332–1343 (2019)
19. Jurtz, V., Paul, S., Andreatta, M., Marcatili, P., Peters, B., Nielsen, M.: Netmhcpan-4.0: improved peptide-mhc class i interaction predictions integrating eluted ligand and peptide binding affinity data. J. Immunol. **199**(9), 3360–3368 (2017)
20. Park, S.-J., Kim, Y.-G., Park, H.-J.: Identification of rna pseudoknot-binding ligand that inhibits the- 1 ribosomal frameshifting of sars-coronavirus by structure-based virtual screening. J. Am. Chem. Soc. **133**(26), 10 094–10 100 (2011)
21. Yang, D., Leibowitz, J.L.: The structure and functions of coronavirus genomic 3′ and 5′ ends. Virus Res. **206**, 120–133 (2015)
22. Ton, A.T., Gentile, F., Hsing, M., Ban, F., Cherkasov, A.: Rapid identification of potential inhibitors of sars-cov-2 main protease by deep docking of 1.3 billion compounds. Mol. Inform. **39**(8), 2000028 (2020)
23. Nambiar, A., Heflin, M., Liu, S., Maslov, S., Hopkins, M., Ritz, A.: Transforming the language of life: transformer neural networks for protein prediction tasks. In: Proceedings of the 11th ACM International Conference on Bioinformatics, Computational Biology and Health Informatics, pp. 1–8 (2020)
24. Keshavarzi Arshadi, A., et al.: Artificial intelligence for covid-19 drug discovery and vaccine development. Front. Artif. Intell. **3**, 65 (2020)
25. Bhati, A.P., et al.: Pandemic drugs at pandemic speed: infrastructure for accelerating COVID-19 drug discovery with hybrid machine learning- and physics-based simulations on high performance computers. Interface Focus **11**(6), 20210018 (2021). arXiv:2103.02843 [physics, q-bio]. http://arxiv.org/abs/2103.02843
26. Brace, A., et al.: Coupling streaming ai and hpc ensembles to achieve 100–1000× faster biomolecular simulations. In: 2022 IEEE International Parallel and Distributed Processing Symposium (IPDPS), pp. 806–816. IEEE, 2022
27. Devlin, J., Chang, M.-W., Lee, K., Toutanova, K.: Bert: pre-training of deep bidirectional transformers for language understanding, arXiv preprint arXiv:1810.04805, 2018
28. Shoeybi, M., Patwary, M., Puri, R., LeGresley, P., Casper, J., Catanzaro, B.: Megatron-lm: training multi-billion parameter language models using model parallelism, arXiv preprint arXiv:1909.08053, 2019
29. Li, S., et al.: Pytorch distributed: experiences on accelerating data parallel training, arxiv preprint arXiv:2006.15704, 2020

30. Liu, Z., Cheng, S., Zhou, H., You, Y.: Hanayo: harnessing wave-like pipeline parallelism for enhanced large model training efficiency. In: Proceedings of the International Conference for High Performance Computing, Networking, Storage and Analysis, pp. 1–13 (2023)
31. Narayanan, D., et al.: Efficient large-scale language model training on gpu clusters using megatron-lm. In: Proceedings of the International Conference for High Performance Computing, Networking, Storage and Analysis, pp. 1–15, 2021
32. Chen, L., et al.: Position paper: the landscape and challenges of hpc research and llms, arXiv preprint arXiv:2402.02018, 2024
33. Ajani, S.N., Khobragade, P., Dhone, M., Ganguly, B., Shelke, N., Parati, N.: Advancements in computing: emerging trends in computational science with next-generation computing. Int. J. Intell. Syst. Appl. Eng. **12**(7s), 546–559 (2024). https://ijisae.org/index.php/IJISAE/article/view/3231
34. Floridi, L., Chiriatti, M.: GPT-3: its nature, scope, limits, and consequences. Mind. Mach. **30**, 1–14 (2020)
35. Ye, J., et al.: A Comprehensive Capability Analysis of GPT-3 and GPT-3.5 Series Models, December 2023. arXiv:2303.10420 [cs]. http://arxiv.org/abs/2303.10420
36. Sharma, V., Gupta, G.K., Gupta, M.: Performance benchmarking of gpu and tpu on google colaboratory for convolutional neural network. In: Gao, X.-Z., Kumar, R., Srivastava, S., Soni, B.P. (eds.) Applications of Artificial Intelligence in Engineering, pp. 639–646. Springer Singapore, Singapore (2021)
37. Chen, C., Li, K., Ouyang, A., Li, K.: FlinkCL: an OpenCL-based in-memory computing architecture on heterogeneous CPU-GPU clusters for big data. IEEE Trans. Comput. **67**(12), 1765–1779 (2018). https://doi.org/10.1109/TC.2018.2839719
38. Naikodi, C., Sanjay, H.: Prediction Model of an HPC Application on CPU-GPU Cluster using Machine Learning Techniques, pp. 92–97, March 2020
39. Fujitsu, Specifications of supercomputer fugaku, 2024. https://www.fujitsu.com/global/about/innovation/fugaku/specifications/. Accessed 14 Apr 2024
40. "About fugaku," https://www.r-ccs.riken.jp/en/fugaku/about/, accessed: [insert today's date]
41. "Hpl-ai," 2019, accessed: [insert today's date]. https://icl.bitbucket.io/hpl-ai/
42. Kudo, S., Nitadori, K., Ina, T., Imamura, T.: Prompt report on exa-scale HPL-AI benchmark. In: 2020 IEEE International Conference on Cluster Computing (CLUSTER), pp. 418–419, September 2020. iSSN: 2168-9253. https://ieeexplore.ieee.org/document/9229634
43. Jiang, Z., et al.: HPC AI500: A Benchmark Suite for HPC AI Systems, November 2019. arXiv:1908.02607 [cs]. http://arxiv.org/abs/1908.02607
44. Jiang, Z., et al.: : HPC AI500: representative, repeatable and simple hpc ai benchmarking, arXiv preprint arXiv:2102.12848, 2021
45. Jiang, Z., Luo, C., Gao, W., Wang, L., Zhan, J.: HPC AI500 v3. 0: a scalable HPC AI benchmarking framework. BenchCouncil Trans. Benchmarks Stand. Eval. **2**(4), 100083 (2022)

Facts and Issues of Neural Networks for Numerical Simulation

Imad Kissami[1], Christophe Cérin[2,3](✉), Fayssal Benkhaldoun[2], and Fahd Kalloubi[1,4]

[1] Mohammed VI Polytechnic University, Ben Guerir, Morocco
{imad.kissami,fahd.kalloubi}@um6p.ma
[2] University Sorbonne Paris Nord, Villetaneuse, France
{christophe.cerin,fayssal.benkhaldoun}@univ-paris13.fr
[3] INRIA Datamove team, Villetaneuse, France
[4] LTI Laboratory, Université Chouaib Doukkali, El Jadida, Morocco

Abstract. Deep learning and AI has revolutionized the field of computer science and have now become the dominant approach to solving a wide problem range, from partial differential equations to molecular discovery. Indeed, numerical simulation techniques have traditionally relied on solving physically derived equations using finite differences and adding heuristic models when they become too complex to solve (Turbulence models in fluid mechanics, for example). This class of techniques remains (computationally) (very) expensive; we are talking about hours to days of simulation on petaflop or exaflop machines. The situation is even worse because the simulation must restart when an engineer wants to change the shape. Today, to achieve an accurate prediction of the simulation at low time and cost, the process is simulated based on an artificial neural network. Different artificial neural networks have been recently developed for progressive prediction of the phenomenon under consideration. The chapter overviews the most recent artificial neural networks proposed by the research communities, explains the pros and cons, and shows the limits and yet unsolved problems. The chapter also illustrates the considered challenges, most of the time, through different partial differential equations.

1 Introduction

Faced with the rise of artificial intelligence (AI), governmental authorities assigned public institutions a lead role in developing AI tools for fundamental research, energy technology development, and national security. For instance in France[1] or in the United States[2], the communities wrote numerous reports to alert on the potential of AI for sciences.

[1] https://www.cnrs.fr/fr/le-centre-artificial-intelligence-science-science-artificial-intelligence-aissai.

[2] https://www.anl.gov/sites/www/files/2023-06/AI4SESReport-2023-v6.pdf.

M. Zbakh et al. (Eds.): CloudTech 2023, LNNS 1220, pp. 160–181, 2024.
https://doi.org/10.1007/978-3-031-78698-3_9

In the context of our book chapter proposal, the US report describes, on page 17, surrogate models as "simpler yet faithful" representations of complex, real-life systems, with the models themselves trained on the outputs of other computational models. It recommends launching pilot programs to develop surrogate models of plasma turbulence and the Earth's oceans, among different systems.

Using deep learning to solve governing PDEs has become a central focus in scientific machine learning [22,40,41,61] (SciML), leveraging universal approximation and high expressivity of neural networks. On the one hand, deep neural networks could approximate any high-dimensional function, given sufficient training data in the input. However, such networks do not consider the physical characteristics underlying the problem. Moreover, their level of approximation accuracy is still heavily dependent on careful specifications of the problem geometry and the initial and boundary conditions. Without this preliminary information, the solution is not unique and may lose physical correctness.

In contrast, Neural Networks for solving PDEs leverage governing physical equations in neural network training. These neural networks are designed and trained to satisfy the given training data and the imposed governing equations. In this fashion, training data that do not necessarily need to be numerous and complete guides the evolution of neural networks. These networks may attempt to find an optimal solution without precise knowledge of the boundary conditions, relying solely on the physical constraints of the problem. Therefore, with some knowledge about the physical characteristics of the problem and some form of training data (even sparse and incomplete), neural networks for solving PDEs may be used to find an optimal solution with high fidelity.

Neural networks for solving PDEs allow for addressing a wide range of problems in computational science and represent a pioneering technology leading to the development of new classes of numerical solvers for partial differential equations (PDEs). These neural networks represent a mesh-free alternative to traditional approaches such as CFD, Electromagnetics, and Quantum Mechanics, or new data-driven approaches for model inversion and system identification. Notably, the trained neural network can predict the values on simulation grids of different resolutions without the need to be trained again. In addition, they allow for the exploitation of automatic differentiation (AD) to compute the required derivatives in the partial differential equations, a new class of differentiation techniques widely used to derive neural networks that are assessed to be superior to numerical or symbolic differentiation.

The recognition of neural networks for solving PDEs as a breakthrough in scientific machine learning is tempered by challenges such as unclear convergence behaviors, scalability issues in complex or high-dimensional domains, and the difficulty in selecting optimal neural architectures. Various approaches have been developed to address these challenges, such as Physics-Informed Neural Networks (PINNs) [31,65], Neural Operators (NO), Transformer Networks (TN), and Newton Informed Neural Operator (NINO). These advanced methodologies offer promising solutions to the inherent challenges of neural networks for solving PDEs, improving their efficiency, and broadening their applicability across diverse scientific problems.

However, these advanced methods are not universally applicable to all problems. Each scientific problem requires a tailored approach, considering its unique characteristics and complexities. These enhancements improve upon the foundational Neural Network approach for solving PDEs but require careful consideration of the given situation to determine its suitability and potential.

In the context of PDEs, the chapter's objective is to shed light on scientific and technical issues of neural networks for numerical simulation. Thus, the book chapter organization is as follows. Section 2 puts in place all the introductory material. First, we introduce a brief history of numerical simulation and neural networks. Second, we present the conventional methods versus data-driven methods for numerical methods. Third, we give reminders on relevant neural networks (Ex: Physics Informed Neural Networks) and examples of partial differential equations (Ex: Advection Equation, Burgers' equation, Shallow water equations, Navier-Stokes Equations).

Section 3 introduces different Neural Network architectures for solving PDEs. We discuss their challenges and notice that training the model can be more expensive than the resolution. The model is not reproducible since we train a new one each time we change the domain or initial condition. Finally, we consider a better understanding of the convergence on the theoretical side, otherwise, the methods are not scalable, and not robust.

In Sect. 4 we evocate the performance evaluation problem: machine learning metrics and HPC metrics. Should we mix them or consider them each on their own? At least, we discuss frugal methods and algorithms in the context. We also open our discussion to issues related, mainly to a System view of the ecosystem. Section 5 relates programming use cases available on the Web to illustrate the practicability of the notions. Section 6 concludes the book chapter.

2 Vocabulary and Fundamentals

2.1 A Brief History of Numerical Simulation and Neural Networks

Neural Networks. Neural networks have a long history, beginning with early attempts to understand how the brain functions and behaves. In 300 BC, Aristotle established the notion of Associationism [8], marking the beginning of humans' efforts to understand the brain. In the 19th century, Alexander Bain introduced Neural Groupings as early neural network models, eventually influencing the Hebbian Learning Rule [64]. Donald Hebb is regarded as the father of neural networks after inventing the Hebbian Learning Rule in 1949, which laid the groundwork for current neural networks. McCulloch and Pitts introduced the MCP Model in 1943 [45], regarded as the progenitor of artificial neural models. In 1958, Frank Rosenblatt introduced the first perceptron [54], resembling modern perceptions. Over the years, neural networks have evolved significantly, leading to the emergence of deep learning models such as Convolutional Neural

Networks (CNNs) [21,38], Deep Belief Networks (DBNs) [26] [28], and Recurrent Neural Networks (RNNs) [46,47]. These models have revolutionized various fields, including computer vision, natural language processing, and speech recognition. In recent years, many enhancements were proposed to improve neural networks convergence, accuracy, and training speed, such as Batch Normalization [30], Dropout [58], GELU [25], Transformer [62], etc.

Numerical Simulation. The concept of numerical simulation [59] dates back, to the pioneering work of 18th-century mathematicians such as Euler and Lagrange who paved the way for differential equations and calculus [19] [37]. The advancement of computers in the 20th century with John von Neumanns contributions propelled numerical simulation by introducing algorithms for solving crucial partial differential equations in fields like hydrodynamics and aerodynamics. In the mid-20th century the finite element method (FEM) emerged, transforming engineering analysis by enabling structures to be broken down into smaller elements [18]. Concurrently, the finite difference method (FDM) was developed to approximate derivatives and solve differential equations on a grid [53]. The 1970s marked not only the growth of fluid dynamics (CFD) but also the introduction of the finite volume method (FVM) [1] which focuses on conserving fluxes across control volume boundaries and is especially suitable for CFD applications [4,43]. The availability of simulation software during this era made these computational tools more accessible, across industries [2].

2.2 Conventional Methods Versus Data-Driven Methods

Conventional Methods. The computational modeling process of any physical process consists of two important stages: (1) a mathematical model development, and (2) computer simulation. The first step in the process is to build a mathematical model that accurately reflects the physical reality. This model is built on governing equations that describe various physical phenomena. For example, the Navier-Stokes equations [12] represent fluid dynamics, while the elasticity equations [27] describe solid deformation under external forces. In electromagnetism, Maxwell's equations [29] provide a complete description of electric and magnetic fields and their interactions. Given the complexity of these equations, simplifications are often necessary to make the problems solvable, analytically or numerically. An example is Laplace's equation utilization for potential flow in scenarios where fluid viscosity's impact is negligible, simplifying the Navier-Stokes equations.

After establishing a mathematical model and defining proper boundaries (system behavior at the domain's edges) and initial conditions (starting values for simulations), one's attention commutes to the solution's model, using numerical methods described in Sect. 2.1. Time-stepping techniques [11,33] advance the solution in discrete intervals, and concepts like convergence and stability [57] ensure accurate and reliable results. Solvers and error analysis address the computational aspects, and parallel computing enhances efficiency [7,17], especially for large-scale simulations.

Data-Driven Methods. Data-driven techniques in solving PDEs represent an important evolution from traditional methodologies, including advanced machine learning (ML) algorithms to improve and speed simulation processes [34,36,63]. These novel models live on their ability to process large informational sets, interpreting subtle patterns and underlying physics that typical numerical approaches miss. One major achievement assisted by ML is its ability to discover and correct discretization mistakes in PDE simulations, increasing precision and reducing computing burden.

This method improves error correction but also excels at anticipating complicated fluid dynamics phenomena such as turbulence [69] and fluctuating aerodynamic motions with high accuracy [36]. The incorporation of machine learning (ML) into PDEs has resulted in the development of neural network-based approaches, such as deep learning [9] and [50], allowing for a more accurate representation of fluid flow. This development allows for the seamless inclusion of experimental and observational data directly into simulations, effectively closing the gap between theoretical constructions and actual phenomena. This integration makes simulations evolve in real time, responding to new data, and considerably improving the prediction potential of PDE models.

Furthermore, data-driven procedures offer a solid foundation for traversing the complex geometries and flow conditions that traditional PDE-solving methods may find difficult. Through the lens of data, ML models are effective at interpreting fluid dynamics when traditional physics-based models may require simplification or incur prohibitive computing costs. The seamless integration of data-driven models with existing PDE-solving approaches ushers in a new age of computational fluid dynamics, marked by faster and more cost-effective simulations and much more accuracy. This innovation has important implications for fields such as engineering and environmental research, offering a future in which simulations are more closely matched with real-world dynamics.

2.3 Reminders on Relevant Neural Networks for Solving PDE

Introduction. Neural networks have experienced tremendous success in solving PDE in different fields, such as biology and physics. Function learning techniques are meant to learn the solution function themselves to find the approximate solutions to PDE. Function learning methods use Physics-Informed Neural Network (PINN) techniques. Although the capability of a PINN to solve an ill-posed direct problem, finding the optimal neural architecture is still a great challenge. Many methods have been proposed to tackle the limitations of PINNs, such as operator learning. Figure 1 shows a classification schema of existing Neural Network architectures for solving PDE. In what follows, every architecture will be dissected.

Physics Informed Neural Networks. Physics-Informed Neural Networks (PINNs) [32,44,52,55,68] are a strong tool for tackling issues related to solving PDE issues by putting physical rules directly into the learning process [32,44,52].

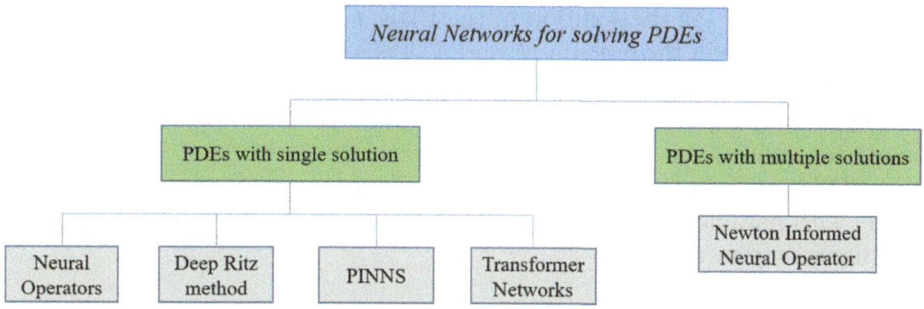

Fig. 1. A classification diagram of Neural Network architectures for solving PDEs.

The PINN process for PDE consists of several fundamental steps. First, the PDE problem should be formulated: both equations to be solved and information to be obtained, compared and interpreted. In particular, physical laws that hold in the system should be specified and encoded in the neural networks' structure.

This commonly involves the architecture of the used neural networks and the activation functions applied. Second, the structure of the neural network is designed: it has to be sophisticated enough to solve a particular problem and deep enough to capture all the effects expected in the fluid motion. Third, the network is trained with the data provided, and the results produced after the network training should be interpreted, as these inference results give information about the fluid dynamics simulation and the simulation's correctness.

Several reasons make PINNs especially convenient for solving PDE. First, the capability of imitating physical laws allows using them in cases when solutions by other numerical methods would be less accurate. Second, in many problems, for example, when information about the flow is difficult to obtain, sparse, or expensive to reproduce, it is common to get such information and use it as training data along with the neural network's predictions about these data in conjunction with the governing laws of fluid motion, ensuring that all solutions derived under such constraints maintain physical consistency. Thus, PINNs solve PDE problems with higher accuracy and productivity. Overall, as the possibilities for PDE simulations grow with the development of data-driven physics-informed, and more comprehensive neural networks, PINNs appear to be an effective tool for solving such problems.

Unlike conventional numerical analysis methods like FEM, PINNs can solve problems where traditional requirements (i.e., governing equations, material properties, geometry, and boundary conditions) can be substituted or augmented with empirical data. PINNs problems are generally formulated as follows [55]:

$$\begin{aligned} \boldsymbol{u}_t + \mathcal{N}[\boldsymbol{u}] &= 0, \quad t \in [0,T], \boldsymbol{x} \in \Omega, \\ \boldsymbol{u}(0,\boldsymbol{x}) &= \boldsymbol{g}(\boldsymbol{x}), \quad \boldsymbol{x} \in \Omega, \\ \mathcal{B}[\boldsymbol{u}] &= 0, \quad t \in [0,T], \boldsymbol{x} \in \partial\Omega, \end{aligned} \tag{1}$$

where $\boldsymbol{u}(t, \boldsymbol{x}) \in \mathbb{R}^{d_u}$ is the solution to the PDE, t denotes time, $\boldsymbol{x}$ is a vector of spatial coordinates in the domain Ω, and $\mathcal{N}[\cdot]$ is a linear or nonlinear differential operator. The function g describes the initial condition (IC) of the PDE, and $\mathcal{B}[\cdot]$ is a boundary operator corresponding to Dirichlet, Neumann, Robin, or periodic boundary conditions.

In PINNs, the solution $\boldsymbol{u}(t, x)$ of the PDE is represented by a neural network $\boldsymbol{u}_\theta(t, x)$ with parameters θ. This model can then be trained by minimizing a loss with three components:

$$\mathcal{L}(\theta) = \lambda_i \mathcal{L}_i(\theta) + \lambda_b \mathcal{L}_b(\theta) + \lambda_f \mathcal{L}_f(\theta). \tag{2}$$

The first two terms consist of a supervised loss which guarantees that the function learned by the neural network satisfies the initial and boundary conditions of the problem. That is, given randomly sampled points $\{\boldsymbol{x}_i^i\}_{i=1}^{N_i}$ from Ω and $\{t_b^i, x_b^i\}_{i=1}^{N_b}$ from $[0, T] \times \partial\Omega$, we have:

$$\mathcal{L}_i(\theta) = \frac{1}{N_i} \sum_{i=1}^{N_i} \left\| \boldsymbol{u}_\theta\left(0, \boldsymbol{x}_i^i\right) - \boldsymbol{g}\left(\boldsymbol{x}_i^i\right) \right\|_2^2, \quad \mathcal{L}_b(\theta) = \frac{1}{N_b} \sum_{i=1}^{N_b} \left\| \mathcal{B}\left[\boldsymbol{u}_\theta\right]\left(t_b^i, \boldsymbol{x}_b^i\right) \right\|_2^2. \tag{3}$$

The third term is the physics-informed objective (i.e., the governing equations), ensuring that the learned function $\boldsymbol{u}_\theta(t, x)$ satisfies the PDE. This is done by minimizing:

$$\mathcal{L}_f(\theta) = \frac{1}{N_f} \sum_{i=1}^{N_f} \left\| \frac{\partial \boldsymbol{u}_\theta}{\partial t}(t, x) + \mathcal{N}\left[\boldsymbol{u}_\theta\right](t, \boldsymbol{x}) \right\|_2^2 \tag{4}$$

where the derivatives of $\boldsymbol{u}_\theta$ are calculated through automatic differentiation. This penalty is enforced on a set $\left\{t_f^i, x_f^i\right\}_{i=1}^{N_f}$ of randomly sampled points from the domain. Furthermore, these losses are weighted by hyper-parameters $\{\lambda_i, \lambda_b, \lambda_f\}$, leading to more flexibility during training. Figure 2 shows a general PINN framework to solve this problem.

PINNs Variant: FI-PINNs (failure-informed PINNs) [20] use adaptive sampling techniques to target under-predicted regions, often known as "failure regions" improving predictability by fine-tuning the model's learning process, particularly in difficult areas. HPINNs (Hard-Constrained Physics-Informed Neural Networks) [42] solve inverse design issues with tight constraints by incorporating hard constraints into a traditional PINN framework utilizing approaches such as the penalty method or augmented Lagrangian.

gPINNs [66] include gradient information from PDE residuals into the loss function to achieve excellent accuracy and training efficiency, especially in steep gradients. Bilevel PINNs [23] tackle PDE-constrained optimization issues using a bilevel optimization module, with the inner loop addressing PDE constraints with PINNs and the outer loop optimizing objective functions, frequently employing sophisticated techniques such as Broyden's hypergradients.

FPINNs (Fractional Physics-Informed Neural Networks) [51] solve issues defined by fractional differential equations and are appropriate for modeling complicated anomalous diffusion processes in materials science and geophysics.

XPINNs (eXtended Physics-Informed Neural Networks) [31] efficiently manage multi-domain problems by partitioning the computational domain into subdomains, each trained with a different neural network. This increases parallelism and computational efficiency while dealing with complex geometries and changing boundary conditions. Bayesian PINNs [65] use Bayesian inference to quantify uncertainty in predictions, offering both point estimates and uncertainty estimation, which is important for applications such as geological simulations and predictive maintenance.

Deep Ritz Method: Deep Ritz method [67] has been proposed for numerically solving variational problems, especially those arising from PDE. The deep Ritz method is nonlinear, adaptive, and tailored to work in high dimensions. Also, the deep Ritz method is reasonably simple and fits well with stochastic gradient descent. However, as stated by the authors, the issue of local minima and saddle points is non-trivial, and there is no consistent conclusion about the convergence rate. Furthermore, treating the essential boundary condition is harder than traditional methods. To tackle the saddle-point problem, the authors in [61] propose a deep double Ritz method combining two neural networks for approximating trial and optimal test functions along a nested double Ritz minimization strategy.

Neural Operators: Neural operators improve the capabilities of ordinary neural networks by learning mappings across function spaces, making them ideal for modeling spatiotemporal processes and PDE. The Fourier Neural Operator (FNO) [39] defines the integral kernel in Fourier space, enabling expressive and efficient systems capable of handling high-resolution issues. FNOs have shown considerable accuracy and computing efficiency gains, especially in fluid dynamics applications. Wavelet Neural Operators (WNOs) [60] use wavelet transformations and neural networks to solve highly nonlinear PDE, using wavelets' spatial and frequency localization features to successfully handle complicated geometry and boundary constraints. Graph Neural Operators (GNOs) [40], such as the Multipole Graph Neural Operator (MGNO) [41], employ graph neural networks to capture spatial relationships and interactions between nodes, efficiently handling long-range interactions through a hierarchical, multi-scale graph structure inspired by the fast multipole method. MGNOs have demonstrated remarkable performance in problems such as Darcy flow and Burgers' equation, achieving great accuracy and computing efficiency.

Physics-Informed Transformer Networks: Physics-Informed Transformer Networks (PITs) [16] combine the benefits of transformer designs and physics-based learning techniques. These networks use attention processes to record dependencies across different portions of the input space, making them ideal for dealing with irregular grids and large-scale challenges. PITs use data and physics-based constraints to provide strong solutions to time-varying PDE and other complicated problems, allowing for better generalization to new domains and efficient modeling of dynamic systems.

Newton Informed Neural Operator: Classical neural network methods for solving nonlinear PDE, such as PINN, Deep Ritz methods, and DeepONet, are limited when multiple solutions are inherent in the nonlinear problem, due to an ill-posed problem. Newton informed neural operator [22] combines classical Newton methods in a single learning process while requiring fewer supervised data points compared to existing neural network methods.

2.4 Examples of Partial Differential Equations

PDEs are mathematical equations involving functions over multiple variables and their partial derivatives. They are fundamental in describing numerous physical phenomena such as fluid flow, heat transfer, and wave propagation. Below are some key examples of PDEs commonly encountered in scientific and engineering problems:

Navier-Stokes Equations. The Navier-Stokes equations describe the motion of fluid substances such as liquids and gases. These equations express the momentum and mass conservation principles and are fundamental in fluid dynamics. The general form of the Navier-Stokes equations for an incompressible flow is:

$$\frac{\partial \mathbf{u}}{\partial t} + (\mathbf{u} \cdot \nabla)\mathbf{u} = -\nabla p + \nu \nabla^2 \mathbf{u} + \mathbf{f} \tag{5}$$

$$\nabla \cdot \mathbf{u} = 0 \tag{6}$$

where $\mathbf{u}$ is the velocity field, p is the pressure field, ν is the kinematic viscosity, and $\mathbf{f}$ represents external forces.

Burgers' Equation. Burgers' equation is a fundamental partial differential equation from fluid mechanics. It is used to model various types of wave phenomena and is given by:

$$\frac{\partial u}{\partial t} + u\frac{\partial u}{\partial x} = \nu \frac{\partial^2 u}{\partial x^2} \tag{7}$$

where u is the velocity field and ν is the viscosity.

Laplace's Equation. Laplace's equation is a second-order partial differential equation often used in fluid dynamics, electromagnetism, and other fields to describe the behavior of scalar fields like electric potential or fluid velocity potential. It is given by:

$$\nabla^2 \phi = 0 \tag{8}$$

where ϕ is the scalar potential function. In Cartesian coordinates, for a function $\phi(x, y, z)$, the equation expands to:

$$\frac{\partial^2 \phi}{\partial x^2} + \frac{\partial^2 \phi}{\partial y^2} + \frac{\partial^2 \phi}{\partial z^2} = 0 \tag{9}$$

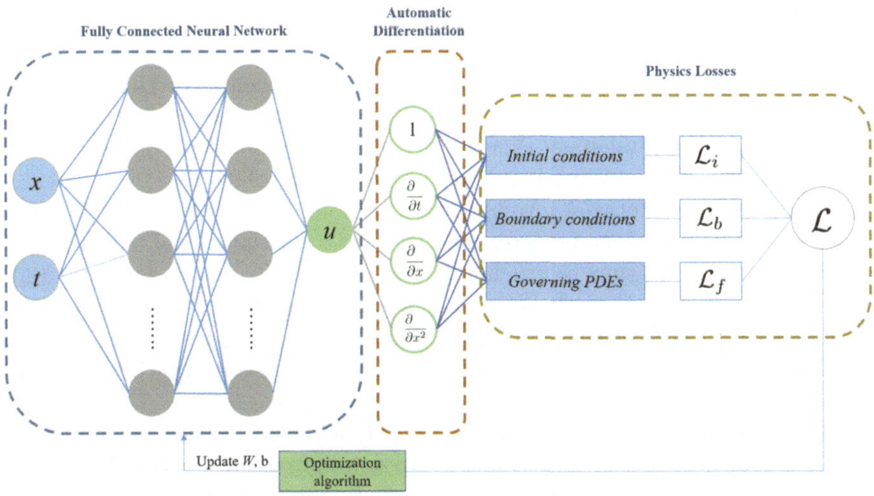

Fig. 2. A general framework to solve the forward problem of nonlinear PDEs [68].

Inverse Problems Involving Nonlinear PDEs. Inverse problems involve determining unknown parameters or inputs of a PDE from known outputs or observations. For example, in fluid dynamics, one might seek to determine the source term **f** in the Navier-Stokes equations from observed velocity and pressure fields. These problems are often formulated as optimization problems, where the objective is to minimize the difference between the observed data and the model predictions, subject to the constraints imposed by the PDEs.

3 Challenges Associated with Neural Networks for Solving PDEs

3.1 Limitations of Neural Networks for Solving PDEs

Neural Networks have demonstrated high potential in solving challenging tasks involving PDEs. However, they have several limitations that should be mentioned as well, including:

- Divergence Behavior: Unlike conventional numerical procedures, we also need to consider these neural networks have a well-defined convergence behavior. Determining how accurately a neural network will handle a given problem is difficult. The primary reason for this difficulty is that there can be multiple local minima that the optimization procedure might get stuck in. The unknown variables in Neural Networks for solving PDEs are the weights of the neural networks. They are determined using an iterative solver that is less robust than the conventional numerical solvers.

- Scalability: These Neural Networks for solving can be difficult to investigate when considering high-dimensional or complicated domain spaces. With the rise in the complexity of the problem, the requirements for computational resources and memory also become more stringent. It means that it is often too expensive to implement Neural Networks for simulating extensive real-world problems demanding highly accurate spatial solutions.
- Optimal Neural Network Architecture: Before a given task is properly explored, it is necessary to determine the design of the neural network. The network's configuration is crucial as performance increases with the right number of neurons in each layer. Besides, the use of the correct activation functions is also essential. This point may take a long time to investigate and develop the optimal neural network, and it is a computationally costly process.
- Data-Gathering Issues: Even if the neural network handles incomplete or limited information the data quality and volume remain crucial. Practically speaking, collecting enough data that sufficiently describes a given problem is always problematic. The same issue arises when implementing a task using a neural network. The size and quality of the data impact the performance of the network.
- The Necessity of Retraining: A neural network requires re-training in most situations every time a new problem is encountered, or the problem is transformed, and the domain of the issue is modified.
- No Complete Reliance on Physical Information Ensures Applicability: Finally, it is impossible to state that including the physical laws of the objects investigated in the neural network's learning ensures perfect application. Whether a given solution provided by the networks is physically correct can depend heavily on the approximation accuracy. For the correct problem, it is necessary to ensure that the problem's geometry, initial conditions, and boundary conditions are precisely defined.

3.2 Understanding of the Convergence of Neural Networks

In neural networks, convergence is defined based on the values of the training loss function. If training loss function values follow a descending trend (assuming we are minimizing the loss function, which is the standard) then it converges.

It is important to study the convergence behavior of neural networks in terms of solving partial differential equations (PDEs), especially as they concern the PDE. In this sense, convergence refers to the ability of a neural network to provide successively better approximations to the true solution of a PDE as training goes on. This point emphasizes why ensuring and understanding convergence is necessary for neural-network-based techniques.

Convergence in Physics-Informed Neural Networks (PINNs). Several factors determine how PINNs converge including choice of neural network architecture, quality and quantity of training data, and optimization algorithms used.

The PINN model seeks to minimize a loss function that combines data and physical knowledge about the problem by combining them, for example through addition or averaging. A typical loss function includes terms that enforce initial/boundary conditions and governing equations.

- Optimization Landscape: The optimization landscape of PINNs can become complex. Trapping the optimization procedure, there are multiple local minima in this landscape. Adaptive learning rates, advanced gradient descent methods, and regularization techniques are utilized to navigate this landscape more effectively and promote convergence towards a global or near-global minimum.
- Loss Function Composition: Composition of the loss function is vital. PINNs use a composite loss function including data-driven terms and physics-based terms. The balance between these terms significantly affects the convergence behavior and is often regulated through hyperparameters. The slow convergence rate and sub-optimal solutions may occur due to improper weighting.
- Gradient Pathologies: Convergence of PINNs can be hindered by gradient pathologies such as vanishing or exploding gradients. These problems can be alleviated and training process stability improved through advanced activation functions like rectified linear units (ReLU), scaled exponential linear units (SELU), and batch normalization techniques.

Convergence in Neural Operators. Fourier Neural Operators and Wavelet Neural Operators are some of the neural operators that learn mapping across function spaces. This leads to challenging convergence situations. They must accurately represent intricate functional relationships which necessitates careful design and training.

- Spectral Methods: Spectral methods, such as Fourier and Wavelet Neural Operators, have been used in this invention. These can improve convergence by efficiently capturing the dominant modes of a solution. Nonetheless, these neural operators should be trained extensively to converge on different problem domains and scales.
- Transfer Learning: Convergence can also be improved through transfer learning whereby neural operator models are pre-trained on related tasks before being retrained with task-specific datasets. It is possible to speed up convergence and enhance the generalization abilities of neural operators using this strategy.

Convergence in Transformer Networks. The transformer networks applied to numerics simulations feature attention mechanisms that help handle cross-dependencies on input spaces. The complexity of the attention and scale of input data can affect their convergence behavior.

- Attention Mechanisms: Long-range dependencies in data are effectively modeled through the self-attention mechanisms of transformer networks, which

may lead to more robust convergence. However, this is not without challenges as it could be expensive to compute and require large datasets for training.
- Scalability and Efficiency: A double-edged sword is termed scalability in transformer networks. While they are effective with massive problems, ensuring convergence in high-dimensional spaces calls for efficient training methods such as gradient clipping and distributed training techniques.

Convergence in Newton Informed Neural Operators. A combination of classical Newton methods and neural network architectures is known as the Newton Informed Neural Operators. The aim is to enhance convergence for nonlinear PDEs with multiple solutions.

- Hybrid Optimization: The integration of Newton's method offers a reliable framework for managing multiple solutions with improved convergence rates. It involves fewer supervised data points and exploits strengths from classical numerical methods and neural networks.
- Iterative Refinement: Newton-informed iterative refinement techniques help converge to accurate solutions by progressively improving the approximation with each iteration within the framework. This method is particularly effective for problems with complex solution landscapes.

The convergence of neural network-based numerical simulations is dependent upon the following: the architecture of the network, the design of the loss function, the optimization algorithm, and features of the problem domain. Neural operators, PINNs, Transformer Networks, and Newton Informed N-operators are important neural network approaches but further studies are necessary to understand and optimize their convergence behavior. Overcoming these challenges will enhance the dependability and relevance of artificial neural networks when dealing with complex scientific and engineering problems that involve numerical computations.

4 Discussion

4.1 Performance Metrics

The evaluation of a numerical simulation program is usually based on the program's execution time for a given problem size. A scalability test is generally carried out by varying the problem size and continuing to measure the execution time. However, the implementation we want to evaluate the performance often requires p processors. The scalability test consists in varying the number of cores on a certain scale (e.g. 2, 4, 8, 16, 32, 64 cores) and measuring the execution time, for a given data size. In this way, we can determine the execution time in each case, and see how the curve varies from case to case. For example, doubling the number of cores does not necessarily mean halving the execution time.

In addition, scalability testing can also be combined with measurements of:

- response time;
- requests per second, transactions per second;
- performance related to the number of users on the system;
- CPU and RAM usage during tests;
- network utilization - number of data items sent and received.

All the previous metrics are quantitative. In PDE modeling, mesh generation is a complicated yet important process in complex numerical computation. In a fluid system, mesh breaks down the 3D model into thousands of smaller flow domains where the governing equations can be discretized and resolved for each domain. The quality of the mesh generated plays an important role in ensuring the accuracy and stability of the simulation results.

Since meshing is one of the bottlenecks in the simulation process, ensuring the mesh is of high quality is important for fast and accurate analysis. For mesh quality assessment, it is necessary to understand the criteria that dictate this analysis. PDE solvers can help engineers in this analysis through mesh quality evaluation and error estimation.

In fluid system simulation, the challenge starts with the system geometry. Mesh enables discretization of the geometry ensuring all relevant flow features are captured uniquely in each cell. This can, however, be challenging given the domain's complexity, the geometric model's complexity, and the type of mesh desired.

Regarding the good properties we may expect for a simulation result, we refer to [49]. This tutorial explains, for instance, the notions of errors and uncertainty.

Performance metrics in machine learning present a different aspect. First, loss functions show a measure of model performance. They're used to train a machine learning model (using some optimization like Gradient Descent) and are usually differentiable in the model's parameters. Metrics, which are different from loss functions, are used to monitor and measure the performance of a model (during training and testing) and don't need to be differentiable. For the regression problems, we need a metric based on calculating some distance between predicted and ground truth.

For instance, to evaluate Regression models, we consider the following metrics: Mean Absolute Error (MAE), Mean Squared Error (MSE), Root Mean Squared Error (RMSE), R^2 (R-squared). To evaluate Classification models, the community considers the following metrics: Accuracy, Confusion Matrix (not a metric but fundamental to others), Precision and Recall, F1-score, and AU-ROC. The definitions of all these metrics can be found through online research.

We observe a great variety in performance metrics, hence the difficulties in comparing experimental results, one produced by a classical PDE simulation, and one through PINNs. This point needs to be clarified and is still an open question. The general question, here, is to decide if a technique is better, in terms of the quality of the results, than another one, and what are the "risks" of using one.

4.2 Frugal Methods and Algorithms

Nowadays, sustainability concerns are becoming more important than in the past because the climatic crisis is challenging. Sustainability includes energy efficiency, preservation of natural resources, and utilizing multiple energy sources as needed in computing devices and infrastructure. Paper [13] questions the IA faced to the environmental crisis. In the form of an open reflection on the goods and evils of AI, the article suggests a drastic change in the AI paradigm, more in phase with the vital obligation to design a more resilient society. Moreover, we know that improving the design of AI algorithms is possible, to get a better memory footprint. For instance, in [14], Tri Dao devised a memory-efficient framework for building attention mechanisms which are the key concept for transformers.

To complement that previous section, we may ask if we also have performance metrics and tools to estimate the environmental impact of using this or this technique. Researchers have to pay attention to the algorithms they devise or even the underlying brick used in libraries, to be sure they are frugal.

Regarding the measure of environmental indicators, people have also developed Python libraries such as CodeCarbon[3], CarbonTraker[4], and ImpactTracker [3]. CodeCarbon estimates and tracks carbon emissions from your compute engine and quantifies and analyzes their impact. CarbonTracker is a tool for tracking and predicting energy consumption and carbon footprint of training deep learning models. ImpactTracker creates a leaderboard for energy-efficient reinforcement learning algorithms to incentivize responsible research in this area as an example for other areas of ML.

4.3 A System View of the Ecosystem

Remember that the algorithms and tools presented in the last two subsections are always part of a larger context. Indeed the ecosystem of neural networks for transforming the ways of doing numerical simulations forms a system i.e., an organized framework or method, and among it a computing platform. In [15] authors gathered papers to discuss advances in High Performance Computing for Artificial Intelligence. This special issue concerns load-balancing application schedulers for CPU-GPU systems, hardware accelerators, and fast neural network training on a cluster of GPUs specifically dedicated to recognition with high accuracy.

Moreover, in [48] the authors presented an ongoing work to design a novel direct-time Graph Neural Networks (GNN) architecture for irregular meshes. It consists of a succession of increasing graph sizes connected by spline convolutions. Finally, in [56] authors introduced Melissa, a system concerned with large-scale ensemble runs typically consisting of executing thousands of physical simulation instances according to a range of different input parameters. These ensemble

---

[3] https://github.com/mlco2/codecarbon.

[4] https://github.com/lfwa/carbontracker.

runs enable sensitivity analyses, deep surrogate pieces of training, reinforcement learning, and data assimilation, but they rely on volumes of data that are too large to store. Melissa is a file-avoiding, fault-tolerant, and elastic framework, generalized to perform ensemble runs such as large-scale sensitivity analysis and large-scale deep surrogate training on supercomputers.

In the same vein, INRIA in France has had the opportunity to launch initiatives as early as 2019. The "HPC - AI - BigData Convergence Days (Conv'2019)"[5] workshop was an opportunity to examine the convergence between "Artificial Intelligence, Big Data, and High-Performance Computing". This workshop is a pioneer regarding the discussions we're having in 2024. The second event, which has just taken place in 2024, is entitled "GAP: Grenoble Artificial Intelligence for Physical Sciences"[6] (presentations soon available to all). The titles of the talks are evocative of the subjects covered. Among these, we'd like to mention "Inferring effective state variables and Dynamics from Data", "Simulation-based Inference for the physical sciences", "An Overview of operator learning", and "Efficient Training of Deep Learning Models" by Julia Gusak, Research Scientist, INRIA, Topal team, focusing on "Neural Ordinary Differential Equations (Neural ODEs)".

Last, INRIA also introduced DeepPhysX[7], a development environment designed to facilitate dialogue between digital technology and artificial intelligence algorithms such as neural networks.

5 Use Cases and Available Resources on the Web

This section presents various programming use cases on the web to illustrate the practicability of the discussed notions. These examples help to understand the real-world applications of the concepts covered in this chapter.

5.1 Solving the Problem with Neural Networks

1. Physics-Informed Neural Networks (PINNs):
   - PINNs[8]: Offers Tensorflow and Pytorch examples and tutorials on using PINNs for examples for solving the Burgers' and Helmholtz equations. The companion tutorial of the original PINN paper [52] is also available online[9].
2. Neural Operators:
   - FNO[10]: Implements neural operators that learn mappings between function spaces, useful for high-dimensional PDEs and spatiotemporal processes. It includes examples for solving the Darcy equation.

[5] https://project.inria.fr/conv2019/.
[6] https://discord.gg/PNUhG88SNC.
[7] https://www.inria.fr/en/combining-numerical-simulation-artificial-intelligence.
[8] https://github.com/omniscientoctopus/Physics-Informed-Neural-Networks.
[9] https://maziarraissi.github.io/PINNs/.
[10] https://github.com/neuraloperator/neuraloperator.

- WNO[11]: Uses wavelet transformations and neural networks to solve highly nonlinear PDEs, handling complex geometries and boundary conditions. It includes examples for 1-D Burger's equation, 2-D Allen-Cahn equation, 2-D Darcy equation, and 2-D Navier-Stokes equation.

3. Physics-Informed Transformer Networks (PITNs):
   - PITNs[12]: Combines transformer architectures with physics-informed learning to solve PDEs, including examples for 1D reaction, 1D wave, convection, and Navier-Stokes equations.

5.2 Solving the Problem with Conventional Methods

1. Finite Element Method (FEM)
   - FEniCS[13] [5]: An open-source computing platform for solving PDEs using FEM.
   - FreeFEM++[14] [24]: open-source PDE solver that uses the FEM for numerics. It provides a high-level programming language for the description of PDEs.
2. Finite Difference Method (FDM)
   - FiPy[15]: A Python library for solving PDEs using finite volume and finite difference methods.
3. Finite Volume Method (FVM)
   - OpenFOAM[16] [10]: An open-source C++ toolbox for the development of customized numerical solvers, and pre-/post-processing utilities for solving continuum mechanics problems, including CFD.
   - Manapy[17] [35]: A Python library for solving PDEs using finite volume methods using JIT compilation. It includes 2D and 3D examples using hybrid meshes.
4. Multigrid Methods
   - PyAMG[18] [6]: A Python library for Algebraic Multigrid (AMG) solvers.

6 Conclusion

This paper gives an exhaustive overview of the recent developments in neural networks to solve partial differential equations (PDEs) with a special focus on different methods such as Physics-Informed Neural Networks (PINNs), Neural Operators, Transformers, and Newton Informed Neural Operators.

---

[11] https://github.com/simon596/Wavelet-Neural-Operator.
[12] https://github.com/AdityaLab/pinnsformer.
[13] https://github.com/FEniCS.
[14] https://github.com/FreeFem.
[15] https://github.com/usnistgov/fipy.
[16] https://github.com/OpenFOAM.
[17] https://github.com/imadki/manapy.
[18] https://github.com/pyamg/pyamg.

Recently, Physics-Informed Neural Networks (PINNs) have been developed to enforce physical laws in training neural networks and improve their ability to generalize and solve PDEs more accurately. PINNs have since evolved to include many of their derivatives including Failure-Informed PINNs, Hard-Constrained PINNs, and Gradient-Enhanced PINNs; all of which target specific challenges, enhancing the performance of PINNs in different settings.

Neural Operators are a class of neural networks that learn to map between function spaces, which makes methods such as the Fourier Neural Operators and the Wavelet Neural Operators extremely powerful at modeling spatiotemporal processes and solving complex PDEs. These operators have shown to be highly accurate, and relatively computationally cheap, especially for problems of fluid dynamics and more generally high-dimensional ones.

Physics-Informed Transformers: Transformer Networks, developed for modeling natural language, are now being repurposed through Physics-Informed Transformers for numerical simulations. It uses attention mechanisms to capture dependencies from across the input space, which makes it appropriate for irregular grids and large problems.

By integrating classical Newton methods within the framework of a neural network, the Newton Informed Neural Operator represents an entirely new approach. In addition to distinguishing among several inherent solutions, this mixed method requires less supervised data than current-day neural network techniques.

Though we have come a long way, several challenges persist. The main issue remains the convergence, and stability, especially in the higher-dimensional spaces. Further work is necessary to scale these methods to large-scale problems and to design the networks for optimal efficiency. The quality and quantity of training data are some of the most influential factors regarding the model's performance.

In the future, further exploration should be done into better training algorithms, and the use of advances in hardware and hybrid approaches that bring together neural networks' strengths with traditional numerical methods. Therefore, customizing neural network architectures to individual science problems will be vital for maximizing efficiency.

References

1. Aboussi, W., Ziggaf, M., Kissami, I., Boubekeur, M.: A highly efficient finite volume method with a diffusion control parameter for hyperbolic problems. Math. Comput. Simul. (2023)
2. Anderson, J.D., Wendt, J.: Computational Fluid Dynamics, vol. 206. Springer, Cham (1995)
3. Anthony, L.F.W., Kanding, B., Selvan, R.: CarbonTracker: tracking and predicting the carbon footprint of training deep learning models (2020). https://doi.org/10.48550/ARXIV.2007.03051, https://arxiv.org/abs/2007.03051
4. Baliga, B., Patankar, S.: A new finite-element formulation for convection-diffusion problems. Numer. Heat Transf. 3(4), 393–409 (1980)

5. Baratta, I.A., et al.: DOLFINx: the next generation FEniCS problem solving environment. Preprint (2023). https://doi.org/10.5281/zenodo.10447666
6. Bell, N., Olson, L.N., Schroder, J.: PyAMG: algebraic multigrid solvers in python. J. Open Sour. Softw. **7**(72), 4142 (2022)
7. Benkhaldoun, F., Cérin, C., Kissami, I., Saad, W.: Challenges of translating HPC codes to workflows for heterogeneous and dynamic environments. In: 2017 International Conference on High Performance Computing & Simulation (HPCS), pp. 858–863. IEEE (2017)
8. Burnham, W.H.: Memory, historically and experimentally considered. I. An historical sketch of the older conceptions of memory. Am. J. Psychol. **2**(1), 39–90 (1888)
9. Calzolari, G., Liu, W.: Deep learning to replace, improve, or aid CFD analysis in built environment applications: a review. Build. Environ. **206**, 108315 (2021)
10. Chen, G., Xiong, Q., Morris, P.J., Paterson, E.G., Sergeev, A., Wang, Y.: Openfoam for computational fluid dynamics. Not. AMS **61**(4), 354–363 (2014)
11. Chen, J., Nakao, J., Qiu, J.M., Yang, Y.: A high-order Eulerian-Lagrangian Runge-Kutta finite volume (EL-RK-FV) method for scalar conservation laws. arXiv preprint arXiv:2405.09835 (2024)
12. Constantin, P., Foiaş, C.: Navier-Stokes Equations. University of Chicago press (1988)
13. Couillet, R., Trystram, D., Ménissier, T.: The submerged part of the AI-ceberg [perspectives]. IEEE Signal Process. Mag. **39**(5), 10–17 (2022). https://doi.org/10.1109/MSP.2022.3182938
14. Dao, T.: FlashAttention-2: faster attention with better parallelism and work partitioning (2023)
15. Dias de Assuncao, M., Rocha Rodrigues, E., Raffin, B.: Preface - special issue advances on high performance computing for artificial intelligence. J. Parallel Distrib. Comput. **156**, 131 (2021). https://doi.org/10.1016/j.jpdc.2021.06.002, https://www.sciencedirect.com/science/article/pii/S0743731521001313
16. Dos Santos, F., Akhound-Sadegh, T., Ravanbakhsh, S.: Physics-informed transformer networks. In: The Symbiosis of Deep Learning and Differential Equations III (2023)
17. Elmisaoui, S., Kissami, I., Ghidaglia, J.M.: High-performance computing to accelerate large-scale computational fluid dynamics simulations: a comprehensive study. In: International Conference on Advanced Intelligent Systems for Sustainable Development, pp. 352–360. Springer (2023)
18. Enkiewicz, O.C.Z.: Introductory lectures on the finite element method. Mech. Solids (1972)
19. Euler, L.: Introductio in analysin infinitorum, vol. 2 (1748)
20. Gao, Z., Yan, L., Zhou, T.: Failure-informed adaptive sampling for PINNs. SIAM J. Sci. Comput. **45**(4), A1971–A1994 (2023)
21. Gu, J., et al.: Recent advances in convolutional neural networks. Pattern Recogn. **77**, 354–377 (2018)
22. Hao, W., Liu, X., Yang, Y.: Newton informed neural operator for computing multiple solutions of nonlinear partials differential equations (2024)
23. Hao, Z., Ying, C., Su, H., Zhu, J., Song, J., Cheng, Z.: Bi-level physics-informed neural networks for PDE constrained optimization using Broyden's hypergradients. arXiv preprint arXiv:2209.07075 (2022)
24. Hecht, F., Pironneau, O., Le Hyaric, A., Ohtsuka, K.: FreeFem++ manual. Laboratoire Jacques Louis Lions (2005)

25. Hendrycks, D., Gimpel, K.: Gaussian error linear units (GELUs). arXiv preprint arXiv:1606.08415 (2016)
26. Hinton, G.E.: Deep belief networks. Scholarpedia **4**(5), 5947 (2009)
27. Hong, H.K., Chen, J.T.: Derivations of integral equations of elasticity. J. Eng. Mech. **114**(6), 1028–1044 (1988)
28. Hua, Y., Guo, J., Zhao, H.: Deep belief networks and deep learning. In: Proceedings of 2015 International Conference on Intelligent Computing and Internet of Things, pp. 1–4. IEEE (2015)
29. Huray, P.G.: Maxwell's Equations. Wiley, Hoboken (2009)
30. Ioffe, S., Szegedy, C.: Batch normalization: accelerating deep network training by reducing internal covariate shift. In: Proceedings of the 32nd International Conference on International Conference on Machine Learning, ICML 2015, vol. 37, pp. 448-456. JMLR.org (2015)
31. Jagtap, A.D., Karniadakis, G.E.: Extended physics-informed neural networks (XPINNs): a generalized space-time domain decomposition based deep learning framework for nonlinear partial differential equations. Commun. Comput. Phys. **28**(5) (2020)
32. Jin, X., Cai, S., Li, H., Karniadakis, G.E.: NSFnets (Navier-Stokes flow nets): physics-informed neural networks for the incompressible Navier-Stokes equations. J. Comput. Phys. **426**, 109951 (2021)
33. Kassam, A.K., Trefethen, L.N.: Fourth-order time-stepping for stiff PDEs. SIAM J. Sci. Comput. **26**(4), 1214–1233 (2005)
34. Kiener, A., Langer, S., Bekemeyer, P.: Data-driven correction of coarse grid CFD simulations. Comput. Fluids 105971 (2023)
35. Kissami, I.: Manapy: An MPI-based python framework for solving Poisson's equation using finite volume on unstructured-grid. In: AIP Conference Proceedings, vol. 3034. AIP Publishing (2024). https://doi.org/10.1063/5.0194750
36. Kou, J., Zhang, W.: Data-driven modeling for unsteady aerodynamics and aeroelasticity. Prog. Aerosp. Sci. **125**, 100725 (2021)
37. Lagrange, J.L.: Théorie des fonctions analytiques (Paris). Oeuvres de Lagrange IX (1797)
38. Li, Z., Liu, F., Yang, W., Peng, S., Zhou, J.: A survey of convolutional neural networks: analysis, applications, and prospects. IEEE Trans. Neural Netw. Learn. Syst. (2021)
39. Li, Z., et al.: Fourier neural operator for parametric partial differential equations (2020). arXiv preprint arXiv:2010.08895 (2010)
40. Li, Z., et al.: Neural operator: graph kernel network for partial differential equations. arXiv preprint arXiv:2003.03485 (2020)
41. Li, Z., et al.: Multipole graph neural operator for parametric partial differential equations. Adv. Neural. Inf. Process. Syst. **33**, 6755–6766 (2020)
42. Lu, L., Pestourie, R., Yao, W., Wang, Z., Verdugo, F., Johnson, S.G.: Physics-informed neural networks with hard constraints for inverse design. SIAM J. Sci. Comput. **43**(6), B1105–B1132 (2021)
43. Maazioui, S., Kissami, I., Benkhaldoun, F., Ouazar, D.: Numerical study of viscoplastic flows using a multigrid initialization algorithm. Algorithms **16**(1), 50 (2023)
44. Mao, Z., Jagtap, A.D., Karniadakis, G.E.: Physics-informed neural networks for high-speed flows. Comput. Methods Appl. Mech. Eng. **360**, 112789 (2020)
45. McCulloch, W.S., Pitts, W.: A logical calculus of the ideas immanent in nervous activity. Bull. Math. Biophys. **5**, 115–133 (1943)

46. Medsker, L., Jain, L.C.: Recurrent Neural Networks: Design and Applications. CRC Press, Boca Raton (1999)
47. Medsker, L.R., Jain, L.: Recurrent neural networks. Design Appl. **5**(64–67), 2 (2001)
48. Meyer, L.T., Pottier, L., Ribés, A., Raffin, B.: Deep surrogate for direct time fluid dynamics. CoRR abs/2112.10296 (2021). https://arxiv.org/abs/2112.10296
49. NASA: Tutorial on CFD verification and validation. https://www.grc.nasa.gov/www/wind/valid/tutorial/tutorial.html
50. Obiols-Sales, O., Vishnu, A., Malaya, N., Chandramowliswharan, A.: CFDNet: a deep learning-based accelerator for fluid simulations. In: Proceedings of the 34th ACM International Conference on Supercomputing, pp. 1–12 (2020)
51. Pang, G., Lu, L., Karniadakis, G.E.: fPINNs: fractional physics-informed neural networks. SIAM J. Sci. Comput. **41**(4), A2603–A2626 (2019)
52. Raissi, M., Perdikaris, P., Karniadakis, G.E.: Physics-informed neural networks: a deep learning framework for solving forward and inverse problems involving non-linear partial differential equations. J. Comput. Phys. **378**, 686–707 (2019)
53. Richardson, L.: The approximate solution of various boundary problems by surface integration combined with freehand graphs. Proc. Phys. Soc. London **23**(1), 75 (1910)
54. Rosenblatt, F.: The perceptron: a probabilistic model for information storage and organization in the brain. Psychol. Rev. **65**(6), 386 (1958)
55. Santos, F.D., Akhound-Sadegh, T., Ravanbakhsh, S.: Physics-informed transformer networks. In: The Symbiosis of Deep Learning and Differential Equations III (2023). https://openreview.net/forum?id=zu80h9YryU
56. Schouler, M., et al.: Melissa: coordinating large-scale ensemble runs for deep learning and sensitivity analyses. J. Open Source Softw. **8**(87), 5291 (2023). https://doi.org/10.21105/JOSS.05291
57. Smith, G.D.: Numerical Solution of Partial Differential Equations: Finite Difference Methods. Oxford University Press, Oxford (1985)
58. Srivastava, N., Hinton, G., Krizhevsky, A., Sutskever, I., Salakhutdinov, R.: Dropout: a simple way to prevent neural networks from overfitting. J. Mach. Learn. Res. **15**(56), 1929–1958 (2014). http://jmlr.org/papers/v15/srivastava14a.html
59. Steinhauser, M.O.: Fundamentals of numerical simulation. In: Steinhauser, M.O. (ed.) Computational Multiscale Modeling of Fluids and Solids, pp. 185–225. Springer, Cham (2022). https://doi.org/10.1007/978-3-030-98954-5_4
60. Tripura, T., Chakraborty, S.: Wavelet neural operator: a neural operator for parametric partial differential equations. arXiv preprint arXiv:2205.02191 (2022)
61. Uriarte, C., Pardo, D., Muga, I., Muñoz-Matute, J.: A deep double Ritz method (D2RM) for solving partial differential equations using neural networks. Comput. Methods Appl. Mech. Eng. **405**, 115892 (2023). https://doi.org/10.1016/j.cma.2023.115892, https://www.sciencedirect.com/science/article/pii/S0045782523000154
62. Vaswani, A., et al.: Attention is all you need. In: Proceedings of the 31st International Conference on Neural Information Processing Systems, NIPS 2017, pp. 6000–6010. Curran Associates Inc., Red Hook (2017)
63. Wang, J.X., Xiao, H.: Data-driven CFD modeling of turbulent flows through complex structures. Int. J. Heat Fluid Flow **62**, 138–149 (2016)
64. Wilkes, A.L., Wade, N.J.: Bain on neural networks. Brain Cogn. **33**(3), 295–305 (1997)

65. Yang, L., Meng, X., Karniadakis, G.E.: B-PINNs: Bayesian physics-informed neural networks for forward and inverse PDE problems with noisy data. J. Comput. Phys. **425**, 109913 (2021)
66. Yu, J., Lu, L., Meng, X., Karniadakis, G.E.: Gradient-enhanced physics-informed neural networks for forward and inverse PDE problems. Comput. Methods Appl. Mech. Eng. **393**, 114823 (2022)
67. Yu, W.E.B.: The deep Ritz method: a deep learning-based numerical algorithm for solving variational problems. Commun. Math. Stat. **6**, 1–12 (2018)
68. Yuan, L., Ni, Y.Q., Deng, X.Y., Hao, S.: A-PINN: auxiliary physics informed neural networks for forward and inverse problems of nonlinear integro-differential equations. J. Comput. Phys. **462**, 111260 (2022). https://doi.org/10.1016/j.jcp.2022.111260, https://www.sciencedirect.com/science/article/pii/S0021999122003229
69. Zhu, Y., Dinh, N.: A data-driven approach for turbulence modeling. arXiv preprint arXiv:2005.00426 (2020)

Scalable Deep Learning for Industry 4.0: Speedup with Distributed Deep Learning and Environmental Sustainability Considerations

Jean-Sébastien Lerat[1,2](✉) and Sidi Ahmed Mahmoudi[2]

[1] Computer Science and Management Department, Haute école en Hainaut, Mons, Belgium

[2] Science and Technology Department, University of Mons, Mons, Belgium

{Jean-Sebastien.Lerat,Sidi.Mahmoudi}@umons.ac.be

Abstract. This paper presents a solution for leveraging High-Performance Computing (HPC) infrastructures, and investigates the integration of distributed deep learning (DDL) techniques to address Industry 4.0 challenges across three distinct applications: intrusion detection with multi-layer perceptron, defect identification with convolutional neural networks, and predictive maintenance with recurrent neural network. Experimental results, underscore the scalability and efficiency of the proposed DDL approach. Notably, computations are accelerated by up to 46 times.

In addition to performance metrics, this research places significant emphasis on environmental sustainability. Detailed examination of energy consumption patterns on the HPC infrastructure aims to minimize the carbon footprint associated with deep learning processes. This dual focus on efficiency and sustainability positions the approach as a holistic and responsible solution for Industry 4.0 applications.

The practical insights enhance the efficiency of deploying DDL in HPC infrastructure. Additionally, they highlight the significance of eco-friendly AI practices for ethical and environmentally sustainable technological progress.

1 Introduction

In the rapidly evolving landscape of modern manufacturing, Industry 4.0 stands as a beacon of technological advancement, promising unprecedented levels of efficiency, productivity, and connectivity. At its core, Industry 4.0 harnesses the power of digital technologies to create intelligent, interconnected systems that revolutionize traditional industrial processes. Central to this paradigm shift is the integration of advanced artificial intelligence (AI) techniques, particularly deep learning, which enables machines to learn from vast amounts of data and make informed decisions autonomously.

The effective application of deep learning in Industry 4.0 scenarios often demands significant computational resources due to the scale, complexity of

M. Zbakh et al. (Eds.): CloudTech 2023, LNNS 1220, pp. 182–204, 2024.
https://doi.org/10.1007/978-3-031-78698-3_10

involved datasets and deep neural networks. To address this challenge, High-Performance Computing (HPC) infrastructure plays a crucial role in enabling the rapid and efficient processing of large-scale data. By leveraging HPC infrastructure, manufacturers can accelerate deep learning computations, thereby facilitating real-time decision-making and enhancing operational efficiency.

However, as Industry 4.0 continues to reshape the industrial landscape, it also presents a myriad of challenges that demand innovative solutions. Traditional deep learning approaches, while powerful, often encounter limitations when applied to the scale and complexity of Industry 4.0 datasets. Addressing these challenges requires a paradigm shift towards distributed deep learning, a technique that leverages the collective computational power of multiple devices to train complex models on massive datasets efficiently.

This chapter explores the integration of distributed deep learning techniques as a solution to key challenges within the Industry 4.0 framework. Through a series of case studies, we investigate the application of distributed deep learning across three distinct industrial scenarios:

Network Intrusion Detection The utilization of multi-layer perceptrons (MLPs) to safeguard Industry 4.0 networks by detecting and identifying intrusion attempts on the MQTT protocol, a critical element of industrial cybersecurity.

Predictive Maintenance The use of Long Short-Term Memory (LSTM) networks to predict potential equipment failures, enabling proactive maintenance strategies that minimize downtime and optimize asset utilization.

Defect Identification The application of Convolutional Neural Networks (CNNs) for identifing defects in images of printed circuit boards, enhancing quality control processes in manufacturing environments.

Moreover, our study extends beyond these applications, incorporating various types of neural networks including multilayer perceptrons, convolutional neural networks, and recurrent neural networks (RNNs). By exploring the efficacy of these different architectures, we provide a comprehensive understanding of their capabilities and limitations in addressing the multifaceted challenges encountered in Industry 4.0 settings.

Experimental results derived from analyses conducted on HPC systems equipped with CPUs and GPUs, underscore the scalability and efficiency of distributed deep learning. Notably, our approach accelerates are accelerated by up to 43 times, demonstrating its potential to drive real-time decision-making and improve operational outcomes.

Furthermore, our research places a significant emphasis on environmental sustainability, recognizing the importance of responsible AI practices in the era of Industry 4.0. By meticulously examining energy consumption patterns on HPC infrastructure, we aim to minimize the carbon footprint associated with deep learning processes. This dual focus on efficiency and sustainability positions the approach as a holistic and responsible solution for Industry 4.0 applications.

In addition to providing practical insights for optimizing the implementation of distributed deep learning in diverse Industry 4.0 contexts, this chapter underscores the importance of embracing environmentally conscious AI practices. By aligning efficiency with sustainability, we strive to contribute towards a future where technology not only enhances industrial capabilities but also fosters responsible and sustainable growth.

The remainder of the paper is organized as follows. In Sect. 2, we provide a comprehensive overview of deep learning in the context of Industry 4.0, discussing its significance and potential applications. The focus of Sect. 3 presents three Industry 4.0 use cases from the state of the art including the implementation of deep learning models for network intrusion detection, defect identification, and predictive maintenance. Subsequently, in Sect. 4, we detail the methods employed in our study, including the development of a measurement tool. Section 5 presents our experimental results and analysis, where we investigate speedup techniques for data loading and processing, examine GPU utilization, and consider environmental sustainability implications. Finally, in Sect. 6, we conclude our findings, discuss their implications, and suggest avenues for future research.

2 Background

Industry 4.0, often referred to as the Fourth Industrial Revolution, represents a transformative era in manufacturing characterized by the integration of digital technologies into industrial processes. Building upon the foundation laid by previous industrial revolutions, Industry 4.0 leverages advancements in artificial intelligence, robotics, the Internet of Things (IoT), and big data analytics to create smart, interconnected systems. These systems enable unprecedented levels of automation, customization, and efficiency, fundamentally reshaping the way goods are produced and services are delivered.

2.1 Deep Learning in Industry 4.0

Deep learning algorithms, particularly neural networks, excel at learning intricate patterns and relationships within complex datasets, enabling machines to perceive, interpret, and make decisions autonomously. In the context of Industry 4.0, deep learning finds applications across a wide range of domains, including predictive maintenance, quality control, supply chain optimization, and process automation. By leveraging deep learning, manufacturers can unlock new levels of efficiency, productivity, and innovation, driving competitive advantage in an increasingly digitalized world.

While deep learning holds immense promise for Industry 4.0 applications, it is not without its challenges. Traditional deep learning approaches often struggle to cope with the scale and complexity of industrial datasets. Issues such as limited computational resources, long training times, and difficulties in deploying models to edge devices pose significant hurdles to the widespread adoption of deep learning in industrial settings.

Moreover, the centralized nature of traditional deep learning architectures may not be well-suited to the distributed and decentralized nature of Industry 4.0 systems. As data volumes continue to grow exponentially and real-time decision-making becomes increasingly critical, there is a pressing need for scalable, efficient, and decentralized approaches to deep learning.

2.2 Data and Model Parallelism

Deep learning models often require extensive computational resources to train on large datasets and complex architectures. To efficiently utilize these resources, researchers and practitioners employ parallel computing techniques, which can broadly be categorized into data parallelism and model parallelism.

2.2.1 Data Parallelism

Data parallelism involves distributing the training data across multiple computational units, such as GPUs or CPU cores, and replicating the model on each unit. Each unit independently computes the gradients using a subset of the data, and then these gradients are aggregated and used to update the model parameters. The key idea behind data parallelism is that while each unit processes different data, they all share the same model parameters. This allows for efficient scaling of training to larger datasets and enables faster convergence by effectively utilizing the computational power of multiple devices.

In data parallelism, the training process is synchronized across devices to ensure that all units update their parameters based on the same global gradient. Techniques such as synchronous gradient updates, where gradients are aggregated and applied simultaneously across all devices, and asynchronous updates, where each device updates its parameters independently and periodically synchronizes with the global parameters, are commonly used to achieve synchronization.

Despite its effectiveness in scaling training to larger datasets, data parallelism may suffer from communication overhead when synchronizing gradients across devices, especially when the model size or the number of devices is large. Additionally, the memory requirements of data parallelism can be substantial, as each device must store a copy of the entire model.

2.2.2 Model Parallelism

Model parallelism, on the other hand, involves partitioning the model itself across multiple computational units. Instead of replicating the entire model on each device, different parts of the model are assigned to different devices, and computations are distributed accordingly. This allows for training larger models that may not fit entirely into the memory of a single device and enables more fine-grained control over resource utilization.

In model parallelism, communication between devices occurs not to synchronize gradients but to exchange intermediate activations or parameters between different parts of the model. This communication overhead can be mitigated by carefully designing the partitioning strategy and minimizing the amount of information exchanged between devices.

Model parallelism is particularly useful for training deep neural networks with intricate architectures, such as those containing recurrent or attention mechanisms, where the computational load is not evenly distributed across the layers of the model. By partitioning the model, researchers can exploit parallelism at a finer granularity and achieve better resource utilization.

2.3 Related Work

The integration of edge computing and cloud computing aligns with the requirements of Industry 4.0 [9]. However, distributed deep learning (DDL) presents challenges due to network communication between compute nodes, which can impede the DDL process [13]. Alternatively, containerization offers a streamlined approach, eliminating the need for complex local infrastructure deployment [7]. While previous methods advocate for using TensorFlow within containers [6], our study opts for PyTorch with MPI, given its superior performance [12]. Furthermore, cloud-based services tailored for deep learning tasks have been explored as an alternative, offering potential energy savings and latency reduction [3]. However, such solutions may entail significant costs and compatibility challenges [5].

Previous studies have demonstrated various methods to reduce network traffic in DDL [13]. However, these methods often compromise accuracy. In contrast, our approach leverages MPI for load distribution, achieving accelerated computation with low network communication, surpassing the state-of-the-art Horovod framework [10]. Future endeavors will focus on scaling distribution and investigating the impact of data location on DDL speedup.

Our research emphasizes a holistic approach to addressing the challenges of DDL deployment in Industry 4.0 contexts, combining efficient computational strategies with considerations for environmental sustainability. Through the integration of containerization, MPI, and PyTorch, our methodology offers a practical solution for accelerating deep learning tasks while minimizing the carbon footprint. As the landscape of Industry 4.0 continues to evolve, our findings pave the way for optimized deep learning implementations that meet the demands of modern manufacturing environments.

Our work makes several contributions to the field of accelerating deep learning in the context of Industry 4.0. We introduce a novel Measure Tool designed to comprehensively collect information on CPU status, memory usage, running processes, disk I/O, network I/O, and GPU utilization simultaneously. This tool fills a crucial gap in existing monitoring solutions by providing a unified platform for comprehensive performance monitoring across multiple system components. Additionally, we propose a comprehensive approach that addresses both data loading and computational workload challenges, offering insights into optimizing mini-batch size, implementing data parallelization techniques, and leveraging GPU acceleration with model parallelism. We also present three distinct use cases covering a broad spectrum of applications, each with tailored neural network architectures and meta-parameters. Overall, our contributions aim to advance the efficiency and effectiveness of deep learning applications in Industry 4.0 environments.

3 Industry 4.0 Use Cases

Three distinct Industry 4.0 use cases are considered, varying in neural network architectures, meta-parameters, as well as data volume and type. This comprehensive approach allows us to cover a broad spectrum of applications. This section presents the data and architectures of the three use cases: MLP-MQTT, CNN-PCB, and LSTM-Machines. These architectures are designed based on two parameters: l, representing the number of hidden layers, and s, denoting the size of those layers. Specific values for these parameters are outlined for each use case.

Finally, these networks are trained using a data split of 70% for training, 10% for validation, and the remaining for testing the final outcome. The physical infrastructure details are provided in Appendix, consisting of two physical machines running the Proxmox operating system, each equipped with two GPUs.

3.1 Network Intrusion Detection Using Multi-layer Perceptrons

A multilayer perceptron is employed to detect intrusion [1] in MQTT communication. The MQTT-IoT-IDS2020 dataset [8] originates from a simulated environment and comprises four types of attacks along with a fifth class for normal network traffic. The number of instances for each class is reported in Table 1.

Table 1. Distribution of images in the different classes of the PCB dataset.

Classes	Number of safe instances	Number of unsafe instances
Bruteforce	70,980,732	19,895,852
Aggressive Scan	70,768	40,624
UDP Scan	210,819	22,436
SSH Bruteforce (Sparta)	90,980,732	39,895,852
Normal	1,056,230	0

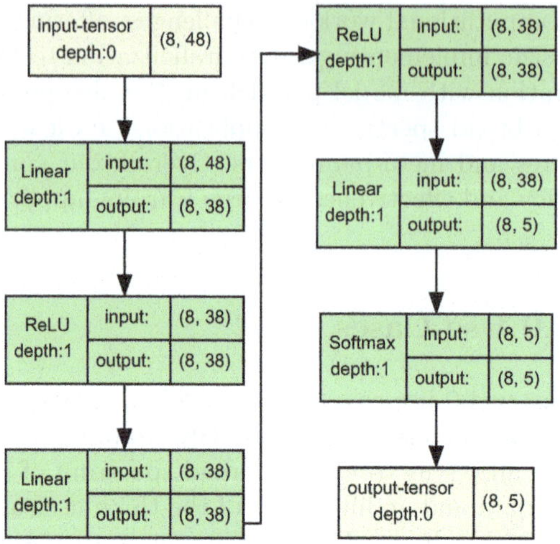

Fig. 1. Schema of the multilayer perceptron used in the MLP-MQTT use case with a mini-batch of 8, $l = 1$ layers of inner (hidden) layer, and $s = 38$ neurons in each inner layer.

The dataset is preprocessed following the methods outlined in the original paper to retain relevant data and transform values (e.g., text) into numerical representations. Fields such as mqtt_flag_uname, mqtt_flag_passwd, mqtt_flag_retain, mqtt_flag_qos, mqtt_flag_willflag, mqtt_flag_clean, and mqtt_flag_reserved are removed due to their predominantly null values, along with the timestamp field as time is not utilized in this model. Missing values are replaced with the median value for each field. TCP and IP fields are label encoded, while the protocol field and the class are one-hot encoded. Finally, the ip_len field is scaled to the $[0, 1]$ range, resulting in 53 fields including the one-hot encoding of classes.

The neural network (Fig. 1) comprises an initial dense layer with a ReLU activation function and 48 neurons. Subsequently, there are l hidden dense layers with the ReLU activation function, each containing s neurons. A final dense layer with the Softmax activation function provides a probability for each class. This network is trained using the cross-entropy loss function and the ADAM optimizer.

Cybersecurity is a fundamental component of Industry 4.0, where industrial systems are increasingly interconnected through communication protocols like MQTT. Ensuring the security of these networks against intrusions is crucial for maintaining the integrity and reliability of automated industrial processes. Utilizing MLP for intrusion detection allows for real-time monitoring and protection of networks, thereby enhancing the overall security of connected industrial systems. This protection is essential to prevent unauthorized access and potential disruptions to critical manufacturing and production processes.

This use-case scenario directly supports Industry 4.0 by providing robust cybersecurity measures. Effective intrusion detection systems enable secure data exchanges and operations across connected devices and systems, ensuring that the advanced interconnected infrastructures of Industry 4.0 remain safe from cyber threats. This, in turn, allows industries to confidently deploy and scale their smart manufacturing solutions without compromising security.

3.2 Defect Identification Using Convolutional Neural Networks

The objective is to identify the type of defect on printed circuit boards (PCB) [2] based on a well-balanced dataset spread over the 6 classes shown in Table 2. To increase the size of the dataset up to 5906 PCBs, each image is reused with a random crop offset by 5 to 10 pixels. All images have three channels (red, green, and blue) with a size of 64 × 64 pixels.

The convolutional neural network designed to predict the defects is inspired by DenseNet [4]. The model is composed of l dense blocks, each constituted of 6 layers with s as the multiplicative factor of bottleneck layers. The growth rate is 32. Figure 2 illustrates such a neural network with the default values for parameters l and s.

The training process computes the loss with cross-entropy and updates the parameters in mini-batch based on stochastic gradient descent. The momentum is 0.9, the initial learning rate is 0.01, and decay of 0.1 every 7 epochs.

Table 2. Distribution of images in the different classes of the PCB dataset.

Classes	Number of pictures
Missing hole	497
Mouse bite	492
Open circuit	482
Short	491
Spur	488
Spurious copper	503

Automated quality control is a cornerstone of Industry 4.0, where manufacturing processes must be both fast and precise. Identifying defects on PCBs using CNNs allows for automatic detection of imperfections, thus enhancing the quality of final products and reducing costs associated with returns and repairs. This automation of quality control processes speeds up production and increases product reliability, which is essential for maintaining high standards in smart manufacturing environments.

This use-case scenario supports Industry 4.0 by integrating advanced image processing and machine learning techniques into the production line. Leveraging CNNs for defect detection enables manufacturers to achieve higher levels of

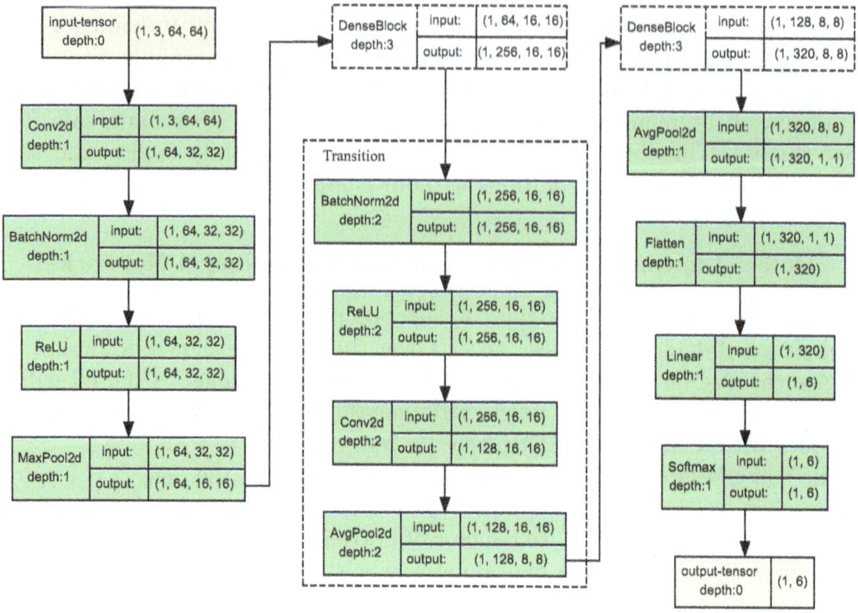

Fig. 2. Schema of the convolutional neural network used in the CNN-PCB use case with a mini-batch of 8 images, $l = 2$ layers of dense block, and $s = 4$ the multiplicative factor of bottleneck layers.

accuracy and consistency in quality control, which is critical for the high-speed, high-precision demands of Industry 4.0. The ability to automatically detect and address defects ensures that products meet stringent quality standards, thereby improving overall production efficiency and reliability.

3.3 Predictive Maintenance with Long Short-Term Memory Networks

A neural network that combines a convolutional layer and Long Short-Term Memory (LSTM) can predict which part of industrial machines will fail [11]. The ability to predict failure is important in Industry 4.0 because a technician can plan and replace the part before it fails.

The dataset contains information about the failure history, maintenance history, error conditions, machine features, and telemetry (i.e., sensor values) of 100 machines from January 1, 2015, to January 1, 2016. These data are organized as regular time series with a measurement frequency of every hour. Therefore, there are 876,000 instances of values, including 3919 failures in the dataset. Failures are caused by one of the four parts of machines, resulting in five one-hot encoded classes, including the no-failure case.

The preprocessing starts with min-max normalization scaled to the $[0, 1]$ range for all features except time. To aid in predicting rare failures of part i, a

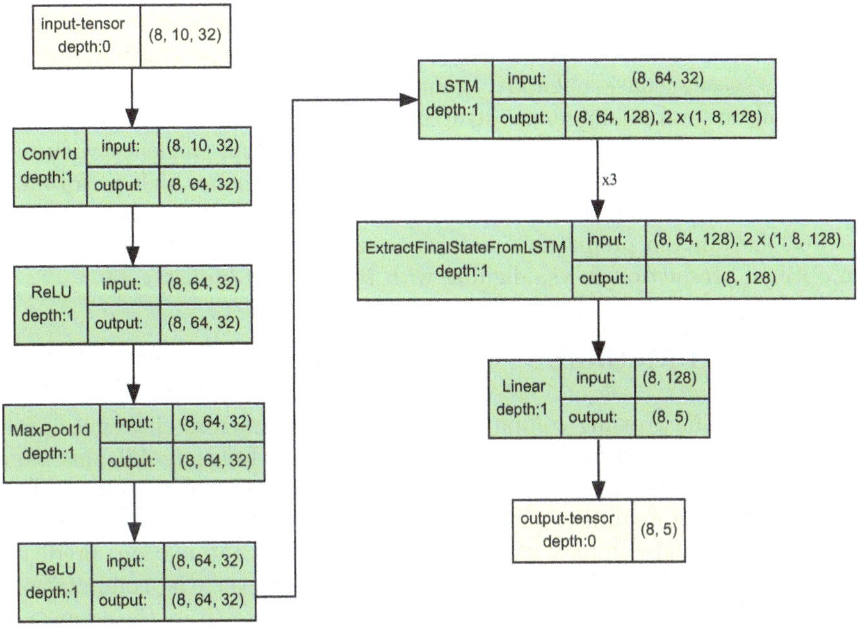

Fig. 3. Schema of the long short-term memory neural network used in the LSTM-Machines use case with a mini-batch of 8, $l = 1$ LSTM layer, and $s = 128$ hidden parameters in each LSTM layer.

time window of 24 h is used to label all considered instances as a failure of part i, which is based on a business use case. Additionally, the number of errors caused by each part of the machine within the window is appended to the instances. A lag feature of three hours is also computed (mean, standard deviation, minimum, and maximum) on telemetry values to capture regular trends. Other features include the machine identifier, the model, the age, and the number of days since each part of the machines was replaced. After preprocessing, each instance contains 37 features, including the one-hot encoding of the class.

The neural network consists of a 1D convolutional layer and l LSTM layers, followed by a final dense layer that receives output from the last LSTM layer to produce a prediction as a one-hot encoding. Each LSTM layer has s hidden parameters. Such a model is illustrated in Fig. 3. The ExtractFinalStateFromLSTM is a wrapper to only keep the output of LSTM and pass them to the final dense layer. This architecture is trained with the maximum absolute error and the ADAM optimizer.

Predictive maintenance is essential for optimizing the utilization of industrial equipment and minimizing unplanned downtime. Using LSTM networks to predict potential machine failures allows companies to plan maintenance activities before a breakdown occurs, thereby reducing repair costs and increasing machine availability. This approach extends the lifespan of equipment and enhances the overall efficiency of industrial operations, which is a key aspect of Industry 4.0.

This use-case scenario enhances Industry 4.0 by incorporating advanced predictive analytics into maintenance strategies. Predictive maintenance enabled by LSTM networks ensures that machinery operates at peak efficiency with minimal downtime. This capability is crucial for the continuous, high-performance operation of smart factories. By proactively managing maintenance schedules based on data-driven predictions, industries can prevent disruptions and maintain optimal production flows, aligning with the goals of Industry 4.0.

4 Proposed Method

In response to the growing computational and environmental challenges posed by the deployment of deep learning techniques within Industry 4.0, this paper introduces a comprehensive and innovative approach designed to enhance both efficiency and sustainability. Industry 4.0 is characterized by its reliance on large-scale data and complex neural networks to drive automation, predictive maintenance, defect detection, and cybersecurity. However, the computational demands and environmental impact of these processes necessitate the adoption of advanced strategies to optimize performance and minimize the carbon footprint.

The core of our proposed method involves the integration of High-Performance Computing (HPC) infrastructures with distributed deep learning (DDL) techniques. By harnessing the collective computational power of multiple devices, this approach aims to significantly accelerate the training and deployment of deep learning models across various industrial applications.

4.1 Accelerating Deep Learning

Our method begins with the utilization of HPC systems equipped with CPUs and GPUs, designed to handle the extensive data processing and computational requirements of Industry 4.0 scenarios. These systems enable the efficient processing of datasets and the training of deep neural networks, facilitating real-time decision-making and operational efficiency.

To fully exploit the capabilities of these HPC systems, we implement distributed deep learning techniques. Data parallelism and model parallelism are employed to distribute the computational workload across multiple devices. Data parallelism involves dividing the training data into smaller subsets, allowing each device to process a subset independently. Model parallelism, on the other hand, partitions the neural network model itself across multiple devices, enabling the training of large models that may not fit into the memory of a single device.

To accelerate the training of deep learning models in the context of Industry 4.0, two time-consuming aspects are considered: data loading and the computational workload processed by the computing device. Initially, our focus lies on optimizing the mini-batch size and implementing data parallelization techniques to expedite data processing. Subsequently, attention shifts towards leveraging GPU acceleration, enabling model parallelism to further enhance computational efficiency. In this phase, we explore computational distribution using MPI and NCCL libraries within the PyTorch deep learning framework. To gather necessary information for analysis, we have developed a dedicated tool.

4.2 Measure Tool

To our knowledge, there is currently no tool available that can comprehensively collect information on CPU status, memory usage, running processes, disk I/O, network I/O, and GPU utilization simultaneously. Therefore, one of our primary contributions is the development of a C++ tool designed specifically for gathering these diverse metrics. This tool addresses a significant gap in existing monitoring solutions by providing a unified platform for comprehensive performance monitoring across multiple system components. By offering detailed insights into the operational dynamics of both CPU and GPU resources, our tool aims to enhance the efficiency and effectiveness of system management and optimization efforts.

Each second, the tool cooperates with the operaring system and the NVIDIA Management Library (NVML) in order to collect information about:

- The sequence represented by timestamp.
- The GPUs: temperature, the power consumption, the maximum power consumption, the clock, the maximum clock, the clock rate, the available mem, the maximum amount of memory, the compute utilization, the memory utilization.
- The network for both the input/output: bytes, packets, errors, rejects, and collisions.
- The CPU: time spent in user/kernel space, priority, the i/o wait time (software interrupt), and idle time.
- The process: number of process, threads, time and duration of read/write operations (file, network), amount of allocated RAM.
- The deep learning task: capture output of the deep learning task (e.g. accuracy, loss).

4.3 Footprint

Normalized consumption is a metric used to evaluate the energy efficiency of different parallelization methods by accounting for the speedup achieved. By dividing the total energy consumption by the speedup factor, we obtain a measure that allows for a fair comparison between methods, irrespective of the actual training time. This approach is particularly useful in identifying the most energy-efficient configuration when considering both power consumption and computational speed. To evaluate the foorprint, we propose these steps:

1. **Measure Power Consumption** Calculate the total power consumed by each configuration during the training process.
2. **Normalize Consumption.** Divide the total power consumption by the speedup factor to get the normalized consumption. This adjusts for the differences in processing time, making it possible to compare the energy efficiency of each method.
3. **Convert to Carbon Footprint.** Use standard conversion factors to translate the normalized energy consumption into equivalent carbon emissions. This involves multiplying the energy consumption (in kWh) by the carbon intensity of the electricity used (e.g., 1 kWh corresponds on average in Europe to 0.45 kg CO2).

For example, if a particular configuration consumes 100 kWh and achieves a speedup factor of 10, the normalized consumption would be 100 kWh / 10 = 10 kWh. To find the carbon footprint, we multiply the normalized consumption by the CO2 conversion factor: 10 kWh * 0.45 kg CO2/kWh = 4.5 kg CO2.

By comparing the normalized carbon footprints of different configurations, industry can identify the methods that offer the best trade-offs between computational efficiency and environmental impact. This approach enables to choose the most sustainable and efficient configuration for a particular deep learning tasks in Industry 4.0 applications.

4.4 Managing New Applications for Efficiency and Effectiveness

When a new application of the Deep Learning process is submitted, authors can manage it to meet efficiency and effectiveness for both power consumption and processing time by following a structured approach:

1. Initial Profiling: Profile the new application's computational requirements and data characteristics. This involves understanding the model complexity, data size, and processing demands.
2. Select Appropriate Parallelism Technique: Based on the profiling results, choose the most suitable parallelism technique. For most cases, distributed data parallelism is preferred due to its balanced performance. However, if the application involves complex models with high data loading times, consider model parallelism.
3. Deploy and Monitor: Implement the chosen parallelism method on HPC infrastructure and use the developed measurement tool to monitor CPU and GPU utilization, memory usage, and power consumption in real-time.
4. Optimize and Adjust: Analyze the monitoring data to identify bottlenecks or inefficiencies. Adjust batch sizes, data partitioning strategies, and parallelism configurations as needed to optimize both processing time and power consumption.
5. Normalize and Compare: Normalize the energy consumption by the achieved speedup to compare against baseline configurations and ensure that the solution is both efficient and environmentally sustainable.

6. Iterative Improvement: Continuously refine the approach based on feedback and performance metrics, ensuring that the new application meets the desired efficiency and effectiveness goals.

By applying this methodical process, authors can effectively manage new deep learning applications to achieve optimal performance and sustainability.

5 Experimental Results and Analysis

This section focuses on optimizing the training process for deep learning applications, particularly focusing on data loading and parallelization techniques. Our study aims to accelerate the training of deep learning models while considering factors such as batch size, parallelization strategies, and energy consumption.

Initially, we explore the impact of batch size on training time across three distinct use cases. Subsequently, we investigate the efficacy of parallelization techniques, specifically focusing on data parallelism and its impact on training time across both CPU and GPU architectures. To quantify the effectiveness of parallelization, we introduce a speedup factor computed as the ratio of training time between sequential and parallelized versions. Furthermore, we delve into the energy consumption aspects associated with parallelization, highlighting the environmental implications of deep learning processes.

5.1 Speedup Data Loading

The first step of our proposal to speedup the training of the deep learning use cases is to focus on data loading. The infrastructure has enough memory to support different batch sizes. The mini-batch training is common in deep learning, and a common batch size that provide good learning curve is 32. A smaller batch size will increase the number of steps to train a good deep learning model, as the well-known stochastic gradient descent. A larger batch size can be damaging to the learning process because it may tend towards a local extremum. Nevertheless, our use cases are not sensitive to the batch size.

Figure 4 shows the time required to train the three use cases with different batch sizes. The curves have different heights but a similar shape. The smallest batch size is 8, which requires more time than a bigger size. At a batch size of 32, the learning process seems to speed up the process, then an increase in time is observed at a size of 64. Then the curve decreases with a gentle slope towards a time similar to the batch size of 32. The exception is the MLP-MQTT case where a bigger batch size keeps accelerating to a batch size of 128. This use case has a lot of instances compared to the other use cases. Processing on a GPU does not always mean training faster because it depends on the model's complexity. Indeed, using a GPU requires that the CPU transfers data from the RAM memory to the GPU memory and again from GPU to CPU to gather results.

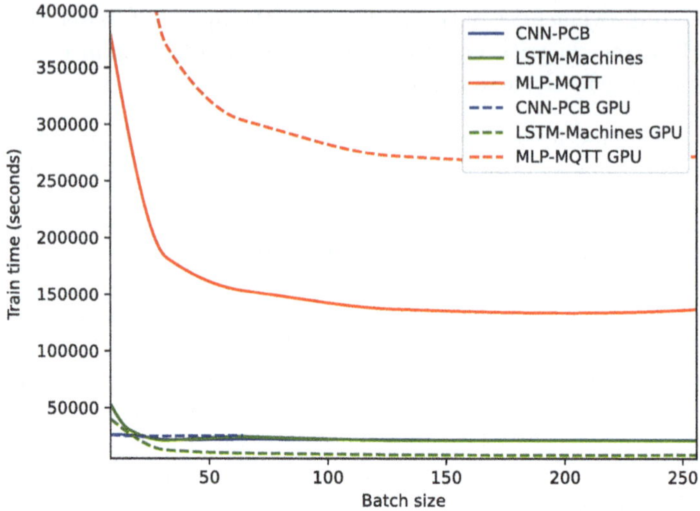

Fig. 4. The y-axis represents the time in seconds, while the x-axis denotes the batch size. Each curve corresponds to a different use case: CNN-PCB is depicted in blue, LSTM-Machine in green, and MLP-MQTT in red. Solid lines represent CPU training, whereas dashed lines indicate GPU training. The interpolation method used for the curves is the Akima spline. (Color figure online)

The batch size enables speeding up the data loading, but the whole training process requires applying forward and backward propagations. During this time, another batch could be loaded to train another copy of the model in parallel. Figure 5 reports the time required with the sequential versions and the data parallelism versions on the CPU and the GPU. Because each machine in our infrastructure has two GPUs, there is only one version of data parallelism on the GPU. As shown, the sequential versions on the CPU and GPU seem equivalent but are faster with data parallelism. The number of processes involved in data parallelism shows that a speedup occurs, but too many processes lead to a slowdown. For instance, the case of LSTM-Machines is faster trained with 2 CPU processes, but time dramatically increases with 4 processes. This behavior also occurs in other use cases but with a higher number of processes.

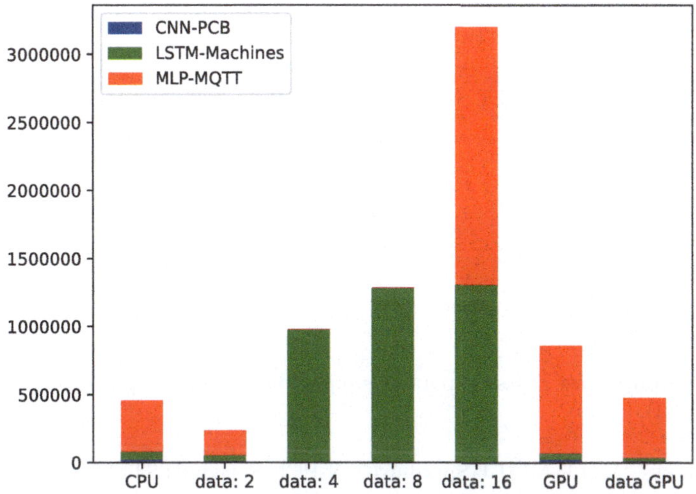

Fig. 5. Time required to train the uses cases over 100 epochs. The y-axis denotes time in seconds, while the x-axis contains labels for the parallelization methods. The three use cases-CNN-PCB, LSTM-Machine, and MLP-MQTT-are distinguished by blue, green, and red respectively. The labels specify the number of processes executed to achieve parallelization, except on GPUs, which are limited by the number of available GPUs. For CPU and GPU labels, they correspond to the sequential version on these devices.

5.2 Speedup Processing

In our methodology, we compute the speedup factor of a method to accelerate the training process as the ratio of the time required for training between the sequential version and a parallelization on the same device. We also distribute the training between the machines in the infrastructure using Message Passing Interface (MPI) and NVIDIA Collective Communications Library (NCCL). The analysis of the experimental results summarized in Table 3 for simple models (few inner layers $l = 1$, except for CNN-PCB where $l = 2$) and in Table 4 for complex models (several inner layers $l = 64$) highlight the effectiveness of different parallelization methods for various use cases and model complexities. The primary metric of interest is the speedup factor, which measures the acceleration in training time achieved by a parallelization method compared to a sequential version on the same device.

Table 3. Speedup of use cases for parallelism and distributed deep learning on simple models ($l = 2$ except for CCN-PCB $l = 2$).

Device	Method	MLP-MQTT Speedup	CNN-PCB Speedup	LSTM-Machines Speedup
CPU	Data	2.062	6.903	1.386
CPU	DDL	5.274	7.206	8.511
GPU	Data	1.799	1.958	2.264
GPU	Model	0.727	1.020	0.898
GPU	DDL-Data	3.564	3.825	2.476
GPU	DDL-Model	1.631	1.966	1.315

Table 4. Speedup of use cases for parallelism and distributed deep learning on complex models ($l = 64$).

Device	Method	MLP-MQTT Speedup	CNN-PCB Speedup	LSTM-Machines Speedup
CPU	Data	1.972	16.01	1.262
CPU	DDL	46.727	13.09	43.337
GPU	Data	2.741	1.27	1.077
GPU	Model	1.052	5.97	0.963
GPU	DDL-Data	22.412	4.28	20.492
GPU	DDL-Model	15.254	2.73	11.077

In Table 3, which summarizes the results for simple models (with few inner layers), the highest speedup factors are achieved by the distributed data parallelism method on both CPUs and GPUs. For the CPU, distributed data parallelism consistently shows significant speedup across all use cases, demonstrating its effectiveness in distributing the computational load among multiple machines. For the GPU, the distributed data parallelism method also achieves high speedup factors, indicating efficient utilization of multiple GPUs to accelerate training.

However, there is an exception in the CNN-PCB use case with a complex model, as shown in Table 4. In this case, the best configuration is not the distributed data parallelism but rather the model parallelism on a single machine. This anomaly is due to the balance between data loading and processing times. In the single machine model parallelism configuration, the time to load the next mini-batch of data is closely aligned with the processing time for an iteration on the GPU. This efficient overlap minimizes idle time and maximizes the GPU utilization. Conversely, the distributed version involves additional time for network communication, which does not significantly improve the speedup due to the overhead of data transfer between machines.

In summary, the optimal configuration depends on the use case and the complexity of the model. Distributed data parallelism is typically the best solution for both CPUs and GPUs, providing the highest speedup factors by effectively distributing the computational load. However, for specific complex models like the

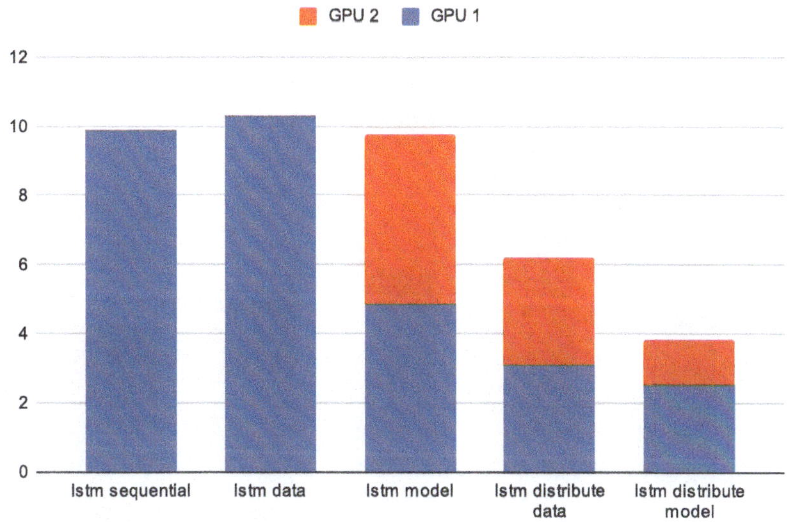

Fig. 6. Average GPU memory utilization (percentage) of the case LSTM-Machines. For the distributed version, the values represent a single machine. GPU1 is distinguished by the color blue, and GPU2 by the color red. (Color figure online)

CNN-PCB, single machine model parallelism can be more advantageous due to better synchronization of data loading and processing, minimizing idle times and avoiding network communication overhead.

5.3 GPU Utilization

The proposed tool can gather metrics on how GPU memory is utilized, the level of GPU compute usage, and the energy requirements of the GPU. These metrics are respectively reported in Fig. 6, Fig. 7, and Fig. 8. For clarity, only the LSTM-Machine case is reported here, as we are particularly interested in the differences between parallelism methods. The results for the other use cases (CNN-PCB and MLP-MQTT) are quite similar, exhibiting the same behavior with only minor variations in values.

In the absence of the distributed version, the underlying libraries used by the tool are unable to differentiate between GPUs because the hardware uses a scalable link interface. This limitation applies only to the memory utilization and compute utilization metrics for the data parallelism method.

Both the sequential and data parallelism methods similarly utilize GPU memory and computational power. Consequently, power consumption is equivalent, but slightly higher for the data parallelism method since it also uses the second GPU.

The data distribution on GPUs results in lower utilization of memory and compute units per machine compared to single-machine data parallelism. This is due to the fact that the reported metrics represent average utilization, and in the

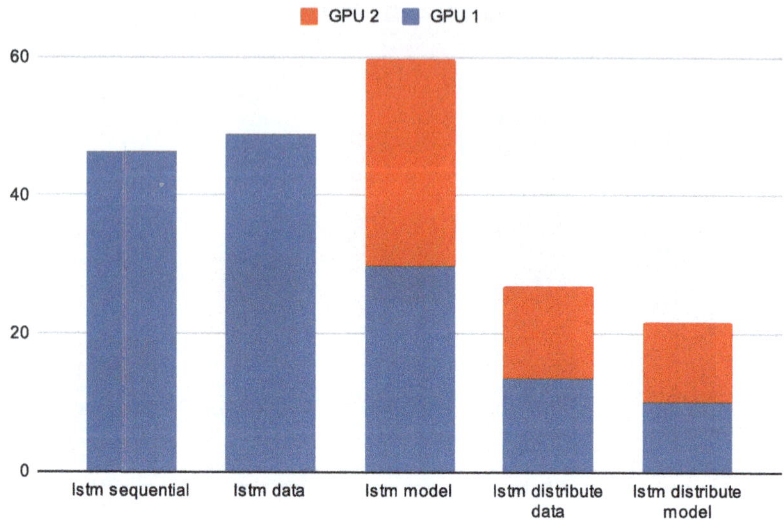

Fig. 7. Average GPU compute utilization (percentage) of the case LSTM-Machines. For the distributed version, the values represent a single machine. GPU1 is distinguished by the color blue, and GPU2 by the color red. (Color figure online)

distributed version, there is a cost of network communication which consumes additional time. However, despite this, the distributed version can still speed up the training of the use cases. Interestingly, the distributed version consumes approximately 20 W less on the GPUs.

The impact of distributed model parallelism compared to model parallelism on a single machine is greater than that of distributed data parallelism compared to data parallelism on a single machine. Indeed, distributed model parallelism consumes less than half the memory and compute units of the single-machine version. However, power consumption is similar to both the single-machine version and the distributed data parallelism version.

5.4 Environmental Sustainability Considerations

In the pursuit of technological innovation, it is imperative to not only optimize for efficiency and performance but also to consider the environmental impact of our actions. As the adoption of deep learning technologies proliferates across various domains, including Industry 4.0, the energy consumption associated with training and inference processes has become a growing concern. To address the environmental sustainability considerations in distributed deep learning, we propose to mitigate the carbon footprint associated with deep learning in Industry 4.0 contexts. Based on Sect. 5.3, we propose to normalize the global consumption (i.e. not only the GPUs but all computers involved in the training) by the speedup factor. This is relevant because the parallelization method speeds up the training and can consume more on a smaller time period. This normalization scale the global

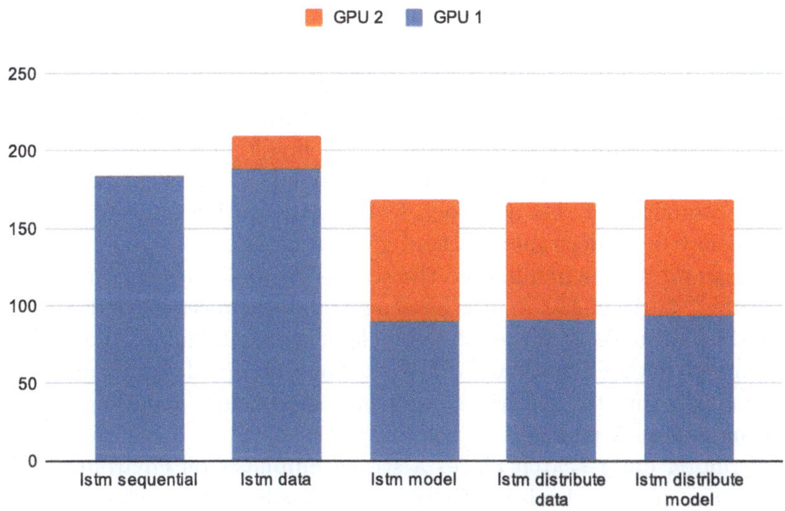

Fig. 8. Average GPU power consumption (watt) of the case LSTM-Machines. For the distributed version, the values represent a single machine. GPU1 is distinguished by the color blue, and GPU2 by the color red. (Color figure online)

consumption to compare value as if they needs as much time as the sequential version. Because the speedup factor is more interesting on complex models (see Sect. 5.2), we report the values in Table 5 for the LSTM-Machines case with the complex model ($l = 64$). Clearly, the most ecological setup is to use distributed data parallelism on CPU. Nevertheless, the speedup of the distributed deep parallelism method on the GPU has a high speedup for triple energy consumption. The speedup baseline is the sequential version on the GPU which is faster than the baseline of the sequential version of the CPU. Consequently, the distributed data parallelism method on GPU is faster than the variant on the CPU. It can be a good trade-off for ecological and technical constraints.

Table 5. Speedup of LSTM-Machines case and power consumption (watt) on the complex model ($l = 64$).

Device	Method	Speedup	Global Consumption	Normalized Consumption
CPU	Data	1.262	198.500	157.290
CPU	DDL	43.337	442.000	10.199
GPU	Data	1.077	331.345	307.656
GPU	Model	0.963	289.165	300.275
GPU	DDL-Data	20.492	574.751	28.047
GPU	DDL-Model	11.077	578.388	52.215

6 Conclusion

Our research represents a significant step towards addressing the challenges and leveraging the opportunities presented by the integration of deep learning techniques within the Industry 4.0 framework. By investigating the application of distributed deep learning across various industrial scenarios, we have demonstrated the potential of this approach to drive innovation and enhance operational efficiency in modern manufacturing environments.

Experimental results highlighted the scalability and efficiency of distributed deep learning, particularly in accelerating training processes and facilitating real-time decision-making. Through meticulous analysis conducted on HPC systems equipped with CPUs and GPUs, we have observed significant speedup factors, underscoring the effectiveness of our approach in handling large-scale datasets and complex models.

Furthermore, our research places a strong emphasis on environmental sustainability, recognizing the importance of responsible AI practices in mitigating the carbon footprint associated with deep learning processes. By examining energy consumption patterns and proposing normalization techniques, we aim to minimize the environmental impact of our methodologies while maximizing efficiency.

It is evident from our findings that the integration of distributed deep learning techniques offers a holistic and responsible solution for addressing the computational challenges of Industry 4.0, while also contributing towards a more sustainable future. By aligning efficiency with sustainability, we pave the way for technological advancements that not only enhance industrial capabilities but also foster responsible growth and environmental stewardship.

Moving forward, we advocate for continued research and innovation in the field of distributed deep learning, with a focus on refining methodologies, optimizing resource utilization, and advancing environmental sustainability practices. By leveraging the power of AI technologies responsibly, we can propel Industry 4.0 towards greater heights of productivity, efficiency, and environmental consciousness, ensuring a brighter and more sustainable future for generations to come.

Appendix

Each physical computer is composed of the following devices:

- **Graphic card** *2 EVGA GeForce RTX 2080 Ti with NVLink*
- **Power supply** *Seasonic Prime 1300 W Gold*
- **Case** *Corsaire Obsidian 750D Airflow*
- **CPU** *AMD 2950X RYZEN THREADRIPPER*
- **CPU Cooling system** *Watercooling Enermax LIQTECH TR4 II 360 RGB*
- **Additional case FAN** *be quiet! Silent Wings 3 120 mm PWM High-Speed* with *SilverStone FF122*
- **Additional device** *DVD ASUS DRW-24D5MT*

- **Motherboard** *ASRock Fatal1ty X399 Professional Gaming*
- **RAM** *Corsair Vengeance LPX Series Low Profile 128 GB (8×16 GB) DDR4 2933Mhz*
- **Storage**
  - *Samsung SSD 970 PRO M.2 PCIe NVMe 1 To*
  - *Samsung SSD 860 EVO 4 To*

The operating system is an Ubuntu Proxmox VE 8.0 with Python 3.10 and specific packages reported in Table 6.

Table 6. Speedup of LSTM-Machines case and power consumption (watt) on the complex model ($l = 64$).

Package name	Package version
certifi	2020.6.20
cryptography	3.4.8
decorator	4.4.2
netifaces	0.11.0
networkx	3.1
numpy	1.21.5
pandas	2.0.0
Pillow	9.0.1
protobuf	4.22.1
PyJWT	2.3.0
pyOpenSSL	21.0.0
python-dateutil	2.8.2
pytorch-triton	2.1.0+e6216047b8
scikit-learn	1.2.2
scipy	1.8.0
torch	2.1.0a0+git0af3203
torchvision	0.16.0a0+0d75d9e

References

1. Khan, M., et al.: A deep learning-based intrusion detection system for MQTT enabled IoT. Sensors. **21** (2021). https://www.mdpi.com/1424-8220/21/21/7016
2. Huang, W., Wei, P.: A PCB dataset for defects detection and classification. ArXiv Preprint arXiv:1901.08204 (2019)
3. Wu, H., Zhang, Z., Guan, C., Wolter, K., Xu, M.: Collaborate edge and cloud computing with distributed deep learning for smart city internet of things. IEEE Internet Things J. **7**, 8099–8110 (2020)

4. Huang, G., Liu, Z., Van Der Maaten, L., Weinberger, K.: Densely connected convolutional networks. In: Proceedings of The IEEE Conference on Computer Vision and Pattern Recognition, pp. 4700-4708 (2017)
5. Boag, S., et al.: Dependability in a multi-tenant multi-framework deep learning as-a-service platform. In: 2018 48th Annual IEEE/IFIP International Conference on Dependable Systems and Networks Workshops (DSN-W), pp. 43–46 (2018)
6. Brayford, D., Vallecorsa, S., Atanasov, A., Baruffa, F., Riviera, W.: Deploying AI frameworks on secure HPC systems with containers. In: 2019 IEEE High Performance Extreme Computing Conference (HPEC), pp. 1–6 (2019)
7. Doukha, R., Mahmoudi, S., Zbakh, M., Manneback, P.: Deployment of containerized deep learning applications in the cloud. In: 2020 5th International Conference on Cloud Computing and Artificial Intelligence: Technologies and Applications (CloudTech), pp. 1–6 (2020)
8. Hindy, H., Bayne, E., Bures, M., Atkinson, R., Tachtatzis, C., Bellekens, X.: Machine learning based IoT intrusion detection system: an MQTT case study (MQTT-IoT-IDS2020 dataset). In: Selected Papers From the 12th International Networking Conference, pp. 73–84 (2021)
9. Teerapittayanon, S., McDanel, B., Kung, H.: Distributed deep neural networks over the cloud, the edge and end devices. In: 2017 IEEE 37th International Conference on Distributed Computing Systems (ICDCS), pp. 328–339 (2017)
10. Sergeev, A., Del Balso, M.: Horovod: fast and easy distributed deep learning in TensorFlow. ArXiv Preprint arXiv:1802.05799 (2018)
11. Wahid, A., Breslin, J., Intizar, M.: Prediction of machine failure in industry 4.0: a hybrid CNN-LSTM framework. Appl. Sci. **12**, 4221 (2022)
12. Lerat, J., Mahmoudi, S., Mahmoudi, S.: Single node deep learning frameworks: comparative study and CPU/GPU performance analysis. Concurrency and Computation: Practice and Experience, pp. e6730 (2021)
13. Sattler, F., Wiedemann, S., Müller, K., Samek, W.: Sparse binary compression: towards distributed deep learning with minimal communication. In: 2019 International Joint Conference on Neural Networks (IJCNN), pp. 1–8 (2019)

Multi-domain Dataset for Moroccan Arabic Dialect Sentiment Analysis in Social Networks

Sara El Ouahabi[1](✉), Safâa El Ouahabi[2], and El Wardani Dadi[1]

[1] SOVIA team, LSA, National School of Applied Sciences Hoceima, Abdelmalek Essaadi University, Tangier, Morocco
saraelouahabi18@gmail.com

[2] Laboratory of Applied Mathematics and Information Systems, Polydisciplinary Faculty of Nador, Mohammed First University, Oujda, Morocco

Abstract. This chapter is focused on the Sentiment Analysis for Moroccan Arabic dialect text, which is collected from social media platforms. The focus will be to achieve close-to-optimal accuracies for a large dataset of 70,332 entries by using different techniques for feature extraction and machine learning approaches. In doing so, and considering the linguistic complexities in nature and paucity of labeled data for the Arabic dialect context—indeed presenting unique challenges—a host of approaches will be investigated within this chapter to drive sentiment analysis accuracy: TF-IDF, Bag of Words (BOW), Word2Vec, and FastText; also including all Machine learning classification algorithms. One critical part of this study involves the utilization of stemming through Tashaphyne (ArabicLightStemmer), which reduces word variations and improves classification. The findings will contribute to better advancement of sentiment analysis of Moroccan Arabic dialect, offering ample insight useful for social media analytics and linguistic research applications and setting a foundation for future tasks in similar linguistic and computational contexts.

Keywords: Moroccan Arabic dialect · sentiment analysis · dataset · social media · machine learning · feature extraction · stemming

1 Introduction

Sentiment analysis, also known as opinion mining or emotion AI, is the process of analyzing text to determine its positive, negative, or neutral sentiment. It finds diverse applications across multiple sectors, encompassing business, politics, and healthcare. For instance, businesses utilize it to scrutinize and interpret customer feedback, politicians employ it to monitor public sentiment, and healthcare organizations utilize it to assess patient satisfaction. In essence, the advancement of sentiment analysis has significantly enhanced our capacity to extract and scrutinize subjective information from textual data, furnishing invaluable insights and facilitating more informed decision-making across various domains.

Social media sentiment analysis is a powerful tool that helps understand how an audience perceives a brand and provides a wealth of data. Consumers actively tag and

M. Zbakh et al. (Eds.): CloudTech 2023, LNNS 1220, pp. 205–235, 2024.
https://doi.org/10.1007/978-3-031-78698-3_11

discuss brands, allowing businesses to tap into real-time insights. By monitoring social sentiment, it is possible to quickly identify negative trends or feedback that could escalate into a crisis.

The social media dataset is of utmost importance in the realm of sentiment analysis. It serves as a vital resource, containing comments or texts paired with sentiment labels, which are employed for training sentiment analysis models. These models acquire the ability to recognize linguistic characteristics and patterns linked to different sentiments, such as positivity, negativity, or neutrality. The accuracy of the model hinges on the diversity and representativeness of the data, with challenges arising from issues like data quality, informality, and noise.

In recent years, sentiment analysis has garnered significant attention, with numerous systems and applications developed for English and other languages. However, limited research has focused on the Moroccan Arabic dialect. Numerous research efforts have focused on sentiment analysis for different Arabic dialects across various Arabic-speaking regions, as documented in studies [1–18]. However, in the context of Morocco, only a limited number of studies have explored sentiment analysis for Darija, as indicated by references [19–29]. Various tools and techniques exist today for automatic sentiment classification, broadly categorized into lexicon-based, machine learning-based, and hybrid approaches.

The task of sentiment analysis in the Arabic language is faced with challenges compared to the English language. The main features that expose these challenges are rich morphologies, dialectal variations, and code-switching. Arabic lacks labeled data. Arabic is a rich morphological language, where words can have plenty of forms, governed by tense, gender, number, and other grammatical aspects. This increases the complexity of accurately analyzing text. In addition to that, it has many dialects to the Arabic language, each with their unique vocabularies and expressions, which further complicates sentiment analysis. Code switching, on the other hand, is a practice of using two or more languages/dialects interchangeably in one conversation and is quite a common practice in the Arabic-speaking regions, which makes the process complicated. The final limitation is related to the lack of good labeled data for sentiment analysis in the Arabic language, which will lead to poor effectiveness of machine learning models.

In our study, we introduce a novel addition to Arabic language resources, comprising an extensive collection of Moroccan Arabic dialect data sourced from diverse social networks and meticulously annotated. To the best of our knowledge [30–49], this dataset stands as the most extensive Moroccan dataset available for sentiment analysis. It stands out for its size (70,332), quality, variety (covering a wide range of domains), and novelty (containing all topics and trends from 2020 to now). This dataset is manually annotated to ensure the validity of data and is extracted from different social media (Facebook, Twitter, YouTube, Instagram, and websites).

Our primary focus is on delving into sentiment analysis for comments in the Moroccan Arabic dialect by leveraging machine learning models. We employ twelve supervised machine learning classifiers to classify sentiment and meticulously assess the influence of different feature extraction models and stemming on the performance of various approaches and models for sentiment analysis derived from dialectal texts.

The stemming is one of the most important preprocessing steps in our sentiment classification pipeline. It relates to the process of reducing words inflected or derived from others to its word stem or root. It is very needed for Arabic, which has very high morphological structure; the same word can take hundreds of forms based on tense, gender, number, etc. In other words, stemming reduces the complexity of text by effectively conflating various forms of a word into a single representation for sentiment models to perform better, keeping variations of a word in one feature for consistency and reliability enhancement of the analysis.

The chapter is organized as follows: In the second section, we address challenges related to sentiment analysis in the Moroccan Arabic dialect. The third section summarizes the main steps followed to achieve the Arabic sentiment analysis system, namely data collection and preprocessing, feature extraction, and system evaluation using various machine learning classifiers with different feature extraction techniques for the N-Gram model. Finally, we conclude our chapter with a conclusion and future works.

2 Challenges of Sentiment Analysis in Arabic Dialect

Sentiment analysis in Moroccan Darija (the Arabic dialect spoken in Morocco) presents several challenges due to its unique linguistic characteristics and the scarcity of available resources.

- Dialectal Variation: Moroccan Darija is a dialect with significant variation across regions and communities. Different cities and even neighborhoods may have distinct vocabulary, idiomatic expressions, and grammatical structures. Building a comprehensive sentiment analysis model that accounts for this variation is challenging.
- Lack of Labeled Data: Unlike English sentiment analysis, where large labeled
- datasets are readily available, there is a scarcity of high-quality labeled data for Moroccan Darija. Creating an annotated dataset for sentiment analysis requires significant effort and expertise.
- Code-Switching: Moroccan Darija speakers often mix Arabic with French, Spanish, and Berber languages. Code-switching occurs frequently in social media posts, making it harder to extract sentiment accurately. Models need to handle this multilingual context effectively.
- Slang and Informal Language: Social media platforms are rich sources of informal language, slang, and emoticons. Moroccan Darija users express themselves using colloquial terms, abbreviations, and creative spellings. Sentiment analysis models must understand these nuances.
- Irony and Sarcasm: Detecting irony, sarcasm, and other nuanced sentiments is challenging even in well-established languages. In Moroccan Darija, where humor and wit are common, identifying such subtleties becomes even more complex.
- Emoticons and Emoji: Moroccan Darija users heavily rely on emoticons and emoji to convey emotions. These visual cues play a significant role in sentiment expression but can be tricky for models to interpret.
- Domain Adaptation: Existing sentiment analysis models trained on standard Arabic or English may not perform well when applied directly to Moroccan Darija. Domain adaptation techniques are necessary to fine-tune models for this specific dialect.

- Data Noise: Social media data is noisy, containing misspellings, abbreviations, and non-standard grammar. Cleaning and preprocessing this data while preserving sentiment information is crucial.

3 Case Study: Moroccan Darija Sentiment Analysis in Social Networks

Sentiment analysis is an application of natural language processing (NLP) technologies that train computer software to understand text in a way similar to humans. The analysis generally goes through several stages before addressing the final result namely data collection, preprocessing, feature extraction training and finally evaluation of the results.

3.1 Data Collection

In our study, we introduce a novel addition to Arabic resources, comprising an extensive collection of Moroccan Arabic dialect data sourced from diverse social networks and meticulously annotated. To the best of our knowledge, this dataset stands as the most extensive Moroccan dataset available for sentiment analysis. It stands out for its size (70 332), quality, variety (covers a lot of domains), and novelty (it contains all topics and trends from 2020 to now). This dataset is manually annotated to ensure the validity of data and it's extracted from different social media (Facebook, Twitter, YouTube, Instagram, and website).

3.2 Data Annotation

We execute the annotation process for our corpus through the following steps:

- **Establish Annotation Guidelines:** We define clear guidelines to direct annotators, specifying the labels to extract from the text. In our case, we focus on identifying the distribution of sentiments (Positive, Negative, and Neutral).
- **Formulate Annotator Guidelines:** We provide detailed instructions to our annotators, outlining the desired labels and expectations.
- **Assemble Annotator Team:** We recruit a team of annotators and allocate the corpus to them for annotation based on the established guidelines.
- **Review and Resolve Discrepancies:** Following annotation, we review the results and address any discrepancies or conflicts among annotators. Manual annotation enables us to ensure data quality and label accuracy. Our dataset encompasses diverse text types, including social media content (Facebook, Twitter, YouTube, Instagram) in Arabic dialect, and accurately represents the distribution of sentiments (Positive, Negative, Neutral (See Fig. 1)).

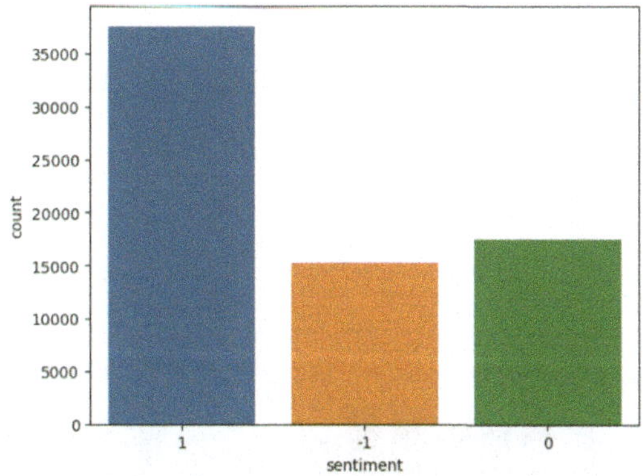

Fig. 1. Distribution of sentiments (Positive, Negative, and Neutral).

Over a span of several months, we gathered a dataset comprising 70,332 comments. These comments originate from diverse social media platforms (See Fig. 2), and are geographically localized within Morocco. Notably, all content is exclusively composed in the Moroccan Arabic dialect. The dataset covers an array of domains, including action, art, comedy, commerce, Covid, cuisine, education, festivals, war, Islamic topics, politics, social issues, and sports (See Fig. 3).

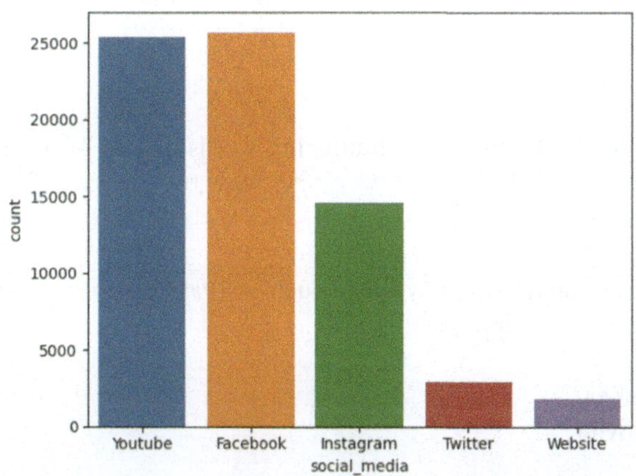

Fig. 2. Distribution of data collected from different types of social networks.

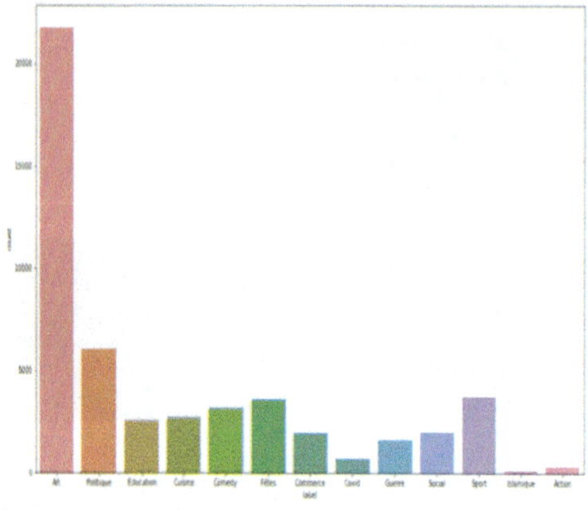

Fig. 3. Distribution of data group by some domains

3.3 Data Preprocessing

We start the data preprocessing by first converting the polarities. Subsequently, we undertake a series of steps outlined below:

- Normalization: This process aims to standardize certain letters within words to a common shape. Example (See Fig. 4 and Fig. 5):

  - Before: "بتسامه" "امي"
  - After: "ابتسامة" "أمي"

- Removal of stop words: Involves eliminating words devoid of meaningful context, such in the example of (Fig. 4) "إن" ,"أنا" ,"و" ,"يا" ,"كانت" ,"هذا" ,"في" ,"وانا". Removed in Fig. 5 after pre-processing.
- Punctuation removal: Aims to eradicate punctuation symbols such as {!"#$&'()*+,،-./:;<=>؟?@[]^_`{|}~}, which do not contribute to our approach.

  - Before: "!" "،" "."
  - After: Removed

- Removal of hashtags: Aims to eliminate phrases starting with the # symbol, as they hold no significance in our approach.
- Removal of numbers and URLs.
- Transformation of emojis into words: Aims to replace emojis with corresponding Arabic words [[حب، مضحك، عادي، محزن]], utilizing a CSV file containing 2639 emojis and their Arabic word equivalents (See Fig. 4 and Fig. 5).

  - Before: "💔" "😢"
  - After: "محزن" "حزين جدا"

- Emoji removal: Aims to eliminate emojis that were not translated in the emoji to word transformation step (See Fig. 4 and Fig. 5).

  – Before: "😭"
  – After: Removed

- Light stemming: Utilizes the affix deletion approach to identify the root of words. In our study, we utilize Tashaphyne (ArabicLightStemmer), which offers stemming and root extraction simultaneously, unlike other stemmers such as Khorja, ISRI, Assem, and Farasa. Tashaphyne is equipped with default prefixes and suffixes and allows for the use of custom prefixes and suffixes, enabling it to handle various aspects and create custom stemmers without necessitating changes to the code.

  – Before: "يتيمة" "بحالي" "رحمتك"(See Fig. 4 and Fig. 5)
  – After: "يتيم" "حال" "رحمة"

- Tokenization: Divides text into tokens, with words often separated by spaces, semicolons, commas, quotation marks, and periods.

  – [رب, رحم, حالف, حال, يتيم, رب, كبر, يتيم, أم, رب, ابتسام, ضحك, رشتن, حنان, أول, يوم, عيد, رمضان, حزن, جدا,الله, رحم, أم, حزن, جدا,الله, رحم, أم,الله, رحم, حزن, جدا,حبيب,شاء, نكون, مكان, آخر,محزن, غفر, له, رحم, جعل, جنة, مثوى, حزن, جدا,رب, رحم, جعل, مثوى, جنة]

يا رب رحمتك تكون حالفه بحالي يتيمة ربيتي وأنا كبرت يتيمة.. كانت أمي ربيتي وابتسامة. ضحكتها. رشتني بأحنانها. وهي بأول أيام عيد رمضان 💔💔 😭😭😥😥
الله يرحمك يا أمي 💔💔
الله يرحمك يا أمي
الله يرحمها 💔💔
حبيبي أنا
إن شاء الله نكون في هذا المكان في الأخرة
💔 الله يغفر له ويرحمه ويجعل الجنة مثواه 🖤 😢
ربي يرحمها. ويجعل مثواها الجنة
الله يرحمها
أمي الله يرحمها
الله يرحمها ويسكنها فسيح الجنان
الله يرحمها ويغفر لها ويسكنها فسيح جنانه
أمي الله يرحمك
ربي يرحمها ويجعل مثواها الجنة
ربي يرحمها ويغفر لها ويرزق أهلها الصبر والسلوان
الله يرحمها ويجعل مثواها الجنة
😭😭😭😭😭😭😭😭😭😭😭😭😭

Fig. 4. Data before preprocessing.

رب رحمة حالف حال يتيم رب أنا كبر يتيم كان أم رب وابتسامة ضحك رشتن بأحنان هي بأول يوم عيد رمضان حزن جدا
الله رحم أم حزن جدا
الله رحم أم
الله رحم حزن جدا
حبيب أنا
إن شاء الله نكون هذا مكان آخر
حزن جدا الله غفر له رحم جعل جنة مثوى حزن جدا
رب رحم وجعل مثوى جنة

Fig. 5. Data after preprocessing.

3.4 Features Extraction

After text preprocessing, the next step is feature extraction, by which the text is converted into vector form. There are a lot of techniques one can use for feature extraction. In this research work, we shall explore the effect of different feature extraction models like TF-IDF, BOW, Word2Vec, and FastText on our data set. We shall also check the effect of stemming to further look upon its working.

- TF-IDF (Term Frequency-Inverse Document Frequency):
  TF-IDF is a statistical measure applied to evaluate the importance of a word in a document relative to a collection of documents (a corpus). It helps in marking important words by reducing the weight of words frequently used. In our dataset, it proves especially effective in capturing the importance of words in the context of Moroccan Arabic dialect, which often encompasses its very unique terms and expressions. It is a valuable sentiment analysis tool due to the way it balances frequency against uniqueness for words.
- Bag of Words (BOW):
  The Bag of Words model is basically a representation of text as an unordered collection of words, disregarding grammar and word order but keeping multiplicity. It is this very simplicity that enables it to be effective as a baseline method for text representation. In the Moroccan Arabic dialect, BOW will help capture the frequency of words, which can be crucial in understanding sentiment, especially when dealing with comments from social media that are mainly repetitive. There have been different sentiment analysis studies where the BOW method has been applied extensively.
- Word2Vec:
  Word2Vec is a neural network model used to represent words within the continuous vector space, capturing relationships between them. The model is particularly nice for dealing with the linguistically rich Moroccan Arabic dialect, and it is rather good at learning associations and contexts from the data. These word vectors in continuous vector space thus capture very subtle differences and meanings important for the correctness of sentiment analysis. A number of studies have also shown that Word2Vec can be quite effective for different NLP tasks, with sentiment analysis being one.
- FastText:
  FastText is also an extension of Word2Vec that incorporates subword information, so it is more efficient for morphologically rich languages like Arabic. It basically breaks

down words into character n-grams in order to understand the internal structure of the words better. Because in the Moroccan Arabic dialect, the shape and writing of words are so much different from one to another, it could be very important. How FastText deals with out-of-vocabulary words and how it efficiently trains large datasets really make this a power tool for our sentiment analysis tasks. Previous work has established the superiority of FastText in capturing linguistic variations and high robustness on very many applications in text classification tasks.

References to Related Work: The techniques of feature extraction listed above have been employed in sentiment analysis processing and other NLP tasks in various works. As such, [Abdelminaam et al., 2021] depicted the capability of TF-IDF and BOW for the sentiment analysis of Arabic dialects. [Mikolov et al., 2013] proposed the method of Word2Vec, reporting great potential for capturing semantic relationships in big corpora. [Bojanowski et al., 2017] has also specified that FastText, to which Word2Vec is compared, surpasses the latter, proving itself suitable for morphologically rich languages by accomplishing tasks with different dialects.

We just attempt to use these methods of feature extraction because we would like to gain at the best rate the linguistic features of Moroccan Arabic. Such an approach allows building a model in a robust way, with efficient performance of sentiment classification on social media comments.

3.5 Sentiment Classification

Sentiment classification is the task of automatically classifying the sentiment expressed in a text. In this work, social media comments are used. Generally, sentiment can be of three types: positive, negative, or neutral, while the major interest of sentiment classification is to make a reasonable decision on the sentiment of the text. We adopted different categories of classification algorithms like linear models, tree-based models, probabilistic models, and others. Such an ordered categorization will help in understanding the different approaches and their effectiveness in carrying out sentiment analysis.

- Linear Model:
  Linear Models are simple and effective algorithms; it assumes the relationship between the input features and output sentiment to be linear. It is favored, of course, for effectiveness in most cases and ease of interpretation. Example of Linear Model applied in this study:

  - Linear SVC: A support vector classifier that finds the hyperplane for best separation between different classes.
  - Ridge Classifier: This is an L2-regularized linear classifier to prevent overfitting.
  - Perceptron: A linear classifier that is updated with weights when a misclassification error happens.

- Tree Models:
  Tree-based models are interpretable and capture non-linear relationships because, in the background, they divide the data into subsets according to the value of features. Examples include:

  - Decision Tree Classifier: It creates a tree structure, where every node has to represent a splitting feature.
  - Random Forest Classifier: Grouping of decision trees, which enhances accuracy by averaging predictions from different trees.

- Probabilistic Models:
  Probabilistic models are developed from the probability theory and are very beneficial in classification problems, where output could be modeled in terms of a distribution over classes. Examples include:

  - MultinomialNB: Naïve Bayes classifier for categorical features.
  - BernoulliNB: Naive Bayes classifier for binary/boolean.

- Other models:
  These models cover a variety of algorithms that do not fall into earlier categories but which work well for sentiment classification.

  - SVC: Support Vector Classifier with a nonlinear kernel.
  - SGD Classifier: A classifier based on the optimization of a loss function by the use of stochastic gradient descent.
  - K-Neighbors Classifier: This classifies by majority voting of its nearest neighbors.
  - Passive Aggressive Classifier: Updates model parameters only on misclassified examples. Very well suited for large-scale learning.
  - Logistic Regression: It is a linear model that uses prediction of probabilities through the logistic function.

To illustrate the impact of stemming, we conducted experiments comparing classifier performance with and without stemming.

Stemming is one of the most important preprocessing steps in our sentiment classification pipeline. In linguistic morphology, stemming is the process for reducing inflected or derived words to their root or base form. This is an important procedure for the Arabic language, since it contains a complex morphology: a lot of forms in word for tenses, genders, and numbers, in addition to other grammatical aspects.

When stemming is applied, we reduce the complexity of the text, effectively conflating different forms of a word into a single representation. This makes sentiment analysis models work better, treating variations of a word as a single feature to enhance the consistency and reliability of the analysis. An example is the word كتبت / يكتب:they have in common the same root كتب(write). Without stemming, this will be treated as two different features, which may dilute the effectiveness of the sentiment classification model. Stemming normalizes both words to their stem form, that is, كتب. So, the model can identify these as the same feature, increasing the accuracy of sentiment classification.

In this study, we choose to employ various feature extraction techniques to train our model. For the classification phase, we split the constructed datasets randomly, allocating 80 to the training subset and the remaining 20 to the test subset. This allows us to evaluate the performance of the developed models.

We utilize a range of algorithms for sentiment classification, including Linear SVC, SVC, MultinomialNB, RidgeClassifier, Perceptron, Bernoulli NB, SGD Classifier, Decision Tree Classifier, Random Forest Classifier, K-Neighbors Classifier, Passive Aggressive Classifier, Logistic Regression. These algorithms are paired with different feature extraction methods and the utilization of a stemmer to enhance the classification process.

3.5.1 Methodology

- Hyperparameters and Settings:
  For each classifier, we used the following hyperparameters and settings during the experiments:

  - **Linear SVC:** Regularization parameter (C) = 1.0, kernel = 'linear'
  - **SVC:** Regularization parameter (C) = 1.0, kernel = 'rbf'
  - **MultinomialNB:** Alpha = 1.0
  - **RidgeClassifier:** Alpha = 1.0
  - **Perceptron:** Max_iter = 1000, tol = 1e−3
  - **BernoulliNB:** Alpha = 1.0
  - **SGD Classifier:** Loss = 'hinge', penalty = 'l2', alpha = 1e−4, max_iter = 1000, tol = 1e−3
  - **Decision Tree Classifier:** Max_depth = None, criterion = 'gini'
  - **Random Forest Classifier:** Number of estimators = 100, criterion = 'gini'
  - **K-Neighbors Classifier:** Number of neighbors = 5, algorithm = 'auto'
  - **Passive Aggressive Classifier:** Max_iter = 1000, tol = 1e−3
  - **Logistic Regression:** Regularization parameter (C) = 1.0, solver = 'lbfgs', max_iter = 1000

- Computational Performance and Evaluation

The computational performance was evaluated using a standard 80/20 split for training and testing datasets. We focused on accuracy, precision, and recall to assess classifier performance. The inclusion of stemming—a crucial preprocessing step—helped reduce the dimensionality of the feature space and improve classifier performance by consolidating different word forms into a single base form.

3.5.2 Experiments and Results

- Before stemming:

  - BOW:

The results presented in Table 1 showcase the performance of various classifiers with different feature extraction techniques, particularly using the bag of words (BOW) model.

Across all n-gram variations, the Linear SVC and Logistic Regression classifiers consistently demonstrate the highest accuracy and precision, ranging from 76% to 77%. Meanwhile, the Random Forest Classifier consistently exhibits lower accuracy and precision scores, hovering around 53%. While some classifiers such as Bernoulli NB and Decision Tree Classifier show lower accuracy rates, others like Linear SVC and Logistic Regression consistently maintain robust performance, indicating their effectiveness in sentiment classification tasks utilizing the bag of words model.

- TFIDF:

As shown in Table 2 The Linear SVC and SVC classifiers consistently exhibit high accuracy and precision scores, ranging from 74% to 76%, across all n-gram variations. Similarly, Logistic Regression also maintains strong performance, particularly in the second and third n-gram variations, with accuracy scores of 75%. Conversely, the Random Forest Classifier consistently shows lower accuracy scores around 53%, suggesting limitations in effectively capturing sentiment using the TF-IDF model. Notably, Decision Tree Classifier exhibits varying precision scores across different n-gram variations, indicating sensitivity to feature representation. Overall, classifiers like Linear SVC, SVC, and Logistic Regression demonstrate effectiveness in sentiment classification tasks when utilizing the TF-IDF feature extraction model, while others show varying degrees of performance inconsistency.

- Word2vec

The results presented reveal the performance of various classifiers utilizing the Word2Vec feature extraction model across different n-gram variations. Notably, the performance of classifiers is relatively inconsistent compared to other feature extraction models.

While some classifiers like Linear SVC and SVC exhibit moderate accuracy scores ranging from 54% to 59%, others such as Bernoulli NB and Passive Aggressive Classifier show lower accuracy scores, indicating challenges in effectively capturing sentiment using the Word2Vec model. Additionally, precision and recall scores also vary across classifiers, suggesting differences in their ability to correctly identify positive, negative, and neutral sentiments. Overall, the results highlight the variability in classifier performance when utilizing the Word2Vec feature extraction model, with some classifiers demonstrating more robust performance than others.

- Fast text

The results illustrate in Table 4 shows the performance of various classifiers utilizing the Fasttext feature extraction model across different n-gram variations. Similar to Word2Vec, the performance of classifiers with Fasttext features exhibits variability. While some classifiers like Linear SVC, RidgeClassifier, and Logistic Regression demonstrate consistent accuracy scores ranging from 54% to 55%, others such as Perceptron, Bernoulli NB, and Passive Aggressive Classifier show lower accuracy scores, suggesting challenges in effectively capturing sentiment using the Fasttext model. Additionally, precision and recall scores vary across classifiers, indicating differences in their ability to correctly classify sentiments. Overall, the results highlight the mixed performance of classifiers when utilizing the Fasttext feature extraction model, with some classifiers displaying more stable performance than others.

– Discussion:

These results provide insights into the performance of various classifiers utilizing different feature extraction techniques before stemming. Across all feature extraction models, classifiers like Linear SVC and Logistic Regression consistently exhibit high accuracy and precision, ranging from 76% to 77% with the bag of words (BOW) model. Meanwhile, classifiers such as Random Forest Classifier consistently show lower accuracy scores around 53%, indicating limitations in effectively capturing sentiment using certain feature extraction models. In contrast, classifiers utilizing TFIDF feature extraction show moderate accuracy scores, with Linear SVC, SVC, and Logistic Regression demonstrating effectiveness. However, the Word2Vec and Fasttext feature extraction models display more inconsistent performance, with varying accuracy scores and challenges in effectively capturing sentiment.

- After stemming

  – TFIDF

These results presented in the Table 5 depict the performance of various classifiers using TFIDF combined with stemming. Across all n-gram variations, classifiers such as SVC, Linear SVC, and Logistic Regression demonstrate relatively high accuracy, precision, and recall scores, ranging from approximately 72% to 77%. These classifiers exhibit robust performance in accurately predicting sentiment with stemming. Conversely, Decision Tree Classifier and Random Forest Classifier consistently display lower accuracy scores, indicating challenges in effectively capturing sentiment using these models. Notably, Random Forest Classifier exhibits particularly low precision scores across all variations, suggesting limitations in correctly identifying sentiments. Overall, these results underscore the effectiveness of certain classifiers, such as SVC, Linear SVC, and Logistic Regression, in sentiment classification tasks with TFIDF and stemming, while highlighting the limitations of others, such as Decision Tree Classifier and Random Forest Classifier. Additionally, K-Neighbors Classifier and Passive Aggressive Classifier demonstrate moderate performance, indicating their suitability for sentiment classification tasks following preprocessing.

– BOW

Table 6 illustrate the performance of various classifiers using the bag of words (BOW) model combined with stemming. Across all n-gram variations, classifiers such as Logistic Regression, Linear SVC, and SGD Classifier consistently demonstrate high accuracy, precision, and recall scores, ranging from approximately 72% to 76%. These classifiers exhibit robust performance in accurately predicting sentiment after stemming. Conversely, Random Forest Classifier consistently displays lower accuracy scores, indicating challenges in effectively capturing sentiment using this model. Notably, Decision Tree Classifier also exhibits consistently low accuracy scores across all variations. Bernoulli NB demonstrates varying degrees of performance inconsistency, particularly showing lower accuracy scores across all n-gram variations. Overall, these results highlight the effectiveness of certain classifiers, such as Logistic Regression, Linear SVC, and SGD

Classifier, in sentiment classification tasks with BOW and stemming, while underscoring the limitations of others, such as Random Forest Classifier and Decision Tree Classifier. Additionally, Multinomial NB, Ridge Classifier, Perceptron, K-Neighbors Classifier, and Passive Aggressive Classifier demonstrate moderate performance, indicating their suitability for sentiment classification tasks.

- Word2vec

The Table 7 illustrate the performance of various classifiers using the Word2Vec model combined with stemming. Across all n-gram variations, the classifiers demonstrate varying levels of accuracy, precision, and recall. Notably, SVC consistently displays relatively higher accuracy scores, ranging from approximately 54% to 54%, across all n-gram variations. Similarly, K-Neighbors Classifier exhibits relatively high accuracy and precision scores, ranging from approximately 56% to 57%, across all ngram variations. Conversely, Passive Aggressive Classifier consistently exhibits lower accuracy scores across all n-gram variations, indicating challenges in effectively capturing sentiment using this model. Additionally, Bernoulli NB demonstrates consistently low accuracy scores, suggesting limitations in its effectiveness for sentiment classification tasks with Word2Vec and stemming. Overall, while some classifiers show moderate performance, such as SVC and K-Neighbors Classifier, others exhibit less stable performance, underscoring the variability in classifier effectiveness when utilizing the Word2Vec model combined with stemming for sentiment classification tasks.

- Fast text

The results of fast text feature extraction and stemming applied to various classification algorithms are outlined in Table 8. Across different algorithms and n-gram sizes, the accuracy ranges from approximately 39% to 57%. SVC consistently demonstrates relatively high precision and recall values, particularly noticeable in n-gram size 2 and 3, suggesting its robustness in handling text classification tasks.

However, Random Forest Classifier shows significant variance in performance across different n-gram sizes, with precision ranging from 28% to 71%. Notably, KNeighbors Classifier achieves the highest accuracy of 57% in n-gram size 3, while Bernoulli NB consistently performs relatively lower compared to other algorithms in terms of accuracy, precision, and recall. These results underscore the importance of algorithm selection and parameter tuning in text classification tasks utilizing fast text feature extraction and stemming techniques.

- **Discussion:**

In the first set of results using TFIDF feature extraction with stemming, classifiers like SVC, Linear SVC, and Logistic Regression consistently perform well, showing accuracy, precision, and recall scores ranging from around 72% to 77%. However, Decision Tree Classifier and Random Forest Classifier consistently lag behind, particularly in accuracy, suggesting difficulty in capturing sentiment effectively. Random Forest Classifier notably struggles with precision across all variations. This underscores the efficacy of SVC, Linear SVC, and Logistic Regression in sentiment classification with TFIDF and stemming, while highlighting the limitations of Decision Tree Classifier and Random Forest Classifier. In contrast, the second set of results using BOW feature extraction with stemming shows similar trends, with Logistic Regression, Linear SVC, and SGD Classifier performing consistently well, while Random Forest Classifier and Decision Tree Classifier struggle. Finally, in the third set of results using Word2Vec feature extraction with stemming, SVC and KNeighbors Classifier exhibit relatively higher accuracy, while Passive Aggressive Classifier consistently performs poorly. Bernoulli NB also consistently shows low accuracy, indicating its limitations.

In comparing the results between feature extraction without stemming and those with stemming, notable differences emerge in classifier performance. In the absence of stemming, classifiers such as Linear SVC and Logistic Regression consistently demonstrate high accuracy and precision, particularly with the bag of words (BOW) model, achieving scores around 76% to 77%. However, Random Forest Classifier consistently lags behind with lower accuracy scores around 53%, indicating challenges in capturing sentiment effectively using certain feature extraction techniques. On the other hand, when stemming is applied, classifiers like SVC, Linear SVC, and Logistic Regression exhibit consistent high performance, with accuracy, precision, and recall scores ranging from 72% to 77%. This contrast highlights the impact of stemming on improving the effectiveness of classifiers, particularly in models like TFIDF, where Random Forest Classifier struggles without stemming but performs better when stemming is employed.

3.5.3 Detailed Analysis of Classifier Performance Before and After Stemming

- Before Stemming:
  - Bag of Words (BOW):
    - High Performers (Linear SVC, Logistic Regression):
      - Characteristics: Both classifiers are robust linear models that work well with high-dimensional sparse data, such as BOW representations. They efficiently separate different classes by finding optimal hyperplanes.
      - Why Effective: The linear nature and regularization techniques help these models handle the complexity and high dimensionality of the text data without overfitting.
    - Low Performers (Random Forest Classifier):
      - Characteristics: As a tree-based model, it tends to overfit when dealing with sparse, high-dimensional data like BOW. It also requires more data to generalize well.
      - Why Less Effective: The BOW model does not capture the relationships between words, which can be a limitation for Random Forests, leading to poor generalization.
  - TFIDF:
    - High Performers (Linear SVC, SVC, Logistic Regression):

  - Characteristics: These classifiers benefit from the TFIDF's ability to highlight important terms while down-weighting common words, improving model precision.
  - Why Effective: The ability to focus on important terms helps these linear models to create more accurate decision boundaries.

- Low Performers (Random Forest Classifier):

  - Characteristics: Similar to BOW, the Random Forest struggles with the high dimensionality and sparsity of TFIDF features.
  - Why Less Effective: TFIDF does not mitigate the overfitting issue inherent to tree-based methods in high-dimensional spaces.

- Word2Vec:

- Moderate Performers (Linear SVC, SVC):

  - Characteristics: These classifiers leverage the dense, continuous representations of words provided by Word2Vec, which can capture semantic similarities.
  - Why Moderately Effective: While Word2Vec embeddings improve understanding of context, the variability in word usage and context might challenge these models.

- Low Performers (Bernoulli NB, Passive Aggressive Classifier):

  - Characteristics: These models might not fully exploit the continuous nature of embeddings, leading to poorer performance.
  - Why Less Effective: The probabilistic and aggressive learning approaches may not align well with the semantic richness of Word2Vec embeddings.

- FastText:

- High Performers (Linear SVC, Ridge Classifier, Logistic Regression):

  - Characteristics: Similar to Word2Vec but also considers subword information, making it effective for morphologically rich languages like Arabic.
  - Why Effective: The ability to capture subword information helps these models to understand the sentiment better, despite the variability in text.

- Low Performers (Perceptron, Bernoulli NB, Passive Aggressive Classifier):

  - Characteristics: These classifiers struggle with the inherent variability and semantic depth of FastText embeddings.

  - Why Less Effective: They fail to leverage the full potential of subword information, leading to lower performance.

- After Stemming:
  - TFIDF:
    - High Performers (SVC, Linear SVC, Logistic Regression):
      - Characteristics: Stemming reduces dimensionality and sparsity by conflating different word forms, which suits these linear models.
      - Why Effective: With reduced feature space, these models can create more robust decision boundaries, enhancing classification performance.
    - Low Performers (Decision Tree Classifier, Random Forest Classifier):
      - Characteristics: Despite stemming, these models still face challenges with high-dimensional data.
      - Why Less Effective: The inherent complexity and need for large datasets to generalize well remain issues.
- Bag of Words (BOW):
  - High Performers (Logistic Regression, Linear SVC, SGD Classifier):
    - Characteristics: These models benefit from the reduction in feature space due to stemming, which simplifies the data structure.
    - Why Effective: The simplified and reduced feature space improves their ability to generalize from the training data.
  - Low Performers (Random Forest Classifier, Decision Tree Classifier):
    - Characteristics: These models continue to struggle with high-dimensional sparse data even after stemming.
    - Why Less Effective: Stemming does not fully mitigate the overfitting tendency of tree-based models in high-dimensional spaces.
  - Word2Vec:
    - Moderate Performers (SVC, K-Neighbors Classifier):
      - Characteristics: These models can better utilize the dense embeddings post-stemming.
      - Why Moderately Effective: Stemming helps reduce noise in the embeddings, leading to better model performance.

- Low Performers (Passive Aggressive Classifier, Bernoulli NB):
  - Characteristics: These classifiers struggle with the dense, semantically rich embeddings.
  - Why Less Effective: The complex nature of embeddings post-stemming is not fully leveraged, leading to poorer performance.

- FastText:
  - High Performers (SVC):
    - Characteristics: SVC can exploit the subword information and the reduced noise due to stemming effectively.
    - Why Effective: The model captures the sentiment nuances better post-stemming.
  - Low Performers (Random Forest Classifier, Bernoulli NB):
    - Characteristics: These classifiers continue to struggle with the high-dimensional data, even with FastText's subword information.
    - Why Less Effective: The complexity of FastText embeddings combined with high dimensionality poses challenges for these models.

We fine-tuned the hyperparameters for each classifier, ensuring optimal performance. For instance, the Linear SVC was set with a regularization parameter $C = 1.0$ and a linear kernel, while the Random Forest Classifier utilized 100 estimators.

3.5.4 Recommendations

Based on the observations from this study, here are some practical recommendations for researchers and practitioners focusing on sentiment analysis for the Moroccan Arabic dialect:

1. Feature Extraction Techniques:
   - Use TFIDF and BOW Models:
     The TFIDF and BOW models are effective techniques for feature extraction in sentiment analysis tasks. In fact, they offer high accuracy and precision values, especially when used in combination with linear models.
   - Leverage Word Embeddings:
     Though effective, consider exploration of word embeddings like Word2Vec and FastText to capture semantic meanings and subword information, especially for languages that are morphologically rich.
2. Selection of Classifier:
   - Linear Models:

Linear SVC, Logistic Regression, and SGD Classifier always perform well with a variety of techniques for feature extraction. They have robustness towards high-dimensional data and give good performance.

- Avoid Tree-Based Models for High-Dimensional Data:
  Tree-based models such as Random Forest and Decision Tree struggle with high-dimensional sparse data. They are definitely not the best models one would look at for feature extraction using BOW and TFIDF.
- Use K-Neighbors Classifier:
- his classifier does a moderate job and can be good to go for, especially when using word embeddings.

3. Use stemming:

   - Stemming significantly improves the classifier's performance by reducing complexity and dimensionality in the feature space. This is especially good for linear models and models that use the feature extraction techniques TFIDF and BOW.
   - Evaluate the impact:
   - one of the best practices, always evaluate the impact of stemming on our particular dataset and model to make sure it enhances performance.

4. Hyperparameter Tuning:
   Optimize the hyperparameters by experimenting with several values and settings of each classifier to find the best possible configuration that maximizes performance. We fine-tuned the hyperparameters for each classifier, ensuring optimal performance. For instance, the Linear SVC was set with a regularization parameter $C = 1.0$ and a linear kernel, while the Random Forest Classifier utilized 100 estimators.
5. Dataset Quality and Representativeness:
   An important part of training an effective sentiment analysis model would require good quality and representative datasets. The dataset should include a variety of topics and various sentiments to enhance model generalization.
6. Computational Aspects:
   The study involved significant computational resources to handle the large dataset and the complexity of feature extraction and classification. The use of efficient algorithms like FastText and Word2Vec, along with powerful linear models, facilitated the processing and analysis of data in a computationally efficient manner.

Such recommendations, when followed, can make the sentiment analysis model effective for the Moroccan Arabic dialect and possibly other morphologically rich languages.

4 Conclusions

In conclusion, this chapter sheds light on the efficacy of sentiment analysis techniques applied to Moroccan Darija comments prior and after to stemming. Through experimentation with various classifiers and feature extraction methods, several insights have emerged. In comparing the results between feature extraction without stemming and those with stemming, notable differences emerge in classifier performance. In the absence of stemming, classifiers such as Linear SVC and Logistic Regression consistently demonstrate high accuracy and precision, particularly with the bag of words (BOW) model, achieving scores around 76% to 77%. However, Random Forest Classifier consistently lags behind with lower accuracy scores around 53%, indicating challenges in capturing sentiment effectively using certain feature extraction techniques. On the other hand, when stemming is applied, classifiers like SVC, Linear SVC, and Logistic Regression exhibit consistent high performance, with accuracy, precision, and recall scores ranging from 72% to 77%. This contrast highlights the impact of stemming on improving the effectiveness of classifiers, particularly in models like TFIDF, where Random Forest Classifier struggles without stemming but performs better when stemming is employed.

The dataset, comprising 70,332 comments with diverse sentiment distributions, poses opportunities for further exploration. Future experiments have been involved refining feature extraction techniques, by assessing the impact of preprocessing steps like stemming. Ultimately, this study provides valuable insights for advancing sentiment analysis in the context of Moroccan Darija, paving the way for improved understanding and interpretation of user sentiments in social media and online discourse.

Future research endeavors could delve into leveraging deep learning techniques for sentiment analysis in Moroccan Darija. Deep learning models, such as recurrent neural networks (RNNs) and convolutional neural networks (CNNs), have shown promise in capturing complex patterns and nuances in textual data. By exploring these advanced architectures, researchers can potentially enhance sentiment analysis accuracy and robustness, particularly in capturing the intricacies of Darija language sentiment expressions.

The methods and findings from this study on sentiment analysis for the Moroccan Arabic dialect demonstrate significant potential for generalization to other languages, particularly those with similar linguistic characteristics. By adapting the stemming techniques, feature extraction methods, and classification models to the specific requirements of the target languages, researchers and practitioners can effectively enhance sentiment analysis capabilities across diverse linguistic landscapes. Further research and experimentation are encouraged to explore and validate these adaptations, contributing to the broader field of natural language processing and sentiment analysis.

Appendix

(See Table 3)

Table 1. Performance Evaluation of Classifiers on Various n-gram Models + Bag-of-Words Approach without Stemming

Classifiers	n-gram	accuracy	precision	recall
Linear SVC	1	76	75	76
SVC	1	73	72	73
Ridge Classifier	1	74	73	74
Perceptron	1	73	74	73
Bernoulli NB	1	65	66	65
SGD Classifier	1	75	75	75
Decision Tree Classifier	1	56	56	56
Random Forest Classifier	1	53	28	53
K-Neighbors Classifier	1	63	70	63
Passive Aggressive Classi fier	1	73	74	73
Logistic Regression	1	76	76	76
Linear SVC	2	76	76	76
SVC	2	73	72	73
Ridge Classifier	2	75	74	75
Perceptron	2	75	75	75
Bernoulli NB	2	59	67	59
SGD Classifier	2	76	75	76
Decision Tree Classifier	2	56	56	56
Random Forest Classifier	2	53	28	53
K-Neighbors Classifier	2	61	72	61
Passive Aggressive Classifier	2	75	75	75
Logistic Regression	2	77	76	77
Linear SVC	3	76	75	76
SVC	3	71	71	71
Ridge Classifier	3	75	74	75
Perceptron	3	75	75	75
Bernoulli NB	3	57	69	57
SGD Classifier	3	75	75	75
Decision Tree Classifier	3	56	56	56
Random Forest Classifier	3	53	28	53
K-Neighbors Classifier	3	61	71	61
Passive Aggressive Classifier	3	75	75	75
Logistic Regression	3	77	76	77

Table 2. Performance Evaluation of Classifiers on Various n-gram Models + TFIDF Approach without Stemming

Classifiers	n-gram	accuracy	precision	recall
Linear SVC	1	74	73	74
SVC	1	76	76	76
Ridge Classifier	1	74	73	74
Perceptron	1	69	70	69
Bernoulli NB	1	70	68	70
SGD Classifier	1	72	71	72
Decision Tree Classifier	1	56	61	56
Random Forest Classifier	1	53	28	53
K-Neighbors Classifier	1	57	69	57
Passive Aggressive Classi-	1	70	71	70
Logistic Regression	1	74	73	74
Linear SVC	2	75	74	75
SVC	2	76	76	76
Ridge Classifier	2	74	74	74
Perceptron	2	72	71	72
Bernoulli NB	2	70	69	70
SGD Classifier	2	72	71	72
Decision Tree Classifier	2	56	58	56
Random Forest Classifier	2	53	28	53
K-Neighbors Classifier	2	58	59	58
Passive Aggressive Classifier	2	70	71	70
Logistic Regression	2	75	74	75
Linear SVC	3	75	74	75
SVC	3	76	76	76
Ridge Classifier	3	75	74	75
Perceptron	3	69	71	69
Bernoulli NB	3	70	69	70
SGD Classifier	3	72	71	72
Decision Tree Classifier	3	56	62	56
Random Forest Classifier	3	53	28	53
K-Neighbors Classifier	3	58	70	58
Passive Aggressive Classifier	3	70	71	70
Logistic Regression	3	75	74	55

Table 3. Performance Evaluation of Classifiers on Various n-gram Models + Word2vec Approach without Stemming

Classifiers	n-gram	accuracy	precision	recall
Linear SVC	1	75	50	55
SVC	1	54	57	54
Ridge Classifier	1	55	50	55
Perceptron	1	49	46	49
Bernoulli NB	1	35	51	35
SGD Classifier	1	55	54	55
Decision Tree Classifier	1	54	57	54
Random Forest Classifier	1	53	41	53
K-Neighbors Classifier	1	59	60	59
Passive Aggressive Classi fier	1	49	53	49
Logistic Regression	1	55	50	55
Linear SVC	2	55	51	55
SVC	2	54	60	54
Ridge Classifier	2	55	50	55
Perceptron	2	55	50	55
Bernoulli NB	2	40	55	40
SGD Classifier	2	36	51	36
Decision Tree Classifier	2	54	51	54
Random Forest Classifier	2	55	55	55
K-Neighbors Classifier	2	53	61	53
Passive Aggressive Classi fier	2	34	54	34
Logistic Regression	2	55	50	55
Linear SVC	3	55	50	55
SVC	3	54	59	54
Ridge Classifier	3	55	50	55
Perceptron	3	50	55	50
Bernoulli NB	3	35	51	35
SGD Classifier	3	55	52	55
Decision Tree Classifier	3	54	49	54
Random Forest Classifier	3	53	42	53
K-Neighbors Classifier	3	59	60	59
Passive Aggressive Classi fier	3	54	53	54
Logistic Regression	3	55	49	55

Table 4. Performance Evaluation of Classifiers on Various n-gram Models + Fast text Approach without Stemming

Classifiers	n-gram	accuracy	precision	recall
Linear SVC	1	55	51	55
SVC	1	54	61	54
Ridge Classifier	1	55	50	55
Perceptron	1	38	58	38
Bernoulli NB	1	35	51	35
SGD Classifier	1	54	49	54
Decision Tree Classifier	1	55	56	55
Random Forest Classifier	1	53	42	53
K-Neighbors Classifier	1	59	60	59
Passive Aggressive Classi fier	1	52	49	52
Logistic Regression	1	55	49	55
Linear SVC	2	55	51	55
SVC	2	54	59	54
Ridge Classifier	2	55	50	55
Perceptron	2	46	55	46
Bernoulli NB	2	36	52	36
SGD Classifier	2	54	52	54
Decision Tree Classifier	2	54	58	54
Random Forest Classifier	2	53	40	53
K-Neighbors Classifier	2	59	59	59
Passive Aggressive Classi fier	2	36	51	36
Logistic Regression	2	55	50	55
Linear SVC	3	55	50	55
SVC	3	54	61	54
Ridge Classifier	3	55	49	55
Perceptron	3	55	50	55
Bernoulli NB	3	36	52	36
SGD Classifier	3	54	49	54
Decision Tree Classifier	3	55	53	55
Random Forest Classifier	3	53	43	53
K-Neighbors Classifier	3	59	59	59
Passive Aggressive Classifier	3	51	48	51
Logistic Regression	3	55	49	55

Table 5. Performance Evaluation of Classifiers on Various n-gram Models + TFIDF Approach with Stemming

Algorithm	N-Gram	Accuracy	Precision	Recall
Linear SVC	1	0.727	0.717	0.727
SVC	1	0.760	0.754	0.760
Ridge Classifier	1	0.713	0.700	0.713
Perceptron	1	0.723	0.712	0.723
Bernoulli NB	1	0.671	0.662	0.671
SGD Classifier	1	0.675	0.659	0.675
Decision Tree Classifier	1	0.717	0.704	0.717
Random Forest Classifier	1	0.549	0.542	0.549
K-Neighbors Classifier	1	0.536	0.287	0.536
Passive Aggressive Classi fier	1	0.626	0.620	0.626
Logistic Regression	1	0.704	0.698	0.704
Linear SVC	1	0.727	0.717	0.727
SVC	2	0.740	0.733	0.740
Ridge Classifier	2	0.766	0.760	0.766
Perceptron	2	0.729	0.718	0.729
Bernoulli NB	2	0.738	0.729	0.738
SGD Classifier	2	0.695	0.687	0.695
Decision Tree Classifier	2	0.691	0.679	0.691
Random Forest Classifier	2	0.726	0.714	0.726
K-Neighbors Classifier	2	0.548	0.527	0.548
Passive Aggressive Classi fier	2	0.536	0.287	0.536
Logistic Regression	2	0.631	0.625	0.631
Linear SVC	2	0.724	0.719	0.724
SVC	2	0.741	0.732	0.741
Ridge Classifier	3	0.740	0.733	0.740
Perceptron	3	0.766	0.760	0.766
Bernoulli NB	3	0.728	0.717	0.728
SGD Classifier	3	0.739	0.731	0.739
Decision Tree Classifier	3	0.694	0.691	0.694
Random Forest Classifier	3	0.690	0.678	0.690
K-Neighbors Classifier	3	0.726	0.715	0.726
Passive Aggressive Classifier	3	0.547	0.556	0.547
Logistic Regression	3	0.536	0.287	0

Table 6. Performance Evaluation of Classifiers on Various n-gram Models + BOW Approach with Stemming

Algorithm	N-Gram	Accuracy	Precision	Recall
Linear SVC	1	0.718	0.706	0.718
SVC	1	0.739	0.730	0.739
Multinomial NB	1	0.722	0.711	0.722
Ridge Classifier	1	0.698	0.687	0.698
Perceptron	1	0.681	0.679	0.681
Bernoulli NB	1	0.673	0.659	0.673
SGD Classifier	1	0.718	0.707	0.718
Decision Tree Classifier	1	0.548	0.587	0.548
Random Forest Classifier	1	0.536	0.287	0.536
K-Neighbors Classifier	1	0.664	0.654	0.664
Passive Aggressive Classifier	1	0.684	0.679	0.684
Logistic Regression	1	0.726	0.715	0.726
MLPClassifier	1	0.703	0.702	0.703
Linear SVC	2	0.754	0.747	0.754
SVC	2	0.740	0.730	0.740
Multinomial NB	2	0.747	0.738	0.747
Ridge Classifier	2	0.740	0.732	0.740
Perceptron	2	0.746	0.740	0.746
Bernoulli NB	2	0.607	0.659	0.607
SGD Classifier	2	0.748	0.739	0.748
Decision Tree Classifier	2	0.548	0.586	0.548
Random Forest Classifier	2	0.536	0.287	0.536
K-Neighbors Classifier	2	0.654	0.647	0.654
Passive Aggressive Classifier	2	0.735	0.731	0.735
Logistic Regression	2	0.762	0.754	0.762
MLPClassifier	2	0.733	0.730	0.733
Linear SVC	3	0.751	0.744	0.751
SVC	3	0.735	0.726	0.735
Multinomial NB	3	0.743	0.737	0.743
Ridge Classifier	3	0.741	0.732	0.741
Perceptron	3	0.753	0.749	0.753
Bernoulli NB	3	0.579	0.683	0.579
SGD Classifier	3	0.748	0.740	0.748
Decision Tree Classifier	3	0.548	0.586	0.548

(continued)

Table 6. *(continued)*

Algorithm	N-Gram	Accuracy	Precision	Recall
Random Forest Classifier	3	0.536	0.287	0.536
K-Neighbors Classifier	3	0.650	0.643	0.650
Passive Aggressive Classifier	3	0.730	0.724	0.730
Logistic Regression	3	0.764	0.757	0.764

Table 7. Performance Evaluation of Classifiers on Various n-gram Models + Word2vec Approach with Stemming

Algorithm	N-Gram	Accuracy	Precision	Recall
Linear SVC	1	0.450	0.536	0.536
SVC	1	0.629	0.540	0.540
Bernoulli NB	1	0.465	0.404	0.404
SGD Classifier	1	0.440	0.534	0.534
Decision Tree Classifier	1	0.591	0.540	0.540
Random Forest Classifier	1	0.287	0.536	0.536
K-Neighbors Classifier	1	0.568	0.566	0.566
Passive Aggressive Classifier	1	0.529	0.335	0.335
Logistic Regression	1	0.457	0.536	0.536
Linear SVC	2	0.463	0.537	0.537
SVC	2	0.633	0.540	0.540
Bernoulli NB	2	0.477	0.402	0.402
SGD Classifier	2	0.452	0.535	0.535
Decision Tree Classifier	2	0.615	0.538	0.538
Random Forest Classifier	2	0.502	0.536	0.536
K-Neighbors Classifier	2	0.568	0.564	0.564
Passive Aggressive Classi fier	2	0.508	0.303	0.303
Logistic Regression	2	0.467	0.536	0.536
Linear SVC	3	0.460	0.537	0.537
SVC	3	0.649	0.540	0.540
Bernoulli NB	3	0.478	0.401	0.401
SGD Classifier	3	0.461	0.536	0.536
Decision Tree Classifier	3	0.638	0.538	0.538
Random Forest Classifier	3	0.502	0.536	0.536
K-Neighbors Classifier	3	0.571	0.567	0.567
Passive Aggressive Classifier	3	0.409	0.530	0.530
Logistic Regression	3	0.458	0.536	0.536

Table 8. Performance Evaluation of Classifiers on Various n-gram Models + Fast text Approach with Stemming

Algorithm	N-Gram	Accuracy	Precision	Recall
Linear SVC	1	0.536	0.453	0.536
SVC	1	0.540	0.645	0.540
Bernoulli NB	1	0.407	0.469	0.407
SGD Classifier	1	0.535	0.437	0.535
Decision Tree Classifier	1	0.538	0.486	0.538
Random Forest Classifier	1	0.536	0.716	0.536
K-Neighbors Classifier	1	0.565	0.568	0.565
Passive Aggressive Classifier	1	0.536	0.504	0.536
Logistic Regression	1	0.535	0.446	0.535
Linear SVC	2	0.536	0.471	0.536
SVC	2	0.541	0.633	0.541
Bernoulli NB	2	0.395	0.467	0.395
SGD Classifier	2	0.536	0.364	0.536
Decision Tree Classifier	2	0.539	0.541	0.539
Random Forest Classifier	2	0.536	0.287	0.536
K-Neighbors Classifier	2	0.561	0.565	0.561
Passive Aggressive Classifier	2	0.462	0.468	0.462
Logistic Regression	2	0.537	0.463	0.537
Linear SVC	3	0.536	0.456	0.536
SVC	3	0.540	0.642	0.540
Bernoulli NB	3	0.400	0.475	0.400
SGD Classifier	3	0.537	0.460	0.537
Decision Tree Classifier	3	0.539	0.518	0.539
Random Forest Classifier	3	0.536	0.502	0.536
K-Neighbors Classifier	3	0.570	0.573	0.570
Passive Aggressive Classifier	3	0.424	0.473	0.424
Logistic Regression	3	0.535	0.443	0.535

References

1. Cherrat, E.M., Ouahi, H., Bekkar, A.: Sentiment analysis from texts written in standard Arabic and Moroccan dialect based on deep learning approaches. Int. J. Comput. Digit. Syst. **15**(1), 1–9 (2024)
2. Abdelhakim, B.A., Mohamed, B.A., Soufyane, A.: Using machine learning and TF-IDF for sentiment analysis in Moroccan dialect: an analytical methodology and comparative study. In: Proceedings of the International Conference on Smart City Applications, pp. 1–8. Springer, Cham (2023)
3. Errami, M., et al.: Sentiment analysis on Moroccan dialect based on ml and social media content detection. In: Proceedings of the International Conference on Digital Technologies and Applications, pp. 1–6. Springer, Cham (2023)

4. Madani, Y., Erritali, M., Bouikhalene, B.: A new sentiment analysis method to detect and analyse sentiments of COVID-19 Moroccan tweets using a recommender approach. Multimed. Tools Appl. **82**(18), 27819–27838 (2023)
5. Lasri, I., Riadsolh, A., Elbelkacemi, M.: Real-time twitter sentiment analysis for moroccan universities using machine learning and big data technologies. Int. J. Emerg. Technol. Learn. 18(5) (2023)
6. El Ouahabi, S., El Ouahabi, S., Dadi, E.W.: Contribution to Moroccan Darija sentiment analysis in social networks. Soc. Netw. Anal. Min. **13**(1), 138 (2023)
7. Mouaad, E., Ouassil, M.A., Rachidi, R., Cherradi, B., Hamida, S., Raihani, A.: Sentiment analysis on moroccan dialect based on ml and social media content detection. Int. J. Adv. Comput. Sci. Appl. **14**(3), 315–325 (2023). https://doi.org/10.14569/IJACSA.2023.0140347
8. Matrane, Y., Benabbou, F., Sael, N.: A systematic literature review of Arabic dialect sentiment analysis. J. King Saud Univ. Comput. Inf. Sci. **35**(6), 101570 (2023). https://doi.org/10.1016/j.jksuci.2023.101570
9. Essefar, K., Baha, H., Mahdaouy, A., El Mekki, A., Berrada, I.: OMCD: offensive Moroccan comments dataset. Lang. Resour. Eval. (2023). https://doi.org/10.1007/s10579-023-09663-2
10. Rachidi, R., Ouassil, M.A., Mouaad, E., Cherradi, B., Hamida, S., Silkan, H.: Classifying toxicity in the Arabic Moroccan dialect on Instagram: a machine and deep learning approach. Indones. J. Electr. Eng. Comput. Sci. **31**(1), 588–598 (2023). https://doi.org/10.11591/ijeecs.v31.i1.pp588-598
11. Slim, A., Melouah, A., Faghihi, U., Sahib, K.: Improving neural machine translation for low resource Algerian dialect by transductive transfer learning strategy. Arab. J. Sci. Eng. **47**(1), 1–8 (2022)
12. Messaoudi, A., et al.: Tunbert: Pretrained contextualized text representation for Tunisian dialect. In: Intelligent Systems and Pattern Recognition, Cham, pp. 278–290 (2022)
13. Brachemi-Meftah, S., Drias, H., Abdessalem, T.: Impact of dimensionality reduction on sentiment analysis of Algerian dialect. In: 2022 IEEE 9th International Conference on Sciences of Electronics, Technologies of Information and Telecommunications (SETIT), pp. 433–440. IEEE (2022)
14. Bourahouat, G., Abourezq, M., Daoudi, N.: Leveraging Moroccan Arabic sentiment analysis using Arabert and Qarib. In: Proceedings of the International Conference on Smart City Applications, pp. 299–310. Springer, Cham (2022). https://doi.org/10.1007/978-3-031-26852-6_29
15. Abdaoui, A., Berrimi, M., Oussalah, M., Moussaoui, A.: Dziribert: pre-trained language model for the Algerian dialect. arXiv preprint arXiv:2109.12346 (2021)
16. Elgezouli, M., Elmadani, K.N., Saeed, M.: SudaBERT: pre-trained encoder representation for Sudanese Arabic dialect. In: Proceedings of the 2020 ICCCEEE, pp. 1–4 (2021). https://doi.org/10.1109/ICCCEEE49695.2021.9429651
17. Garouani, M., Chrita, H., Kharroubi, J.: Sentiment analysis of Moroccan tweets using text mining. In: Proceedings of the International Conference on Digital Technologies and Applications. Springer, Cham (2021)
18. Garouani, M., Kharroubi, J.: MAC: an open and free Moroccan Arabic corpus for sentiment analysis. In: Proceedings of the International Conference on Smart City Applications, pp. 849–858. Springer, Cham (2021). https://doi.org/10.1007/978-3-030-94191-8_68
19. Abdelminaam, D.S., Fahmy, A.A., Abdelrazek, M., El-din Hemdan, E.E., Elngar, A.A., Abd El-Latif, A.A.: ArabicDialects: an efficient framework for Arabic dialects opinion mining on Twitter using optimized deep neural networks. IEEE Access **9**, 97079–97099 (2021)
20. Mihi, S., Benaicha, A., El, A., Arezki, S., Laachfoubi, N.: MSTD: Moroccan sentiment twitter dataset. Int. J. Adv. Comput. Sci. Appl. **11**(10), 363–372 (2020)

21. Habbat, N., Anoun, H., Hassouni, L.: Topic modeling and sentiment analysis with LDA and NMF on Moroccan tweets. In: Proceedings of the Third International Conference on Software and Computer Applications (ICSCA), pp. 147–161. Springer, Cham (2020)
22. Kwaik, K.A., Chatzikyriakidis, S., Dobnik, S., Saad, M., Johansson, R.: An Arabic tweets sentiment analysis dataset (ATSAD) using distant supervision and self-training. In: Proceedings of the 4th Workshop on Open-Source Arabic Corpora and Processing Tools, with a Shared Task on Offensive Language Detection, pp. 1–8 (2020)
23. Moudjari, L., Akli-Astouati, K., Benamara, F.: An Algerian corpus and an annotation platform for opinion and emotion analysis. In: Proceedings of the 12th Language Resources and Evaluation Conference, pp. 1202–1210 (2020)
24. Fourati, C., Messaoudi, A., Haddad, H.: TUNIZI: a Tunisian Arabizi sentiment analysis dataset. In: Proceedings of the 12th Language Resources and Evaluation Conference, pp. 1202–1210 (2020)
25. Abu Kwaik, K., Chatzikyriakidis, S., Dobnik, S., Saad, M., Johansson, R.: LSTM-CNN deep learning model for sentiment analysis of dialectal Arabic. In: Habash, N., Bougares, F., Zaghouani, W. (eds.) Arabic Language Processing: From Theory to Practice, ICALP 2019. LNCS, vol. 11738, pp. 108–121. Springer, Cham (2019)
26. Atoum, J.O., Nouman, M.: Sentiment analysis of Arabic Jordanian dialect tweets. Int. J. Adv. Comput. Sci. Appl. **10**(2), 142–148 (2019)
27. Bettiche, M., AlShamma, S., Koutaibi, M., Bennamoun, M.: Opinion mining in social networks for Algerian dialect. In: Proceedings of the 17th International Conference on Information Processing and Management of Uncertainty in Knowledge-Based Systems. Applications, vol. 854, pp. 629–641. Springer, Cham (2018)
28. Baly, R., Khaddaj, A., Hajj, H., El-Hajj, W., Shaban, K.B.: ArSentDLEV: a multi-topic corpus for target-based sentiment analysis in Arabic. In: Proceedings of the Eleventh International Conference on Language Resources and Evaluation (LREC 2018) (2018)
29. Alnawas, A., Arici, N.: The corpus based approach to sentiment analysis in modern standard Arabic and Arabic dialects: a literature review. Politeknik Dergisi **21**(2), 461–470 (2018)
30. Al-Thubaity, A., Alharbi, M.: A Saudi dialect twitter corpus for sentiment and emotion analysis. In: Proceedings of the 2018 21st Saudi Computer Society National Computer Conference (NCC), pp. 1–6 (2018)
31. Baly, R., Hajj, H., El-Hajj, W., Shaban, K.B.: Comparative evaluation of sentiment analysis methods across arabic dialects. Procedia Comput. Sci. **117**, 266–273 (2017). https://doi.org/10.1016/j.procs.2017.10.118
32. El-Beltagy, S.R., Ali, A., Shoukry, A.: NileTMRG at SemEval-2017 Task 4: Arabic sentiment analysis. In: Proceedings of the 11th International Workshop on Semantic Evaluation (SemEval-2017), pp. 790–795. Association for Computational Linguistics (2017). https://aclanthology.org/S17-2133
33. Itani, M., Roast, C., Al-Khayatt, S.: Corpora for sentiment analysis of Arabic text in social media. In: Proceedings of the 2017 8th International Conference on Information and Communication Systems (ICICS), pp. 64–69 (2017)
34. Al-Twairesh, N., Al-Khalifa, H., Al-Salman, A., Al-Ohali, Y.: AraSenTi-tweet: a corpus for Arabic sentiment analysis of Saudi tweets. Procedia Comput. Sci. **117**, 63–72 (2017)
35. Assiri, A., Emam, A., Al-Dossari, H.: Saudi twitter corpus for sentiment analysis. Int. J. Comput. Inf. Eng. **10**(2), 142–148 (2016)
36. Al-Rubaiee, H., Qiu, R., Li, D.: Identifying Mubasher software products through sentiment analysis of Arabic tweets. In: Proceedings of the 2016 International Conference on Industrial Informatics and Computer Systems (CIICS), pp. 1–6 (2016)
37. Bin Hathlian, N.F., Hafezs, A.M.: Sentiment-subjective analysis framework for Arabic social media posts. In: Proceedings of the 2016 4th Saudi International Conference on Information Technology (Big Data Analysis) (KACSTIT), pp. 1–6 (2016)

38. Sghaier, M.A., Zrigui, M.: Sentiment analysis for Arabic e-commerce websites. In: Proceedings of the 2016 International Conference on Engineering & MIS (ICEMIS), pp. 1–7 (2016)
39. Ibrahim, H.S., Abdou, S.M., Gheith, M.: MIKA: A tagged corpus for modern standard Arabic and colloquial sentiment analysis. In: Proceedings of the 2015 IEEE 2nd International Conference on Recent Trends in Information Systems (ReTIS), pp. 353–358 (2015)
40. Abdul-Mageed, M.: Subjectivity and sentiment analysis of Arabic as a morphologically-rich language. Ph.D. thesis, Indiana University (2015)
41. Duwairi, R.: Sentiment analysis for dialectical Arabic. In: Proceedings of the 2015 IEEE International Conference on Automation and Computing (IACS), pp. 1–6 (2015). https://doi.org/10.1109/IACS.2015.7103221
42. Nabil, M., Aly, M., Atiya, A.: ASTD: Arabic sentiment tweets dataset. In: Proceedings of the 2015 Conference on Empirical Methods in Natural Language Processing, pp. 2515–2519 (2015)
43. Al-Ayyoub, M., Alsmadi, I., Shatnawi, M., Al-Kabi, M.: Lexicon-based sentiment analysis of Arabic tweets. Int. J. Soc. Netw. Min. **2**(2), 101–114 (2015). https://doi.org/10.1504/IJSNM.2015.072280
44. Ben Salamah, J., Elkhli, A.: Microblogging opinion mining approach for Kuwaiti dialect. In: Proceedings of the International Conference on Computer Technology and Information Management, pp. 388–396 (2014)
45. Abdul-Mageed, M., Diab, M., Kübler, S.: SAMAR: subjectivity and sentiment analysis for Arabic social media. Comput. Speech Lang. **28**(1), 20–37 (2014)
46. Abdulla, N.A., Ahmed, N.A., Shehab, M.A., Al-Ayyoub, M.: Arabic sentiment analysis: lexicon-based and corpus-based. In: 2013 IEEE Jordan Conference on Applied Electrical Engineering and Computing Technologies (AEECT), pp. 1–6 (2013)
47. Shoukry, A., Rafea, A.: Sentence-level Arabic sentiment analysis. In: Proceedings of the 2012 International Conference on Collaboration Technologies and Systems (CTS), pp. 546–550 (2012)
48. Akerkar, R., Sajja, P.: Knowledge-Based Systems. Jones & Bartlett Publishers (2009)
49. Duwairi, R., Al-Refai, M., Khasawneh, N.: Stemming versus light stemming as feature selection techniques for Arabic text categorization. In: 2007 Innovations in Information Technologies (IIT), pp. 446–450 (2007). https://doi.org/10.1109/IIT.2007.4430403

On the Challenges of Migrating to Microservices Architectures for Better Cloud Solutions

Hassan Farsi(✉), Driss Allaki, Abdeslam En-Nouaary, and Mohamed Dahchour

Institut National des Postes et Télécommunications, Rabat, Morocco
{farsi.hassan,d.allaki,abdeslam,dahchour}@inpt.ac.ma

Abstract. Driven by the increasing demand for agility and scalability in modern software development, the cloud-native paradigm has emerged as an approach for building and deploying applications. This shift necessitates a migration from traditional monolithic architectures towards microservices. While various strategies and patterns have been suggested to facilitate this migration, there is still a need to investigate their impact on the emergence of anti-patterns and their potential evolution during the migration phases. This chapter highlights challenges and explores research directions to deal with anti-patterns while migrating to microservices. Four main research questions are presented and discussed, providing valuable insights into this topic. This work intends to inspire the community to conduct research on this subject, aiming to improve migration quality and facilitate the creation of efficient microservices architectures.

1 Introduction

The cloud-native paradigm is an approach to software development to create agile and scalable software solutions. This approach promotes the creation of modular designs, allowing for faster development, dynamic scaling, and greater adaptability to changes [1]. To follow the cloud-native paradigm, applications often adopt a microservices architecture. This architectural style breaks applications into a set of small and loosely coupled services [2], promoting various principles of the cloud-native paradigm. For instance, microservices enable greater agility by fostering the creation of small and autonomous development teams, enabling them to work on individual services without disrupting each other. This is achieved through the independent development and deployment of microservices. This leads to faster development since teams can release updates without waiting for others. It also facilitates easier maintainability, as modifications only impact the relevant service. Moreover, it simplifies testability, as each service is tested independently without the need to consider the entire application. Microservices also facilitates dynamic scaling by allowing each service to be independently scaled in response to changing demand. As a result, the *Time-to-Market* (timeframe for a software product, feature, or update to reach the market) is significantly reduced [3].

M. Zbakh et al. (Eds.): CloudTech 2023, LNNS 1220, pp. 236–257, 2024.
https://doi.org/10.1007/978-3-031-78698-3_12

Implementing the cloud-native paradigm requires a migration from monolithic architectures to microservices. This migration needs to be executed incrementally and with careful consideration to ensure that the new system aligns with the principles of the cloud-native paradigm (including agility and scalability) [4], while also retaining the same functionalities of the monolithic system. For instance, poorly defining service boundaries while migrating to microservices may lead to a tightly coupled architecture. This reduces agility by increasing collaboration between teams, delays maintainability as changes may affect multiple services, and, limits scalability by preventing independent scaling. Accordingly, a set of migration strategies and patterns have been recommended to avoid common migration issues and help improve the quality of the resulting microservices architecture.

The quality of the resulting microservices architecture quality can be influenced by various factors, including lack of architectural knowledge, communication challenges, and difficulties in data management. Another important factor is the emergence of anti-patterns and their impact on the overall architecture. Anti-patterns are solutions that may initially appear effective but ultimately result in unfavorable consequences, such as increased complexity, restricted scalability, and prolonged maintenance [5].

This chapter aims to present challenges and explore research directions related to the emergence of anti-patterns while using migration strategies and patterns to transition from monolithic architectures to microservices. We present and analyze four main research inquiries, each addressing a specific aspect of the emergence of anti-patterns during this migration.

The following sections of this chapter are structured as follows. Section 2 presents the migration strategies and patterns, followed by a presentation of both monolithic and microservices anti-patterns. Section 3 offers a detailed description of the research inquiries presented by this work. Section 4 sheds light on the related work, while Sect. 5 serves as a conclusion for this paper.

2 Background

This section presents the elements to understand the problem at hand. It includes an overview of migration strategies and patterns, along with a presentation of anti-patterns in monolithic and microservices architectures. Each of these elements is supported by illustrative examples for a better understanding.

2.1 Strategies and Patterns for Migrating to Microservices

Migration strategies and patterns are a set of actions and good practices that facilitate a safe and incremental transition from monolithic architectures to microservices [6,7]. As they follow an incremental approach, these strategies and patterns mitigate potential risks by tackling issues at an earlier stage, which makes the identification and assessment of emerging problems easier. The adaptability is further enhanced through the ability to make adjustments after each incremental step, which in turn promotes early delivery by providing a functional part of the system during the initial migration phases.

2.1.1 Migration Strategies

Migration strategies are steps that organize the migration to microservices. The goal of these strategies is not necessarily to replace entirely the monolith with microservices; instead, they often involve refactoring certain functionalities from the monolith to microservices in order to address the initial migration triggers (better agility and scalability) [7,8]. If necessary, the new microservices system may also expand by moving all the monolith functionalities to microservices, leading to replacing entirely the monolithic system.

According to [6], there are three main strategies for migrating incrementally to microservices. The first strategy is to **implement new features as services**. This consists of stopping the expansion of the monolith and incorporating new features as new services, enabling the integration of additional features (services) without increasing the complexity of the existing monolith. Figure 1 presents an example of implementing a new notification feature as a service.

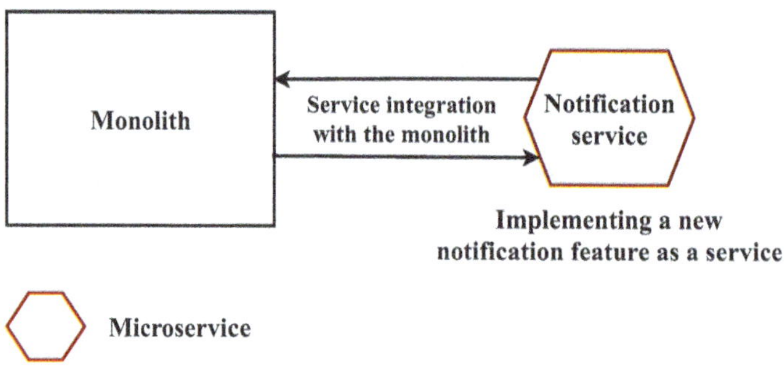

Fig. 1. Implementing a new notification feature as a service.

In the example shown in Fig. 1, the first migration strategy was utilized to create a separate microservice for the new notification feature. It is important to integrate the service with the monolith to enable data access and invoke other existing functionalities. The integration can be implemented by synchronous communication using for example REST *Application Programming Interface (API)* calls, or asynchronous communication via domain events. This may require some adjustments within the monolith, which are facilitated by the patterns described in Sect. 2.1.2.

The second strategy is to **separate the presentation of the monolith from the backend**, which involves splitting the monolith into a presentation layer and a business and data access layer. These layers are typically deployed as separate applications, which communicate through exposed API. Figure 2 presents an example of this strategy.

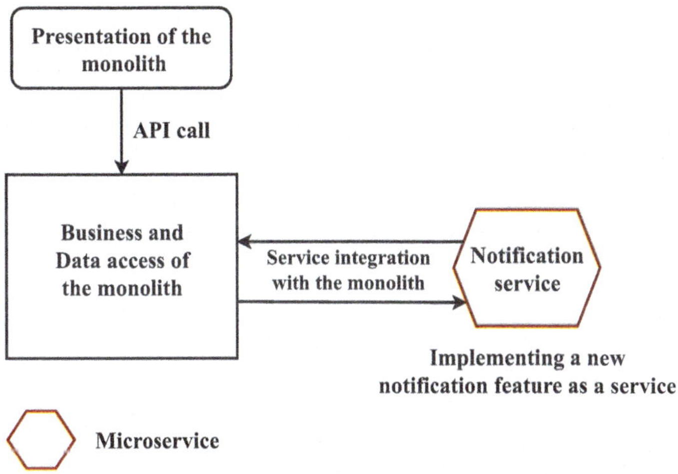

Fig. 2. Separating the presentation and the backend of the monolith.

Figure 2 shows the separation of the monolith's presentation from its business and data access logic. These elements are presented as separate applications, with the monolith's presentation invoking the business and data access functionalities through API requests. This strategy prepares the monolith for service extraction in the third migration strategy by relocating the presentation code to an independent application. Moreover, the API exposed by the business and data access application could be also accessible to the newly created microservices.

The third strategy is to **break up the monolith by extracting functionality into services**, leading to an increase in the number of the created microservices and a reduction in the size of the monolith. The challenge while implementing this strategy is to split the business domain and define the service boundaries. One approach to achieve this is *Domain-Driven Design (DDD)* [9], which is an approach to identify microservice boundaries based on business domains, rather than the technical aspects. Figure 3 presents an example of this strategy.

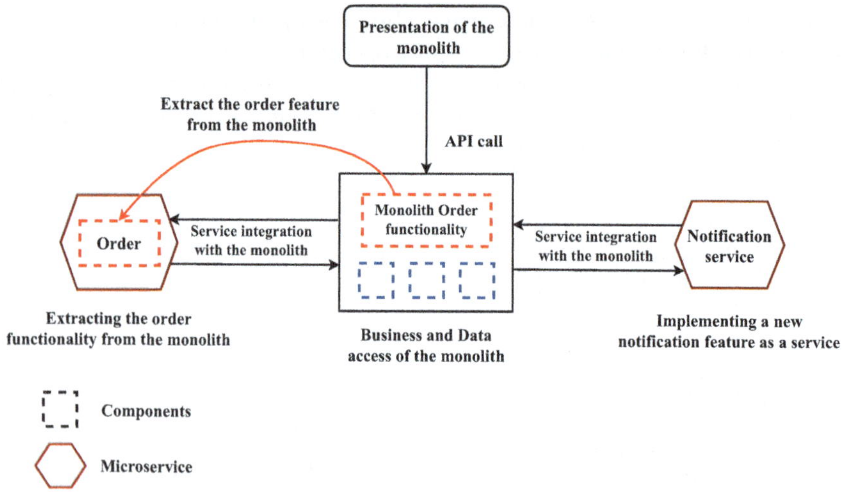

Fig. 3. Breaking up the monolith by extracting the order functionality.

The illustration in Fig. 3 presents the extraction of the order functionality from the monolith and its implementation as a new microservice. The component (one or a set of classes) responsible for the monolith order functionality is retrieved and deployed as separate order microservices. The monolith order functionality remains in place until the new order microservice is tested and validated to offer the same answers as the monolith order functionality.

Implementing the first two migration strategies seems straightforward, however, executing the third strategy (breaking up the monolith by extracting functionality into services) requires careful planning to ensure a smooth and secure transition. For example, dealing poorly with the dependencies related to the functionality to be extracted could cause disturbances in other functionalities that rely on it. As such, a set of recommended migration patterns is available to aid in this matter.

2.1.2 Migration Patterns

Migration patterns are solutions to address common problems that might occur during the migration to microservices [7]. These patterns assist in migration strategies by providing effective solutions to potential challenges that may arise during implementation. It is important to note that these patterns are not relevant in all situations, as each one has its own set of benefits and constraints.

There are two categories of migration patterns known as "*Monolith splitting patterns*" and "Data Decomposition Patterns" [7]. On one hand, the "*Monolith splitting patterns*" provides solutions to extract functionalities from the business and data access logic of the monolith and implement them as new microservices. On the other hand, the "*Data Decomposition Patterns*" deal with the challenges related to data decomposition, since microservices architecture promotes that each microservice owns and controls its data.

Several patterns fall under the category of "*Monolith splitting patterns*". In this work, we present some of these patterns.

Strangler Fig Application Pattern [10]**:** The *Strangler Fig Application* pattern is a technique to rewrite systems incrementally. This pattern is used for migrating monolithic systems to microservices by extracting features gradually and implementing them as new services. Using a proxy, the incoming calls to the monolith that invoke the extracted functionality in the monolith are intercepted and redirected to the newly created services. Figure 4 presents an example of extracting the order functionality into a new order service while redirecting external calls to the service using an HTTP proxy.

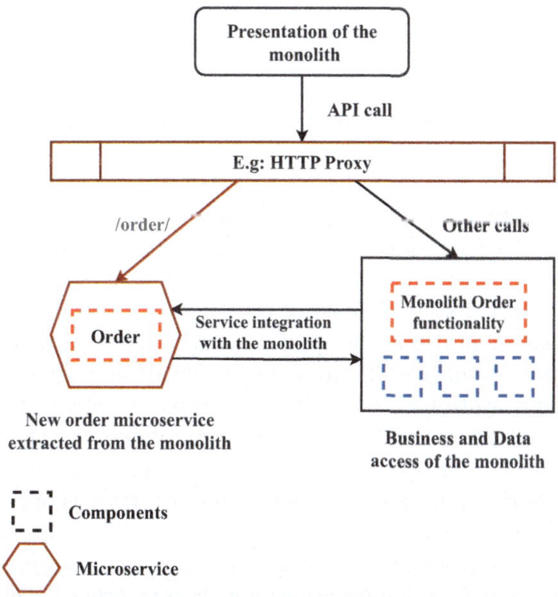

Fig. 4. Overview of the Strangler Fig Application pattern.

The example of Fig. 4 shows the utilization of the *Strangler Fig Application* pattern to extract and implement the component responsible for the ordering functionality as a new microservice. Incoming requests from the monolith's presentation that invoke the ordering operations in the monolith are intercepted and redirected to the new order service via an HTTP proxy.

Employing the *Strangler Fig Application* pattern reduces the risk by enabling the gradual extraction of monolithic features, which allows for an early detection of any problems. Besides, configuring the proxy permits to switch seamless the incoming calls back to the monolith's implemented feature if issues arise within the newly established microservices. This pattern also reduces interruptions as the monolith continues to operate smoothly while the new microservices are being implemented.

The functionality within the monolith remains until the new microservice is tested and confirmed to provide output that matches that of the monolith. Once it is done, the functionality inside the monolith should be removed. However, this task could be disruptive for other developers involved in the codebase, as their components within the monolith could rely on the extracted features (deals with internal interactions). Therefore, the utilization of the *Branch by Abstraction* pattern proves beneficial in minimizing disruptions during such modifications.

Branch by Abstraction Pattern [11,12]**:** The *Branch by Abstraction* is a pattern for implementing system changes without disrupting other teams. It involves using an abstraction to switch smoothly between multiple implementations without disruption to other teams. In the context of migrating to microservices, two implementations are established: one maintaining the existing monolithic functionality, while the other redirecting dependencies from the monolithic system to the newly developed microservice. Figure 5 shows an example of the utilization of the *Branch by Abstraction* pattern to manage the extraction of the order functionality.

In the example of Fig. 5, the *Branch by Abstraction* pattern is implemented through four main steps. The initial step involves creating an *Order abstraction* for the *Monolith order functionality*. The *Monolith order functionality* implements the created *Order abstraction*, while all the dependencies are routed to the same abstraction. The second step consists of creating an *Order service implementation*. This new implementation calls upon the newly established microservice. The *Monolith order functionality* will still be active until the work on the *Order service* and the implementation is done. The third step is to activate the *Order service implementation* instead of the *Monolith order functionality*. This allows the use of the newly created order service, with the flexibility to revert to the *Monolith order functionality* if required. Once the new order service is operational, tested, and ready for use, the *Monolith order functionality* and the *Order abstraction* are no longer in use and subsequently eliminated in the fourth step. As a result, the extraction of the order functionality does not disrupt other development teams, such as those handling the *Invoice* and *Payment* microservices.

To confirm that the new microservice provides the same output as the monolith existing functionality, a comparison between the two implementations' results is required. Instead of switching between the implementations and only activating one at a time, the *Parallel run* pattern enables the utilization of both implementations simultaneously which makes the comparison easier.

Parallel Run Pattern [7]**:** The *Parallel Run* is a pattern that allows both the monolith existing functionality and the microservices implementation to be invoked simultaneously, enabling the comparison of their output for equivalence.

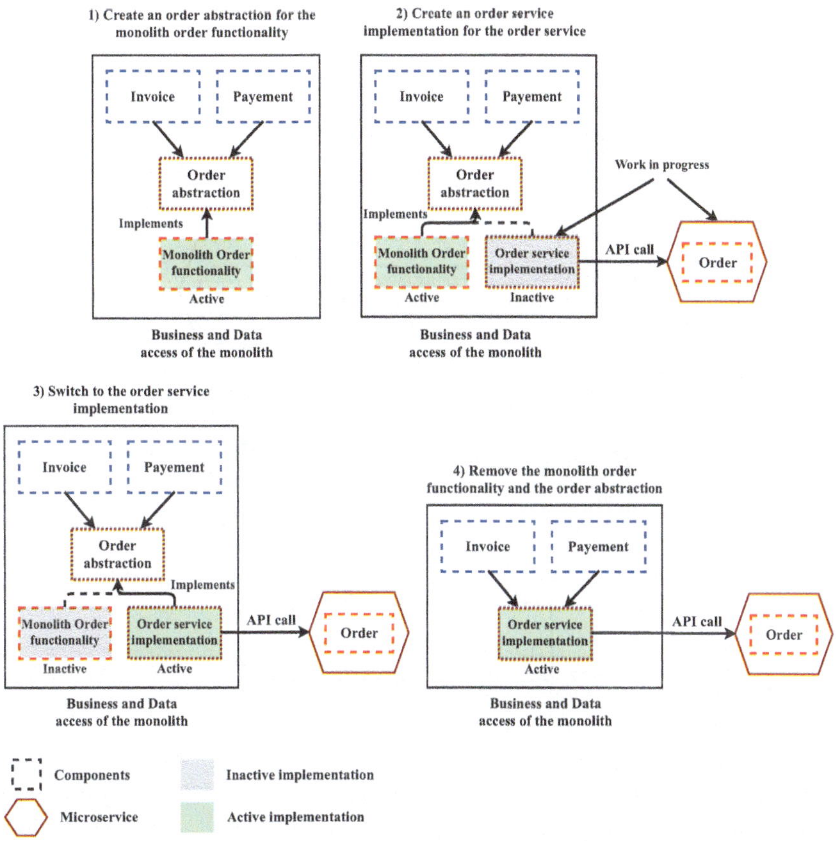

Fig. 5. Overview of the Branch by Abstraction pattern.

The comparison is conducted while taking the functionality within the monolith as the source we trust. Once the microservice demonstrates a level of precision that matches the monolith's existing functionality, it attains the status of being the main implementation and the source we trust. Subsequently, the monolith existing functionality could be removed following the fourth step of the *Branch by Abstraction* pattern. Figure 6 shows an overview of the *Parallel Run* pattern.

Figure 6 example shows an example of employing the *Parallel Run* pattern to invoke at the same time the order functionality in the monolith and the newly created order service. The outputs of handling the order requests from both implementations are saved, enabling a comparison process to evaluate the equivalence of the order microservices functionalities against the monolith order functionality. Currently, the monolithic order functionality is the main source from which the results are generated, making its feedback the primary information conveyed to the presentation layer.

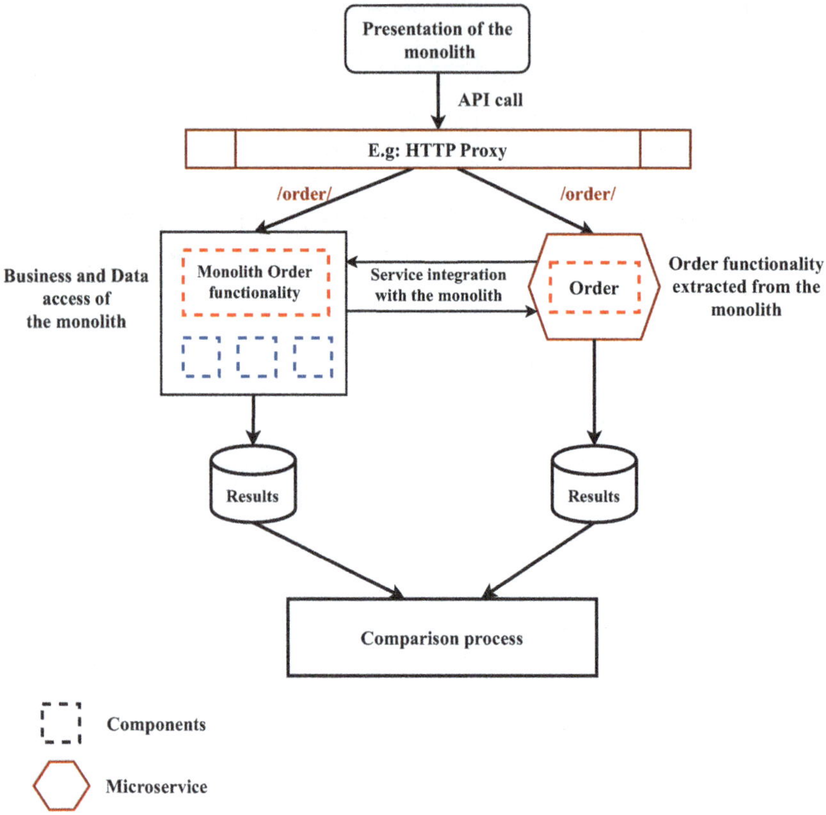

Fig. 6. Overview of the Parallel Run pattern.

Integrating new services with the existing monolith may necessitate certain modifications to the monolith itself. For example, to inform a service about the behaviors within the monolith, adjustments in the monolith are required to send the information to the service. In case the monolith is in highly level of complexity, making direct modifications before simplifying it by extracting functionalities might be a difficult task. In such cases, implementing the *Decorating Collaborator* pattern proves beneficial.

Decorating Collaborator Pattern [7]**:** The *Decorating Collaborator* pattern facilitates the transfer of behaviors from a monolithic system to its associated services without altering the monolith. This is achieved by improving the proxy employed in the *Strangler Fig Application* pattern, which captures the monolith's response to the presentation layer and subsequently triggers a call to the microservices, notifying them of any updates. Figure 7 shows an example of applying the *Decorating Collaborator* pattern with a new loyalty feature implemented as a new microservice.

In the example of Fig. 7, the *Loyalty* service is used to attribute a certain loyalty point when an order is placed successfully. Rather than changing the

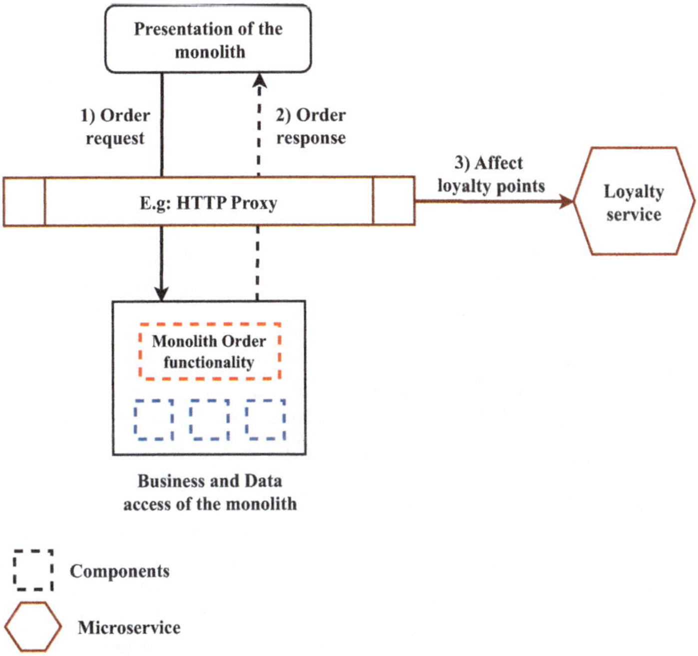

Fig. 7. Example of the Decorating Collaborator pattern.

monolith to update to call the *Loyalty* service, the *Decorating Collaborator* pattern is used to intercept the order response and update the *Loyalty* service via the proxy implemented by the *Strangler Fig Application* pattern. It is important to pay attention to the amount of code integrated within the proxy to prevent increasing its complexity.

There is no specific order to execute the aforementioned patterns. However, a common approach is to begin with the *Strangler Fig Application* pattern followed by the *Branch by Abstraction* patterns to extract the functionality of the monolith. The *Decorating Collaborator* pattern could be used when needed to integrate new microservices without altering the monolith, while the *Parallel Run* pattern is employed during the implementation of each microservice to assess and compare its functionalities with that of the current monolith.

The "Data Decomposition Patterns" category contains various patterns to deal with data decomposition. Some of these patterns are solutions used when both monolith and microservices coexist, enabling the microservices to access and manage the data owned by the monolith. The other patterns in this category focus on efficiently extracting data from the monolith database and allocating it to its corresponding microservice.

Database View Pattern [7,13]**:** The *Database View* is a pattern that enables newly created microservices to retrieve data from the monolith database. This pattern consists of creating a view within the monolith database that contains the data needed by a microservice. Utilizing a view ensures the confidentiality of other content of the database as the microservice is confined to only interacting with the view. Figure 8 shows an example of using the *Database View* pattern where the *Loyalty* service interacts with a view to access the data it needs.

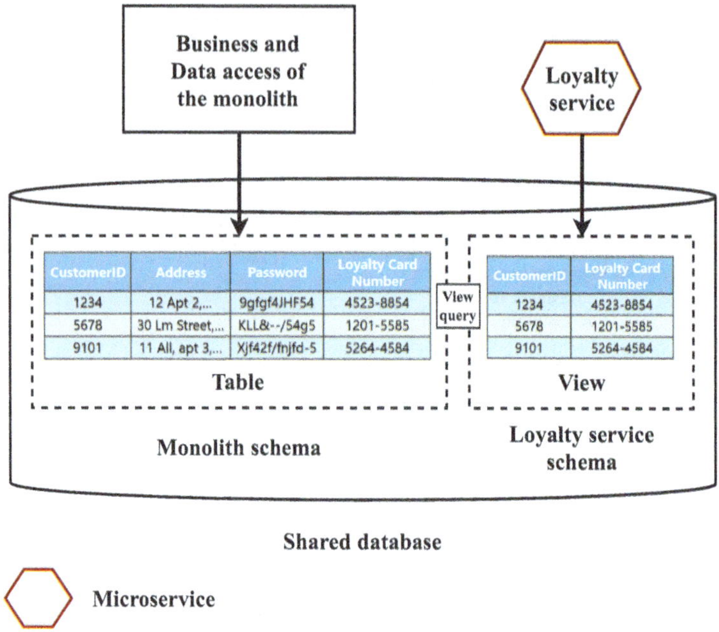

Fig. 8. Example of the Database View pattern.

In the example of Fig. 8, a view has been generated using a view query that includes the data needed for the *Loyalty* service. This particular view is established within a separate schema to streamline the extraction of the *Loyalty* service data to a separate database. The remaining data in the table is protected as the Loyalty service only reads the view in the *Loyalty service schema.*

The generated view is a result of a view query which makes it read-only. This limits its effectiveness as the information contained therein cannot be altered. Moreover, placing the view within the same database engine as other data may impact the accessibility of *Loyalty* service data, since a failure in the database renders all data inaccessible. Therefore, it is important to segregate the data appropriately by employing the *Split Table* pattern.

Split Table Pattern [7]**:** The *Split Table* is a pattern to split a table containing data managed by various services. In shared databases, it's common for multiple sets of data to be stored in a single table. This pattern splits the table (by column) so that each microservice encapsulates, controls, and manages its own data. Figure 9, presents an example of this pattern.

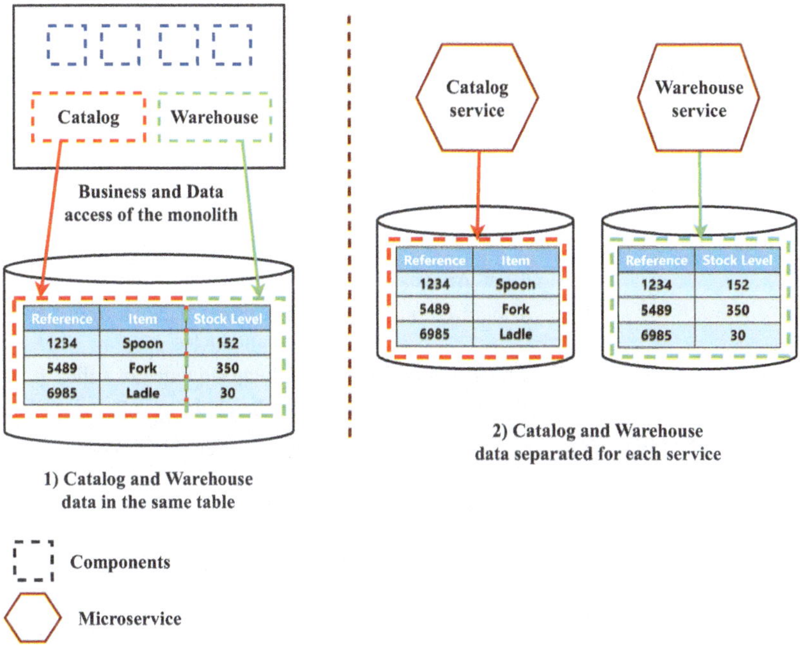

Fig. 9. Example of the Split Table pattern.

The example in Fig. 9 presents a use case where both the Catalog and the Warehouse data are stored within the same table. The *Split Table* pattern is used to separate these data into two separate tables, each residing in a distinct database managed by its corresponding microservice. The challenge with this pattern is to identify the owner of the data in case multiple microservices access the same table.

2.2 Monolithic and Microservices Anti-patterns

Anti-patterns are solutions that initially appear to be good but later prove to be bad for the system [5]. An example of this is the *Blob* anti-pattern [14], in which multiple attributes and functionalities are added to a single class. While this may seem convenient for smaller applications, it can lead to tight coupling, increase complexity, and obstruct reusability and maintenance as the application grows. There are several anti-patterns identified in both monolithic (current architecture) and microservices (architecture to migrate to) [14–16]. Table 1 presents some anti-patterns in both architectures.

The anti-patterns presented in Table 1 are categorized into two sections. On one hand, the *Monolith and microservices anti-patterns* include anti-patterns that are relevant to both monolithic and microservices architectures. For instance, the aforementioned *Blob* anti-pattern could appear in the components of the monolith as in the newly created microservices. The anti-patterns of

Table 1. Monolith and microservices anti-patterns

Architecture	Category	Anti-pattern	Description
Monolith and microservices anti-patterns	Design anti-patterns	The Blob	An object with a lot of responsibilities
		Lava Flow	A remaining obsolete code
		Spaghetti Code	A difficult and non-comprehended code
		Boat Anchor	A piece of software that has no use
		Global Hub	A component with multiple dependencies
		Message Chain	A long sequence of calls
		Cyclic dependencies	Cyclic chain of calls
	Architectural anti-patterns	Stovepipe System	Poorly related concepts or components
		Architecture by Implication	Developing without architecture
		Swiss Army Knife	Classes with overly complex interfaces
		Broken Modularization	Poorly defined modules
Microservices anti-patterns	Design anti-patterns	Wrong Cut	Microservices decomposition based on technical layers
		Megaservice	A microservice with a lot of responsibilities
		Nanoservice	A very small microservice
	Implementation anti-patterns	Shared Libraries	Shared libraries or file among microservices
		Inappropriate Service Intimacy	Microservice connected to the private data of another
		Shared Persistence	Microservices shares one database
		Hardcoded Endpoints	Explicitly specified IP addresses, ports, and endpoints

this section are organized as *Design anti-patterns*, which pertain to the design and organization of code within individual components (inside the monolith or microservices), and *Architectural anti-patterns* which are related to the overall structure of the architecture and the interactions between components. On the other hand, *Microservices anti-patterns* present anti-patterns specific to microservices. These anti-patterns are further divided into two categories: *Design anti-patterns*, which pertain to the design of microservices, and "*Implementation anti-patterns*, which concern low-level coding and executing the design.

The presence of anti-patterns affects the overall quality of the architecture by mitigating agility and scalability. These anti-patterns can increase the complexity of the architecture by introducing additional dependencies. This reduces agility by leading teams to more collaboration, thereby increasing time attributes such as *Time-to-Market* and *Lead time* [3,17]. For example, the cyclic dependencies anti-pattern [18] can increase the level of coupling in the architecture. A high degree of coupling between microservices means that alterations in one component may necessitate adjustments in others, leading to more collaborative efforts among teams. Anti-patterns also make the independent scale of components harder. In a tightly coupled architecture, scaling one microservice can trigger the scaling of other microservices that are needed to cover the functionality required by the users, leading to challenges in resource optimization.

Various factors can contribute to the emergence of anti-patterns in microservices, including insufficient architectural knowledge, communication gaps, lack of collaboration, and inadequate planning and design [14]. In this work, we explore the emergence of anti-patterns in microservices due to the presence of anti-patterns in the existing monolith. This raises concerns about the potential persistence of these anti-patterns during the migration, leading to their manifestation in the microservices architecture.

3 Research Questions

In this section, four research questions related to the potential emergence of microservices anti-patterns from those in the monolith are presented. Each question is structured as a subsection and includes an introduction to the context, the question itself, an elaboration, and the significance it holds.

3.1 Impact of Monolithic Anti-patterns on Microservices

Context: The presence of anti-patterns in microservices is often attributed to poorly designed decisions and lack of communication [14]. However, another important factor could be the monolith itself that is being migrated to microservices. This indicates the potential for anti-patterns to emerge in microservices due to the pre-existing anti-patterns present in the monolithic system.

Research Question: *What is the impact of the monolith's anti-patterns on the emergence of anti-patterns in microservices?* From a wider perspective: *What factors (including anti-patterns) contribute to the emergence of microservices anti-patterns during the migration from a monolith?*

Elaboration: The code-level design anti-patterns (Table 1) in the monolith may persist within the internal code of each microservice. This indicates that the existence of anti-patterns within a microservice's codebase can be attributed to the presence of similar anti-patterns found in the monolithic components encapsulated by the same microservice. For instance, if a component in the monolith has a *Lava Flow* anti-pattern, migrating that component to a microservice would carry over the same anti-pattern. In this case, the anti-pattern only manifests itself in the created microservice. However, further investigation is necessary to determine whether these anti-patterns in the monolith could lead to the emergence of other anti-patterns. This concept can be further generalized by exploring any factor that might lead to the appearance of microservices anti-patterns.

Significance: Exploring this research question enhances understanding the causes behind the anti-patterns' emergence in microservices. As a result, suitable measurement could be implemented to mitigate or eliminate the occurrence of these anti-patterns while migrating; thereby enhancing the migration quality and leading to more agile and scalable microservices architectures.

3.2 Influence of Migration Strategies and Patterns on Monolithic Anti-patterns

Context: To understand the impact of the monolithic anti-patterns on the emergence of anti-patterns in microservices (first question), it is necessary to conduct an investigation into the influence of migration strategies and patterns on these anti-patterns. This entails closely examining how these strategies and patterns affect existing anti-patterns in the monolith, potentially leading to their emergence or elimination in microservices.

Research Question: *In what ways can the migration strategies and patterns influence or potentially transform the anti-patterns present in the monolith, leading to the emergence or elimination of microservices anti-patterns?*

Elaboration: Migration strategies and patterns require altering the monolith to extract functionalities and implement them as microservices. These strategies and patterns may affect the anti-patterns associated with these functionalities, potentially leading to the emergence or elimination of microservices anti-patterns. For example, the *Strangler Fig Application* pattern does not affect the aforementioned design anti-patterns related to the code. Despite migrating the components to a new microservice, the existing anti-patterns continue to persist in the code of the microservice. As such, further investigation is required to fully understand the influence of the other strategies and patterns on the monolithic anti-patterns.

Significance: Understanding this influence can aid in adapting or potentially proposing new patterns to prevent the emergence of microservices anti-patterns, thereby improving migration quality.

3.3 Detecting and Tracking Anti-patterns

Context: To investigate the influence of migration strategies and patterns on anti-patterns within the monolith (second question), it is important to employ detection and tracking tools and methods to identify these anti-patterns before, during, and post each step of the migration to microservices. This enables the detection of anti-patterns within the monolith, following them during migration, and observing their impact on microservices.

Research Question: *How can we detect and track anti-patterns before, during, and post each step of the migration to microservices?*

Elaboration: The first step to understand the impact of the migration on the monolithic anti-patterns entails identifying the presence of these anti-patterns in the monolith. This requires the use of appropriate tools and identification methods for each anti-pattern. Then, after every incremental step of extracting functionalities from the monolith, these tools should be employed to identify and track any alterations in the monolith's anti-patterns, as well as any possible emergence of new anti-patterns in the microservices implementation.

Significance: Efficient detection of anti-patterns helps assist the migration to microservices. By providing a means to track the outcome of each incremental step, these tools and methods aid in identifying any emergent anti-patterns; thereby determining appropriate measures to eliminate them at an early stage of the migration, leading to a smooth and safe transition to microservices.

3.4 Anti-patterns in the Coexistence of the Monolith and the Microservices

Context: The incremental migration approach to microservices allows for the monolith and the newly implemented microservices to coexist. As a result, the two implementations can have anti-patterns at the same time.

Research Question: *How does the existence of anti-patterns in both the monolith and microservices impact the other implementation components and the overall system in cases they coexist?*

Elaboration: Migration strategies and patterns require a gradual extraction of functionalities from the monolith to microservices. This gradual extraction creates a coexistence between the monolith and microservices. It is important to understand the impact of each anti-pattern on the performance of both implementations and the overall system. Any anti-pattern present in the microservices can detrimentally impact not only their performance but also the functionalities within the monolith that are dependent on the microservice's responses.

Significance: Understanding anti-patterns in the coexistence of both monolith and microservices aids in identifying the main source that mitigates the overall system's performance. By applying suitable refactoring measurements, enhancements can be made to the performance, consequently leading to an enhanced user experience.

3.5 Discussion

This work presents challenges and explores research directions regarding the occurrence of anti-patterns when adopting migration strategies and patterns for migrating from monolithic architectures to microservices. It presents four research inquiries, each specifically targeting a different aspect of anti-pattern occurrence during the migration.

The four research questions explore different aspects of anti-patterns while migrating from a monolith to microservices. The first question looks into the impact of the monolithic anti-patterns in the emergence of microservices anti-patterns. This requires exploring the influence of migration strategies and patterns on monolithic anti-patterns which is carried out by the second research question. To investigate this influence, the appropriate set of tools and methods for detecting and tracking anti-patterns is needed in both architectural styles, which is covered by the third research question. During the transition of the migration, a coexistence phase occurs between the monolith and the new microservices. Hence, it is important to understand the impact of anti-patterns from both implementations (monolith and microservices) on each other and the overall system (fourth question).

Answering these inquiries allows for a deeper understanding on the occurrence of anti-patterns in microservices. This enables the creation of preventive measures, including updating existing strategies and practices or implementing new ones, to prevent the occurrence of these anti-patterns and manage them effectively. This ultimately reduces the complexity of the resulting architectures, expedites development and testing, streamlines maintenance, and optimizes scalability. As a result, greater agility is ensured in the resulting architectures allowing for the creation of autonomous development teams and ensuring better software delivery.

4 Related Work

Regarding the works on anti-patterns while migrating to microservices, [19] has presented some strategies, challenges, and anti-patterns associated with migrating to microservices. The study showed some anti-patterns such as "microservices that are difficult to manage and maintain", "lack of consideration for the implications of migration", and using "legacy data stores or existing repository clusters" may occur as bad habits during the migration.

Several researchers have worked on assisting the quality while migrating to microservices. In [20] structural metrics for measuring the maintainability quality of microservice architecture were introduced. These metrics facilitate early assessment, enabling to optimize microservices maintainability while migrating. An examination in [21] explores enhancing quality attributes during the migration to microservices. It shows that across various phases, numerous quality attributes experience improvement, especially through the phase in which microservices are identified. In [22] an approach was proposed to facilitate the migration to microservices by automatically identifying microservices from object-oriented (OO) source code. The quality of the identified microservices is evaluated using metrics that measure both internal and external coupling as well as data dependencies. In [23] a validation framework for microservices decomposed from monolithic applications was provided, assisting in the migration to microservices. This framework utilizes quality attributes such as granularity, coupling, and cohesion for validation. Then, it was validated through testing on five applications, showing its scalability and valuable metrics.

Other works have provided practical guidelines to migrate to microservices. In [8] practical guidance on migrating to microservices was provided, outlining the benefits and challenges associated with this migration. A set of patterns illustrated by real-world examples is presented to overcome prevalent migration issues and reduce associated risks. [24] provides a roadmap and a methodology for the migration to microservices, utilizing patterns for decomposing services, isolating data, and replication. This roadmap includes three stages: a preparation phase, a scaling and enhancement stage, and an optimization stage.

Some works have presented methods and techniques to identify anti-patterns within microservices that could improve the migration quality. For instance, [18,25], and [26] have proposed graph-based approaches to identify a range of anti-patterns such as Hard-Coded Endpoints, Nanoservices, Shared persistence, and Cyclic Dependency. Some of these works have further offered implemented these approaches, providing tools such as MSANose [27] or Arcan [28].

Some works have focused on defining and illustrating the symptoms and consequences of anti-patterns. [14] offers an overview of various anti-patterns that can manifest in both monolithic and single microservices. The anti-patterns are classified into three main categories: Software Development Anti-patterns, Software Architecture Anti-patterns, and Software Management Anti-patterns. Each anti-pattern is defined, highlighting its symptoms, consequences, and recommended refactoring solutions.

Regarding microservices anti-patterns, [15] has proposed a catalog for microservices anti-patterns. These anti-patterns were identified based on reviewing 27 microservices papers and analyzing 67 open-source systems. Then, they were grouped into Design anti-patterns, Implementation anti-patterns, and Deployment anti-patterns. [16] presents a catalog and a taxonomy of 20 common microservices anti-patterns, based on interviewing practitioners. These anti-patterns were identified as technical anti-patterns and organizational anti-patterns.

To the best of our knowledge, there has been no investigation into the anti-patterns' emergence while migrating to microservices. This chapter expands on the findings presented in [29], where an initial discussion on the possible correlation between anti-patterns in monolithic and microservices architectures was introduced. This work provides more detailed explanations, additional analysis, further discussion, and enhanced illustrative examples on transitioning to microservices and its influence on the emergence of anti-patterns.

The research and studies presented in the literature can assist in addressing one or more of the research inquiries outlined in this work. These works typically cover various aspects and elements that pertain to one or more of the presented research questions. Therefore, Table 2 presents a classification of the relevant studies based on the outlined research questions.

Table 2. Classification of related work according to the research questions

Research questions	Relevant work	Description
3.1	[19]	Presents some anti-patterns that may occur when moving from monolithic systems to microservices
3.1	[14–16]	Provide an overview of various anti-patterns, their symptoms, and consequences
3.1	[6,7]	Provide a detailed explanation on migration strategies and patterns
3.2	[19]	Provides practices and recommendations to be adopted while migrating to microservices
3.2	[22]	Proposes an approach to facilitate the migration to microservices
3.2	[8]	Provides guidance and patterns on migrating to microservices
3.2	[24]	Provides a roadmap and a methodology for the migration to microservices, utilizing patterns for decomposing services
3.2	[6,7]	Provide a detailed explanation on migration strategies and patterns
3.3	[20]	Offers metrics for measuring the maintainability. These metrics could be used to detect anti-patterns
3.3	[22]	Proposes metrics to evaluate the quality of the microservices. These metrics could be utilized in detecting anti-patterns
3.3	[23]	Utilizes some quality attributes are utilized to validate microservices decomposition, which could be used to detect the presence of anti-patterns
3.3	[18,25,26]	Propose approaches to detect anti-patterns in microservices
3.3	[27,28]	Present tool to detect anti-patterns
3.3	[14–16],	Provide anti-patterns symptoms that help their detection
3.4	[6,7]	Provide patterns that allow for the coexistence of monolithic and microservices systems

Table 2 presents a classification of related works aligned with the research questions presented in Sect. 3. A selection of resources for research questions 3.1 and 3.3 are highlighted, offering analyses on anti-patterns including their symptoms, consequences, and identification approaches. This facilitates the recognition of anti-patterns within both architectures, followed by the definition of strategies, methodologies, and techniques for detecting and monitoring them during migration. Regarding research question 3.2, the works on migration strategies and patterns are presented. This provides valuable insight into understanding these strategies and patterns, ultimately aiding in the study of their impact on anti-patterns during the migration. Similarly, for research question 3.4, a focus is placed on works that delve into the coexistence phase between monolithic and microservices architectures. This allows for a better understanding of this phase and in turn, the emergence of anti-patterns and their effects on both architectures can be comprehensively analyzed.

5 Conclusion

Ensuring a safe migration from monolithic architectures to microservices requires dealing with the emergence of anti-patterns. Despite multiple factors contributing to the emergence of such anti-patterns, it remains important to explore their possible origination from those present within the monolith. This work presents four research questions, each revolving around a different aspect of the emergence of anti-patterns from those present in the monolith during the migration.

Further investigation into these research questions enhances the understanding of the factors contributing to the emergence of anti-patterns. As a result, through appropriate measurement, the appearance of these anti-patterns could be mitigated and eventually eliminated. This in turn aids in improving the implementation of migration strategies and patterns, resulting in higher migration quality and the development of more agile and scalable microservices architectures.

In future work, we intend to investigate and provide answers to these research questions. Moreover, we intend to further explore this subject to contribute to providing better migration quality.

References

1. Gannon, D., Barga, R., Sundaresan, N.: Cloud-native applications. IEEE Cloud Comput. **4**(5), 16–21 (2017)
2. Lewis, J., Fowler, M.: Microservices: a definition of this new architectural term. MartinFowler. com **25**(14–26), 12 (2014)
3. Singleton, A.: The economics of microservices. IEEE Cloud Comput. **3**(5), 16–20 (2016)
4. Scholl, B., Swanson, T., Jausovec, P.: Cloud native: using containers, functions, and data to build next-generation applications. O'Reilly Media, Inc. (2019)
5. Rising, L.: The Patterns Handbook. Cambridge University Press, Cambridge (1998)
6. Richardson, C.: Microservices Patterns: With Examples in Java. Simon and Schuster (2018)
7. Newman, S.: Monolith to Microservices: Evolutionary Patterns to Transform Your Monolith. O'Reilly Media (2019)
8. Fritzsch, J., Bogner, J., Wagner, S., et al.: Microservices migration in industry: intentions, strategies, and challenges. In: 2019 IEEE International Conference on Software Maintenance and Evolution (ICSME), pp. 481–490. IEEE (2019)

9. Evans, E.: Domain-Driven Design: Tackling Complexity in the Heart of Software. Addison-Wesley Professional (2004)
10. Fowler, M.: Strangler Fig Application (2004). https://martinfowler.com/bliki/StranglerFigApplication. Accessed 7 Mar 2024
11. AWS Prescriptive Guidance Decomposing monoliths into microservices - Branch by abstraction pattern. https://docs.aws.amazon.com/prescriptive-guidance/latest/modernization-decomposing-monoliths/branch-by-abstraction.html. Accessed March 2024
12. Fowler, M.: BranchByAbstraction (2014). https://martinfowler.com/bliki/BranchByAbstraction.html. Accessed March 2024
13. Materialized View pattern. https://learn.microsoft.com/en-us/azure/architecture/patterns/materialized-view. Accessed March 2024
14. Brown, W.H., Malveau, R.C., Mccormick, H.W., et al.: AntiPatterns: Refactoring Software, Architectures, and Projects in Crisis. Wiley (1998)
15. Tighilt, R., Abdellatif, M., Moha, N., et al.: On the study of microservices antipatterns: a catalog proposal. In: Proceedings of the European Conference on Pattern Languages of Programs 2020, pp. 1–13 (2020)
16. Taibi, D., Lenarduzzi, V., Pahl, C.: Microservices anti-patterns: a taxonomy. Microserv.: Sci. Eng. 111–128 (2020)
17. Forsgren, N., Humble, J., et Kim, G.: Accelerate: the science of lean software and DevOps: building and scaling high performing technology organizations. IT Revolution (2018)
18. Farsi, H., Allaki, D., En-Nouaary, A., et al.: A graph-based solution to deal with cyclic dependencies in microservices architecture. In: 2022 9th International Conference on Future Internet of Things and Cloud (FiCloud), pp. 254–259. IEEE (2022)
19. Muraca, F., Pollo-cattaneo, M.F.: Migration from monolithic applications to microservices: a systematic literature mapping on approaches, challenges, and anti-patterns (2023)
20. Hasan, M.H., Osman, M.H., Novia, I.A., et al.: From monolith to microservice: measuring architecture maintainability. Int. J. Adv. Comput. Sci. Appl **14**(5) (2023)
21. Capuano, R., Muccini, H.: A systematic literature review on migration to microservices: a quality attributes perspective. In: 2022 IEEE 19th International Conference on Software Architecture Companion (ICSA-C), pp. 120–123. IEEE (2022)
22. Selmadji, A., Seriai, A.-D., Bouziane, H.L., et al.: Re-architecting OO software into microservices: a quality-centred approach. In: Service-Oriented and Cloud Computing: 7th IFIP WG 2.14 European Conference, ESOCC 2018, Como, Italy, 12–14 September 2018, pp. 65–73. Springer (2018)
23. Cojocaru, M., Uta, A., Oprescu, A.-M.: MicroValid: a validation framework for automatically decomposed microservices. In: 2019 IEEE International Conference on Cloud Computing Technology and Science (CloudCom), pp. 78–86. IEEE (2019)
24. Henry, A., Ridene, Y.: Migrating to microservices. Microserv.: Sci. Eng. 45–72 (2020)
25. Gamage, I.U.P., Perera, I.: Using dependency graph and graph theory concepts to identify anti-patterns in a microservices system: a tool-based approach. In: 2021 Moratuwa Engineering Research Conference (MERCon), pp. 699–704. IEEE (2021)
26. Borges, R., Khan, T.: Algorithm for detecting antipatterns in microservices projects (2019)
27. Walker, A., Das, D., Cerny, T.: Automated code-smell detection in microservices through static analysis: a case study. Appl. Sci. **10**(21), 7800 (2020)

28. Pigazzini, I., Fontana, F.A., Lenarduzzi, V., et al.: Towards microservice smells detection. In: Proceedings of the 3rd International Conference on Technical Debt, pp. 92–97 (2020)
29. Farsi, H., Allaki, D., En-Nouaary, A., et al.: Dealing with anti-patterns when migrating from monoliths to microservices: challenges and research directions. In: 2023 IEEE 6th International Conference on Cloud Computing and Artificial Intelligence: Technologies and Applications (CloudTech), pp. 1–8. IEEE (2023)

Missing Data Imputation Approach for IoT Using Machine Learning

Abderrahim Lachguer[1(✉)], Abderrahmane Sadiq[2,3], Youssef Es-saady[1,3], and Mohamed El Hajji[1,4]

[1] IRF-SIC Laboratory, Ibnou Zohr University, Agadir, Morocco
abderrahim.lachguer@edu.uiz.ac.ma, y.essaady@uiz.ac.ma
[2] ISIMA Laboratory, Polydisciplinary Faculty of Taroudant, Ibnou Zohr University, Agadir, Morocco
a.sadiq@uiz.ac.ma
[3] Polydisciplinary Faculty of Taroudant, Ibnou Zohr University, Agadir, Morocco
[4] Regional Center for Education and Training Professions, Souss Massa, Morocco
m.elhajji@crmefsm.ac.ma

Abstract. Over the last decade, numerous cutting-edge technologies have emerged, triggering unprecedented progress in both physical and cyber systems. Among these technologies, connected devices are increasingly integrated into various environments for monitoring and data collection purposes. Missing data imputation remains one of the main challenges in applications utilizing these devices, leading to imprecision of the derived insights and inefficiency of related systems. Recently, numerous machine learning methods and techniques have been proposed to address this challenge. However, these solutions often rely on the use of historical data or the use of records of other features that are highly correlated with the feature containing the missing value. These approaches often overlook large subsequences of missing values as well as missing values in correlated features. This study proposes a new approach to address the missing data issue, assuming that data are collected in several locations. The approach utilizes historical values of the corresponding feature from other locations where data have been recorded with no missing values. One of the aims of this study is to conduct the data imputation process at the edge level, considering the limitations in processing capabilities of edge devices. Several popular lightweight machine learning and deep learning algorithms have been compared in terms of inference time and accuracy.

Keywords: Missing data · machine learning · IoT · constrained devices · edge computing

1 Introduction

The world nowadays is evolving at an unprecedented rate, with new breakthrough discoveries and technologies on a daily basis. Artificial intelligence (AI) and related technologies have fundamentally reshaped our perception of what a computer can achieve. This progress in software solutions was enabled by earlier advancements in physical

M. Zbakh et al. (Eds.): CloudTech 2023, LNNS 1220, pp. 258–273, 2024.
https://doi.org/10.1007/978-3-031-78698-3_13

architectures and processing capabilities [1]. These developments also served as a catalyst that sparked the development of new small and cost-effective devices, such as microcontrollers, wearables, sensors, and actuators [2]. Together, these advancements, combined with communication technologies and protocols, define the realm of Internet of Things (IoT).

Despite these technological advancements, the persistent issue of missing data continues to be a primary focus of various research studies. Before the revolution of machine learning (ML), the problem of missing data was handled using statistical and interpolation approaches [3]; these approaches are still used for missing data imputation due to their ease of implementation. However, they are prone to performance decrease, particularly when imputed values diverge from the used values for imputation. In contrast, ML techniques can learn complex patterns between variables, leading to more precise and accurate imputation compared to statistical and interpolation approaches [4].

Real-time monitoring of smart environments and modern networked infrastructures is sensitive to data quality, especially regarding missing data. Various factors can lead to data missingness in IoT-based applications. These factors include device failure, manual intervention, synchronization issues, sensor aging, devices running out of power, meteorological extremes, and unstable internet connections [5, 6]. Ignoring or improperly handling a missing value in a real-time IoT application can decrease reliability and lead to severe consequences. Therefore, addressing missing data promptly as it occurs is a crucial step in developing sustainable and highly resilient smart environments.

It is worth noting that imputation and prediction are two approaches that differ both conceptually and methodologically. Imputation typically involves replacing a missing value with a newly generated one. This approach assumes the presence of a record with a missing value and provides access to the entire dataset. Therefore, not only historical values but also subsequent values are utilized for imputation, which helps achieve exceptional precision [7]. In contrast, prediction relies on historical data to estimate a value that has not yet been recorded. The methodology adopted in this study combines these two approaches: we treat the value to be recorded as missing and in need of imputation, which is essentially an act of imputation. However, unlike traditional imputation, no access to future values is allowed, similar to the prediction approach [8].

The cloud-centric paradigm has shown its limits in handling the dramatic increase in the number of connected devices [1]. These devices produce tremendous amounts of data that need to be pushed to the cloud for further processing and storage, yet the bandwidth for data transmission has not kept pace with the growth in data production [9]. These two issues are the main reasons limiting the performance of cloud-based IoT applications. To address these limitations, edge computing brings computing capabilities closer to the end devices where data is being generated [10]. It addresses several issues associated with cloud-centric applications, such as latency, system response time, bandwidth constraints, privacy protection, and single point of failure. The edge server acts as a bridge between sensing devices and the cloud server, as it supports various communication protocols and embeds communication hardware of various types [11].

This study, which is an extension of our previous work [12], addresses missing data imputation in a distributed IoT architecture using lightweight ML algorithms. We implemented the classical approach for data imputation, which involves using historical data from a sensor to impute missing values. Our proposed approach assumes that data are collected in various locations; in our experimentation, we used data collected from one location to impute missing data in another location. We compared the two approaches using different machine learning methods in terms of accuracy and inference time. We also evaluated these approaches by varying multiple parameters, such as the window size of historical values used for imputation and the ratio of missing data.

The main contributions of our study are summarized as follows:

- Proposition and implementation of an approach for missing data imputation in a distributed IoT architecture.
- Evaluation of a lightweight imputation process, designed to be implemented on constrained devices at the edge level, under variable parameters.
- Assessment of the performance of popular machine learning algorithms for regression problems, in terms of accuracy and inference time.

The remainder of this article is organized as follows: Sect. 2 presents the problem statement and motivation. Section 3 discusses related works. Section 4 describes the details of the proposed approach, followed by Sect. 5, which outlines the evaluation methodology. Section 6 presents the results and discussion. Finally, Sect. 7 concludes the paper and offers perspectives on future work.

2 Problem Statement and Motivation

Complete and precise measurements are crucial for real-time or near real-time decision-making in IoT-based applications. Therefore, adopting real-time data imputation or recovery approaches is essential to maintain the operability and reliability of any IoT-based system [4]. Consequently, before implementing any data recovery approach, it is vital to understand the various mechanisms, patterns, and aspects of missing data, along with the existing methods for handling them [13].

2.1 Missing Data Mechanisms and Patterns

It is well established in the literature that missing data mechanisms can be categorized into three classes [14]. Missing Completely at Random (MCAR) refers to the case where the probability of a missing value within a variable is independent of any other data. This means that the probability of a value being missing is the same for all cases of the variable. For example, a temperature sensor might generate missing values due to random transmission problems. Missing at Random (MAR) is another mechanism that occurs when the probability of missing data is not uniform across all values of a variable; however, this can be accounted for using other observed data. For example, a temperature sensor might generate more missing values during the day than at night. Here, the missing values depend on another observed variable, which is the time. Missing Not at Random (MNAR) corresponds to the scenario where the probability of missing data varies within the values of a variable and cannot be explained by other observed information. This means that the probability of missingness is related to unobserved data. For example, a temperature sensor fails to collect data when the temperature exceeds a specific threshold. Here, the missingness is directly related to the unrecorded higher temperatures. Missing data under the assumptions of MCAR and MAR mechanisms are very common in sensing devices and are relatively easy to handle. However, missing data under the MNAR mechanism assumption are uncommon and more complex to handle [15].

Missing data can also be classified into patterns [16]. The main patterns are univariate or multivariate, and monotone or non-monotone. A univariate pattern means that the missingness exists only in one variable, while a multivariate pattern indicates that missingness exists in multiple variables. In contrast, a monotone pattern describes a case where a variable starts missing from a certain point and continues to be missing until the end, while a non-monotone pattern defines a scenario where a variable is missing for one or more consecutive time points but does not remain missing until the end.

2.2 Missing Data Amount and Gaps

Other aspects of missing data include the missing amount and gaps [11]. The missing amount refers to the ratio of missing data in the dataset; this aspect defines the dataset's quality. A missing gap refers to consecutive missing values within a variable. It is another critical aspect of missing data that defines the success or failure of the imputation process. Missing gaps are commonly classified in four categories, depending on the amount of missing data that a gap encompasses. Point missing gap is a gap with a single missing value. A missing gap of more than two values up to 3% of the data is referred to as a short missing gap. A high missing gap is defined as a gap that represents between 3% and 10% of the data. If the amount of missing data contained in a gap exceeds 10% it is termed as a very high missing gap. While point and short missing gaps, which are frequent in datasets, are typically easy to handle and well-addressed in the literature, high missing gaps are less frequent but often complicate the imputation process and are less addressed [15].

2.3 Handling Missing Data

Various approaches are utilized for handling missing data [6]. Listwise deletion, or complete case deletion, is the most popular and straightforward method, which involves removing all records that contain one or more missing values. Pairwise deletion is another method that removes a record only if the missing value it contains is used in the imputation process. While this method preserves other values contained in cases with missing values, it affects the coherence of the analysis since each computation may use a different subset of records. Good model performance can be maintained after the deletion of missing values only if the amount of missingness is less than 10% [4].

Statistical and interpolation-based approaches are widely employed for missing data imputation [17]. The results from these approaches are generally satisfactory. However, machine learning and deep learning-based approaches have been the most studied and widely adopted techniques in recent years. Their ability to learn complex patterns between variables has allowed for unprecedented accuracies. Related models consist of two major parts: training and inference. Both can be executed at the edge level if the model is of low complexity. The first part demands powerful processing capabilities, while the second part does not require as much processing power [18]. Hence, the inference stage, which is the final goal of any ML model, can be integrated into constrained edge devices to enhance autonomy and real-time response without compromising processing power and energy savings [19].

2.4 Motivation

Many deep learning-based solutions have been proposed to address the missing data problem at the edge. However, the complexity of these models often exceeds the implementation capabilities of edge devices. Additionally, high gaps of missing data are frequently excluded or not addressed. A lightweight imputation mechanism, proposed by [14], utilizes a deep learning model that estimates the missing value based on the last three recorded values (a window of historical values). However, the reason for selecting this window size is not well justified. Furthermore, deep learning algorithms, known for their resource intensity, are adopted without considering other machine learning techniques that could deliver comparable results while using fewer resources. Additionally, gaps of missing data were not addressed, and a missing value cannot be imputed if the preceding values are also missing.

In our study, we address the limitations of this approach by exploring the utilization of different window sizes to precisely determine the optimal window size for maximum model precision. Additionally, we tackled the problem of high gaps of missing data by harnessing recorded data from another location. Moreover, we explored the conditions under which the use of deep learning is more beneficial than using traditional machine learning techniques. For this purpose, we evaluated a simple deep learning algorithm, namely MLPR(Multi-Layer Perceptron Regressor) against several machine learning algorithms, specifically LR(Linear Regression), GBR(Gradient Boosting Regressor), and SVR(Support Vector Machine) within the context of a distributed IoT architecture.

3 Related Works

A wide range of imputation techniques based on machine learning for IoT environments have been proposed in the literature, addressing the common problem of missing data generated by IoT sensors. This section includes a non-exhaustive list of recent works, such as the study conducted by Okafor and Delaney [20], which investigates time points and gaps of missingness in low-cost sensor networks. In this study, several machine learning algorithms for missing data imputation were evaluated across three different simulated scenarios. The first scenario addresses data missingness in one variable across various gap sizes. The second scenario introduces data missingness in multiple variables, also across various gap sizes. The final scenario examines missingness at specific time points in multiple variables under various percentages of missing data. The results showed that accuracy decreases as the missing data ratio, or the size of the gaps increases. It has also been established that the Variational Autoencoder (VAE), a deep learning algorithm, exhibits the best accuracy in all the studied scenarios. Agbo et al. [21] investigated data missingness in low-cost environmental sensors as well. They proposed a data clustering technique termed Best Fit Missing Value Imputation (BFMVI). This approach addresses data missingness in a univariate missing pattern. The core principle of BFMVI is its ability to select the most suitable technique to fill in missing values, based on the characteristics of the missing data. According to the study's outcomes, BFMVI exhibited superior accuracy compared to other machine learning algorithms.

In a study conducted by França et al. [14], two models based on the Multi-Layer Perceptron (MLP), designed for implementation on constrained devices, were utilized for missing data imputation. The first model imputes the current missing value using the last three recorded values, while the second model additionally incorporates the value recorded at the same time on the previous day. Both models were assessed in terms of accuracy and inference time; and compared with other machine learning algorithms for missing data imputation. The results demonstrated improved overall performance relative to the other algorithms. Another deep learning algorithm based on MLP was introduced by Pan et al. [4]. The authors used seven datasets from various backgrounds to evaluate their model against other commonly used algorithms for missing data imputation. Additionally, various proportions of data missingness under different missing data mechanisms were utilized in the experiment. The authors concluded that the accuracy of each machine learning model can be higher for some datasets while being less accurate for others under different circumstances.

A missing data imputation process called the Deep Missing Data Protocol (Deep-MDP) was proposed and evaluated by Kok and Ozdemir [5]. This process involves detecting missing values in real-time data generated by three types of sensors: temperature, relative humidity, and light sensors. The authors developed a deep learning model for each type of sensor data, utilizing the Long Short-Term Memory (LSTM) algorithm. These models use available data from two sensors to impute the missing value in the third sensor, harnessing the strong correlation between the different sensors' values. The protocol then transmits the complete data in a compact format to subscribers. Notably, this protocol is implemented in a distributed IoT architecture, with the imputation process performed primarily at the edge devices. Vedavalli and Ch [13] also developed an algorithm based on LSTM that leverages spatial and temporal correlations between IoT nodes for missing data imputation. Initially, the nodes were grouped into clusters based on their spatial-temporal correlation. Whenever a missing value was encountered during data processing, values from devices that belong to the same cluster were used for imputation. The developed algorithm is an extension of the LSTM, referred to as ST-HLSTM (Spatial-Temporal-Hierarchical LSTM), specifically designed to impute missing data given a matrix of recorded values from the relevant cluster. The developed algorithm was assessed and compared with other missing data imputation algorithms, demonstrating superior performance.

4 The Proposed Approach

In a real-time, IoT-based application context, machine learning models for missing data imputation can utilize only historical sensor values and the complexity of the model determines whether the process can be implemented on a constrained device. In this study, we employed a strategy that utilizes recent historical data to estimate the current missing value. This strategy is implemented within our proposed approach, which involves imputing missing data in one sensor based on values collected from another sensor. We compared this method with the classical approach, which imputes missing values within a sensor using its own historical records. Given that larger window sizes increase the model's complexity, our experiment also evaluated various window sizes of historical values to identify the minimum size that yields the best results. To this end, we transformed a time series dataset into a dataset with predictor variables and a target variable. The predictor variables represent the window of the last recorded values, while the target variable represents the value to be imputed.

The implemented strategy is illustrated in Fig. 1 which displays two sequences of records from each dataset at the two stations. This transformation of the data enabled us to feed it into two models, each corresponding to one of the approaches. In the example shown in the figure, to impute the current value of TC, which is the target variable using the local method, we use the last four historical values from the same sensor as predictor variables. To impute the value of TC using the distant method, the model uses the corresponding value from another sensor, recorded at the same time as the missing value, along with the last three recorded values as predictor variables.

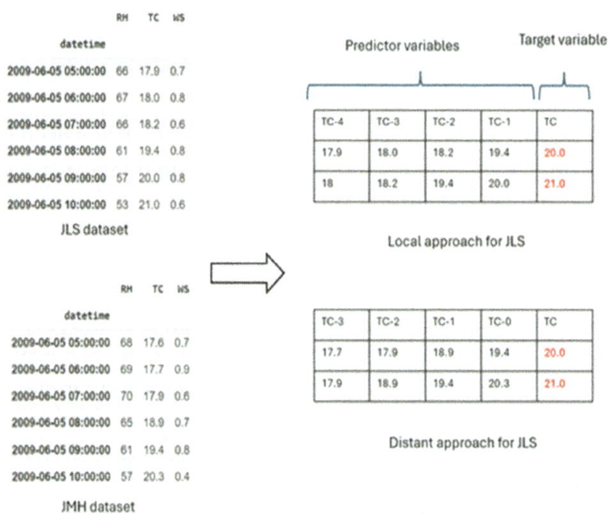

Fig. 1. Data preparation strategy.

The distant approach addresses many challenges associated with data imputation on IoT devices. It handles large missing gaps effectively, since each value is imputed using data from another sensor. This evaluation helps us precisely determine the optimal window size that yields accurate results while ensuring the model remains as simple as possible. Consequently, the model is customized for integration into constrained devices. The accuracy of the model is enhanced because it utilizes not only historical values but also the current value from the other sensor, which has the advantage of being as close as possible to the missing value.

5 Methodology

This study was conducted using two datasets obtained from pre-existing air quality monitoring stations located in Marrakesh, Morocco. The datasets were publicly accessible on the Figshare platform [22] at the time of this study. The two stations, namely Mhamid Station (MHS) and Jamaa Lafna Station (JLS), are located in the neighborhoods of Mhamid and Jamaa Lafna, respectively. The distance between the two stations is approximately 6 km. The datasets included air quality and meteorological measurements, collected hourly from June 1, 2009, to November 28, 2010. The meteorological data measured at these stations covered temperature (TC) in degrees Celsius, relative humidity (RH) in percent, and wind speed (WS) in meters per second. Our study focused on data imputation analysis for these features. Different phases of this study are illustrated in Fig. 2.

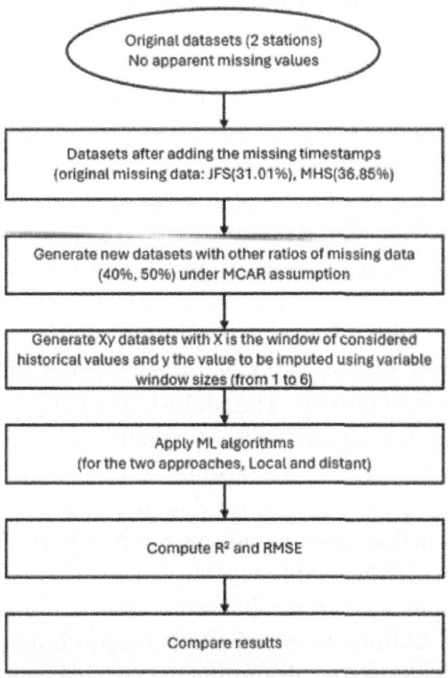

Fig. 2. The main phases of the study.

Initial exploration of the datasets revealed no missing data. However, a thorough examination later showed that a substantial number of records were completely missing. To investigate the true extent of missing data, these missing records were added back to the datasets with 'nan' as their values, corresponding to their respective timestamps. This process enabled us to generate new datasets that explicitly reveal missing data percentages of 31.01% for MHS and 36.85% for JLS, respectively. A visual representation of the number of missing records for each month, for both stations, is provided in Fig. 3; it illustrates the period between June 2009 and October 2010. This process also enabled the computation of missing gaps, which is another significant parameter in assessing data missingness. As illustrated in Table 1, the datapoint missing gap is the most frequent. As the gap size increases, the number of missing gaps decreases. However, some large gaps of missing values have been observed; for example, a gap of more than 2000 consecutive missing values was recorded at JLS.

In similar data imputation studies, various simulated percentages of missing data (between 5% and 50%) have been utilized for evaluation. In our study, we simulated 40% and 50% of missing data, as the datasets already contained relatively high percentages of data missingness. To create datasets with missing data under the assumption of the MCAR mechanism, the corresponding number of values for each percentage was deleted randomly, following a uniform distribution. This procedure allows us to assess the model's performance under extreme conditions of data missingness.

A crucial step in the input data preprocessing involves preparing various types of sub-datasets for model training. The number of these sub-datasets was determined by multiple factors: the originating station (MHS and JLS), the number of predictor variables representing the window size of historical values (ranging from 1 to 6), the percentage of missing values (original missing data percentage, 40%, and 50%), and the features to be analyzed (TC, RH, and WS). This resulted in 216 sub-datasets for analysis ($2 \times 6 \times 3 \times 3 \times 2$). To reduce this number, the evaluation of missing data under different percentages of missingness was omitted for the distant approach, retaining only the original missingness of the datasets. Subsequently, each sub-dataset was split into training and testing sets. The training sets included data recorded between June 1, 2009, and July 31, 2010, while the test sets contained data recorded from August 1, 2010, to November 28, 2010. This split resulted in approximately 77% of the data being allocated for training and 23% for testing.

The implementation of the data imputation approaches utilized various machine learning and deep learning algorithms. The Multi-Layer Perceptron Regressor (MLPR) is considered a deep learning algorithm, as it comprises several layers of neural networks capable of learning complex relationships between variables. It is regarded as a lightweight deep learning algorithm compared to more complex models. The machine learning algorithms used in this study include the Support Vector Regressor (SVR), Gradient Boosting Regressor (GBR), and Linear Regression (LR). These machine learning algorithms are well-documented in the literature; thus, to maintain brevity, we omit detailed explanations. In addition to the mentioned algorithms, a baseline strategy was also utilized. This strategy involves substituting the missing value with the last recorded value (LV) in the local approach, and using the value recorded simultaneously with the missing one in the distant approach.

It is worthwhile to highlight that numerous machine learning algorithms are sensitive to data scale, including some that we implemented in our study. Hence, we applied a standard scaler to the predictor variables of the sub-datasets prior to the training phase. Regarding the performance evaluation of the models, model accuracy and inference time were employed. Model accuracy was computed using two metrics commonly used in literature: the coefficient of determination (R2) and the Root Mean Squared Error (RMSE). To limit the number of figures related to accuracy metrics included in this paper, we omitted the use of RMSE and retained the use of R^2, as the latter is more informative and does not have the interpretability limitations of the former [23]. It is worth mentioning that this case study was implemented using the Scikit-learn library [24] and the Google Colab platform [25].

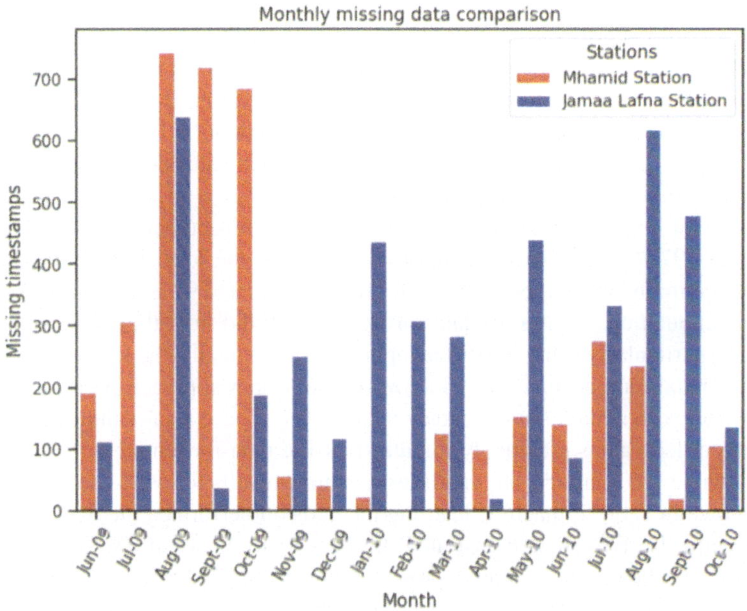

Fig. 3. Monthly missing data comparison.

Table 1. Missing gaps in the datasets.

Gap size	1	2	3	4	5	6–20	20–100	100–1000	>1000
MHS	84	30	21	14	8	22	13	10	0
JLS	42	5	3	7	2	14	12	4	1

6 Results and Discussion

A thorough analysis was conducted to evaluate the impact of window size on historical values under different percentages of missingness. Illustrations in Figs. 4, 5, 6, 7 and 8 show the accuracy of the algorithms and the baseline method under 31%, 40%, and 50% of data missingness for JLS using the local approach. We observed that using a window size of 1, results in lower accuracy, while accuracy slightly improves starting from a window size of 2 in cases of lower data missingness. However, for 50% data missingness, the accuracy decreases when using a window size greater than 2. Table 2 illustrates the varying accuracies of the algorithms used for TC in JLS under 31% of data missingness. The insights provided in the table indicate a slight advantage for MLPR with increasing window sizes, whereas LR shows better accuracy at smaller window sizes.

Figure 9 illustrates a comparison between the distant and local approaches. The illustrations clearly show that the distant approach outperforms the local approach in terms of accuracy. Furthermore, the distant approach can impute missing values using only the last recorded value from another station. Also, the local method achieves better accuracy with just the last two recorded values. These findings are critical for implementing machine learning models for missing data imputation in constrained devices, as they demonstrate that using only one or two historical values can maintain effective imputation accuracy while reducing computational costs.

The distant imputation approach achieved maximum accuracies of 99.4% for temperature and 97.8% for relative humidity. In contrast, the local imputation approach recorded maximum accuracies of 98.8% for temperature and 97% for relative humidity. However, imputation of missing data for wind speed did not attain acceptable accuracy levels, particularly using the distant approach. Specifically, the local approach only achieved a maximum accuracy of 67%, while the distant approach performed significantly worse, with a maximum accuracy of just 20%. These results indicate that using the last recorded values effectively imputes missing data for temperature and relative humidity with high accuracy, but yields less accurate results for wind speed imputation.

As shown in Table 2, machine learning methods outperformed the baseline strategy (LV). Specifically, MLPR demonstrated the best performance in the local approach, while GBR excelled in the distant approach. However, this advantage is only observed when sufficient training data is provided, i.e., when data missingness is below 40% in our case study. In situations with higher data missingness, LR showed superior accuracy compared to the other algorithms (cf. Fig. 10). The assessment in terms of inference time is illustrated in Fig. 11. The findings indicate that LR had the lowest inference time among the machine learning-based methods. Although MLPR and GBR also exhibited reasonable inference times, SVR showed the highest inference time, suggesting that this algorithm may not be well suited for imputation in constrained devices.

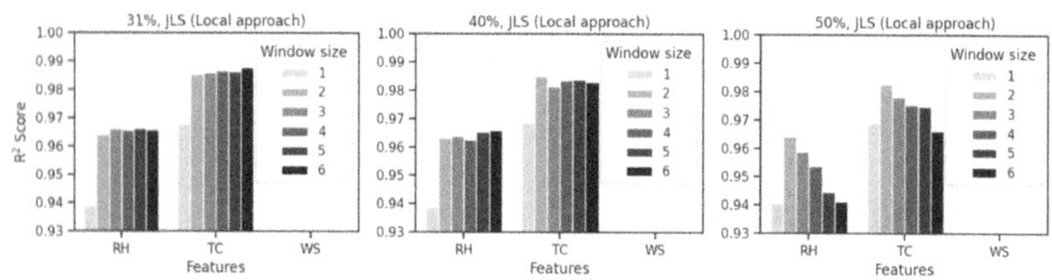

Fig. 4. MLPR accuracy for studied features under various missing data percentages.

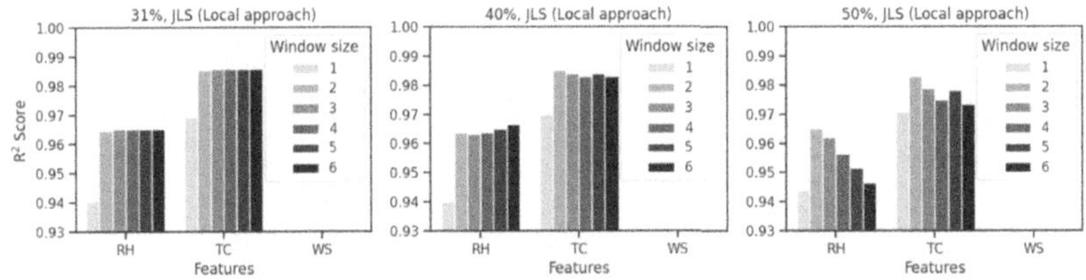

Fig. 5. LR accuracy for studied features under various missing data percentages.

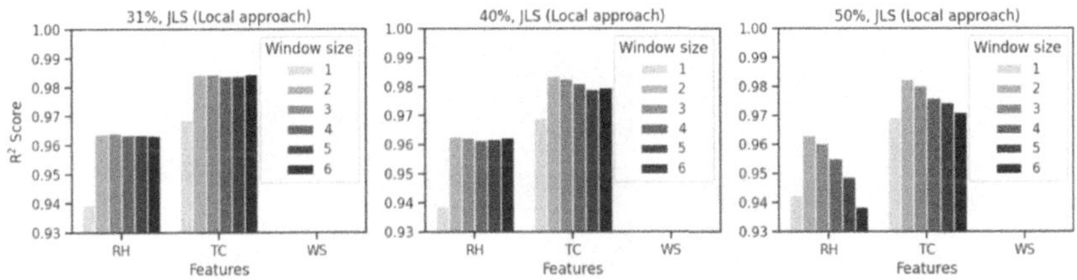

Fig. 6. GBR accuracy for studied features under various missing data percentages.

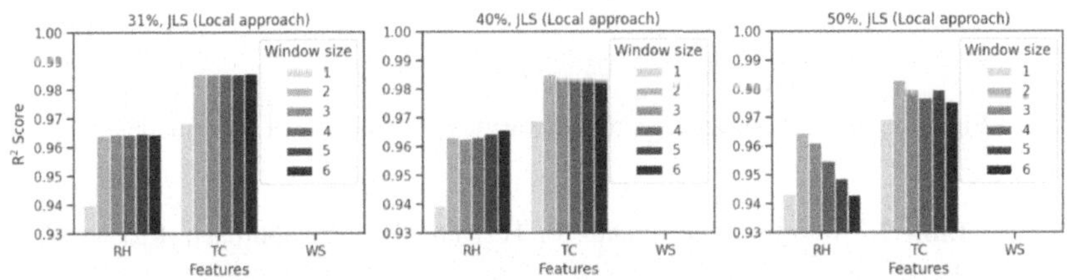

Fig. 7. SVR accuracy for studied features under various missing data percentages.

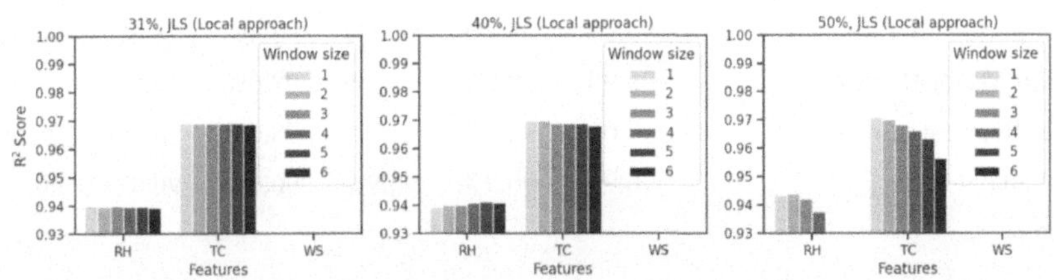

Fig. 8. The baseline method (LV) accuracy.

The present work clearly demonstrates the feasibility of missing data imputation at the edge in constrained devices. The implemented strategy utilizes a window of the last recorded entries to impute the current missing value. Two approaches have been implemented: the classical approach, which computes the missing value based on the last historical data from the same sensor, and our proposed approach, termed the distant approach, which computes the missing value based on data recorded from another distant sensor. When missing data occurs, the edge device can either use the last recorded value from the sensor or use other recorded data of the same feature from another sensor. In

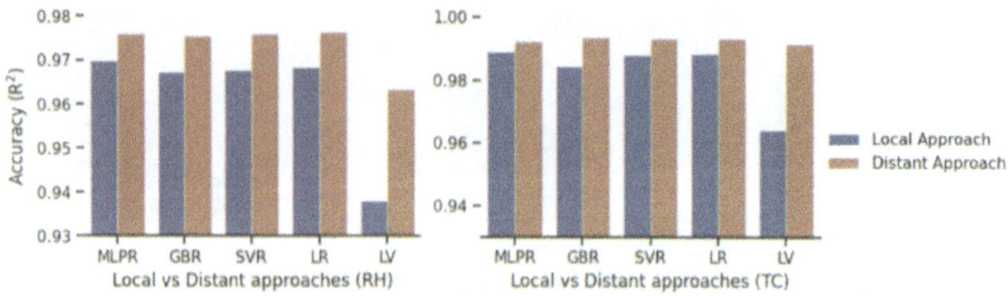

Fig. 9. Comparison of the two approaches under 31% of missing data in JLS.

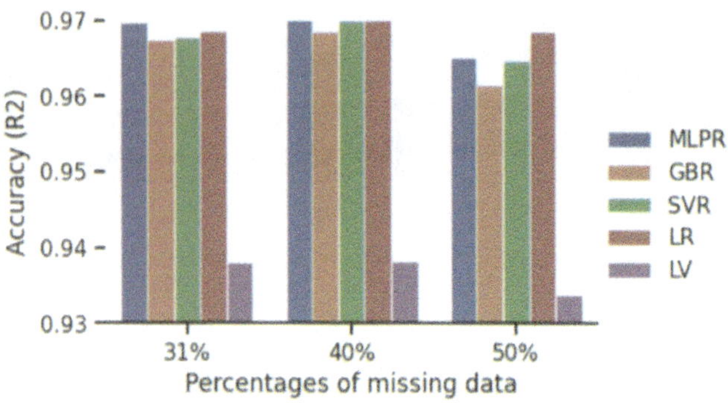

Fig. 10. Algorithms accuracy under various percentages of missing data.

Table 2. Algorithms accuracies for TC using the local approach (R2).

Window Size	1	2	3	4	5	6
Multi-Layer Perceptron Regressor	96,8%	98,6%	98,6%	98,7%	98,6%	98,8%
Gradient Boosting Regressor	96,9%	98,4%	98,5%	98,4%	98,4%	98,5%
Support Vector Regressor	96,9%	98,6%	98,6%	98,6%	98,6%	98,6%
Linear Regression	97,0%	98,6%	98,6%	98,6%	98,6%	98,6%
Last Value	96,9%	96,9%	96,9%	96,9%	96,9%	96,9%

our experimentation, we employed only two sensors for each feature. However, multiple sensors can be utilized to ensure the availability of sufficient historical samples for imputation.

It has been established that accurate missing data imputation can be achieved by using only the latest recorded value with the distant approach, and the last two values with the local approach. Thus, one value for the distant approach and two values for the local approach represent the ideal window sizes that maintain precise results while minimizing resource usage. Following this, various machine learning algorithms have been evaluated to identify those most suitable for missing data imputation in constrained devices. Despite having a higher inference time, MLPR and GBR showed the best accuracy for the local and distant approaches, respectively. These two algorithms are appropriate when the accuracy of the model is of greater importance. However, if the edge device has limited computational capabilities, or if minimizing inference time is a priority, LR would be the optimal choice since it exhibits improved accuracy with minimal inference time.

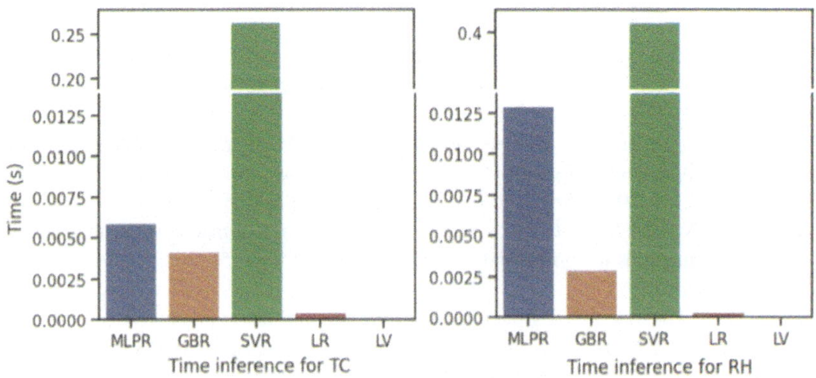

Fig. 11. Inference time assessment.

7 Conclusion and Perspectives

Incomplete data is an inherent sensing issue in IoT. This issue may be attributed to various factors, ranging from network problems to security attacks. Sending raw data containing missing values to the cloud raises further issues, such as increased latency and the risk of single point failure. To address these challenges, further studies are needed to develop new data processing approaches at the edge level, particularly for constrained devices like microcontrollers and single board computers. In this study, we addressed the missing data imputation problem at the network edge in a distributed IoT architecture. We utilized two datasets containing meteorological data collected from two stations within the same city. We implemented a lightweight imputation mechanism that uses the last recorded values for imputation. We evaluated four well-known machine learning algorithms for regression analysis using two different approaches: the classical approach, which uses historical values from the same sensor for imputation, and the proposed approach, which uses the last recorded values from another station. These approaches were assessed under various conditions, including different percentages of missing data and window sizes of historical values. The findings from this study have highlighted the advantages and limitations of the two approaches, as well as the most appropriate machine learning models and parameters for various scenarios.

This work explored numerous aspects of missing data imputation at the edge. However, some limitations remain unexplored and require further investigation. Leveraging correlations between various types of variables, either within the same station or across different stations, is worth exploring. Additionally, harnessing data from multiple sensors simultaneously using simple mechanisms suitable for resource-constrained devices is a promising direction. Furthermore, the approaches tested need to be assessed in real-world scenarios, where data are dynamic, and the processes are executed under more stringent constraints. These limitations and opportunities will be explored in future studies.

References

1. Hajjaji, Y., Boulila, W., Farah, I.R., Romdhani, I., Hussain, A.: Big data and IoT-based applications in smart environments: a systematic review. Comput. Sci. Rev. **39**, 100318 (2021). https://doi.org/10.1016/j.cosrev.2020.100318
2. Al-Fuqaha, A., Guizani, M., Mohammadi, M., Aledhari, M., Ayyash, M.: Internet of things: a survey on enabling technologies, protocols, and applications. IEEE Commun. Surv. Tutorials **17**(4), 2347–2376 (2015). https://doi.org/10.1109/COMST.2015.2444095
3. Kim, J., Lee, Y., Lee, M.H., Hong, S.Y.: A comparative study of machine learning and spatial interpolation methods for predicting house prices. Sustain **14**(15) (2022). https://doi.org/10.3390/su14159056
4. Pan, H., et al.: Discrete missing data imputation using multilayer perceptron and momentum gradient descent †. Sensors **22**(15), 1–23 (2022). https://doi.org/10.3390/s22155645
5. Ibrahim, K.: DeepMDP : a novel deep learning based missing data prediction protocol for IoT," vol. XX, no. c, pp. 1–12 (2020). https://doi.org/10.1109/JIOT.2020.3003922
6. Adhikari, D., et al.: A comprehensive survey on imputation of missing data in internet of things. ACM Comput. Surv. **55**(7) (2022). https://doi.org/10.1145/3533381
7. Hasan, M.K., Alam, M.A., Roy, S., Dutta, A., Jawad, M.T., Das, S.: Missing value imputation affects the performance of machine learning: a review and analysis of the literature (2010–2021). Inf. Med. Unlocked **27**, 100799 (2021). https://doi.org/10.1016/j.imu.2021.100799
8. Li, Q., Tan, H., Wu, Y., Ye, L., Ding, F.: Traffic flow prediction with missing data imputed by tensor completion methods. IEEE Access **8**, 63188–63201 (2020). https://doi.org/10.1109/ACCESS.2020.2984588
9. Čolaković, A., Hadžialić, M.: Internet of things (IoT): a review of enabling technologies, challenges, and open research issues. Comput. Networks **144**, 17–39 (2018). https://doi.org/10.1016/j.comnet.2018.07.017
10. Mocrii, D., Chen, Y., Musilek, P.: IoT-based smart homes: a review of system architecture, software, communications, privacy and security. Internet Things (Netherlands) **1–2**, 81–98 (2018). https://doi.org/10.1016/j.iot.2018.08.009
11. Pazhoohesh, M., Pourmirza, Z., Walker, S.: A comparison of methods for missing data treatment in building sensor data. In: Proceedings of 2019 7th International Conference on Smart Energy Grid Engineering, SEGE 2019, February 2018, pp. 255–259 (2019). https://doi.org/10.1109/SEGE.2019.8859963
12. Lachguer, A., Sadiq, A., Es-Saady, Y., El Hajji, M.: Towards an IoT architecture based on machine learning for missing data prediction on the edge. In: 2023 International Conference on Cloud Computing and Artificial Intelligence: Technologies and Applications, CloudTech 2023, p. 10366054 (2023). https://doi.org/10.1109/CloudTech58737.2023.10366054
13. Vedavalli, P., Ch, D.: A deep learning based data recovery approach for missing and erroneous data of IoT nodes. Sensors **23**(1) (2023). https://doi.org/10.3390/s23010170
14. França, C.M., Couto, R.S., Velloso, P.B.: Missing data imputation in internet of things gateways. Inf. **12**(10), 1–22 (2021). https://doi.org/10.3390/info12100425
15. Ahmed, H.M., Abdulrazak, B., Blanchet, F.G., Aloulou, H., Mokhtari, M.: Long gaps missing IoT sensors time series data imputation: a Bayesian Gaussian approach. IEEE Access **10**, 116107–116119 (2022). https://doi.org/10.1109/access.2022.3218785

16. Bhajantri, L.B., Gangadharaiah, S.: A comprehensive survey on resource management in internet of things. J. Telecommun. Inf. Technol. **2020**(4), 27–43 (2020). https://doi.org/10.26636/JTIT.2020.145220
17. Yen, N.Y., Chang, J.W., Liao, J.Y., Yong, Y.M.: Analysis of interpolation algorithms for the missing values in IoT time series: a case of air quality in Taiwan. J. Supercomput. **76**(8), 6475–6500 (2020). https://doi.org/10.1007/s11227-019-02991-7
18. Laroui, M., Nour, B., Moungla, H., Cherif, M.A., Afifi, H., Guizani, M.: Edge and fog computing for IoT: a survey on current research activities & future directions. Comput. Commun. **180**, 210–231 (2021). https://doi.org/10.1016/J.COMCOM.2021.09.003
19. Dutta, D.L., Bharali, S.: TinyML meets IoT: a comprehensive survey. Internet Things (Netherlands) **16**, 100461 (2021). https://doi.org/10.1016/j.iot.2021.100461
20. Okafor, N.U., Delaney, D.T.: Missing data imputation on IoT sensor networks: implications for on-site sensor calibration. IEEE Sens. J. **21**(20), 22833–22845 (2021). https://doi.org/10.1109/JSEN.2021.3105442
21. Agbo, B., Al-Aqrabi, H., Hill, R., Alsboui, T.: Missing data imputation in the internet of things sensor networks. Futur. Internet **14**(5) (2022). https://doi.org/10.3390/fi14050143
22. Sadiq, A., Elfazziki, A., Ouarzazi, J.: Air quality monitoring stations data - Marrakech - Morocco. https://figshare.com/articles/dataset/Air_quality_monitoring_stations_data_Marrakech_Morocco/3113515?file=4837747, Accessed 02 Mar 2024
23. Chicco, D., Warrens, M.J., Jurman, G.: The coefficient of determination R-squared is more informative than SMAPE, MAE, MAPE, MSE and RMSE in regression analysis evaluation. PeerJ. Comput. Sci. **7**, 1–24 (2021). https://doi.org/10.7717/PEERJ-CS.623
24. Pedregosa, F., et al.: Scikit-learn: machine learning in Python. J. Mach. Learn. Res. **12**, 2825–2830 (2011)
25. Google, Google Colaboratory (2024). https://colab.research.google.com/

Holistic Approach for Enhanced Object Recognition in Complex Environments

Zainab Ouardirhi[1,2](✉), Sidi Ahmed Mahmoudi[1], and Mostapha Zbakh[2]

[1] Computer and Management Engineering Department, FPMS, University of Mons, Mons, Belgium
zainab.ouardirhi@umons.ac.be

[2] Communication Networks Department, ENSIAS, Mohammed V University in Rabat, Mons, Morocco

Abstract. Addressing occlusion in object detection is a critical challenge as the use of smart video surveillance systems (SVS) grows. Current methods often struggle with relative occlusions among objects, which can severely compromise the efficacy of real-world SVS applications. Existing occlusion-handling strategies typically face limitations related to data inflexibility and the accurate discrimination of objects. To overcome these challenges, this paper introduces a sophisticated end-to-end solution that utilizes point cloud density analysis for precise occlusion rate assessment and combines depth information with RGB data to effectively separate occluded objects. This approach leverages advancements in integrating 2D and 3D data, employing cutting-edge network architectures and enhanced preprocessing techniques. Our model demonstrated its effectiveness on the KITTI dataset, particularly in scenarios where occlusion exceeded 20%. It surpassed the performance of YOLO3D by an average improvement of 10% in AP, and CompNet by 5%. This enhancement is indicative of the model's advanced capabilities in discerning and accurately detecting objects, even when they are partially obscured, which is a substantial step forward in the realm of occlusion handling.

Keywords: Deep Learning · Object Detection · Computer vision · SVS · Occlusion handling · Voxelization · Density-Aware

1 Introduction

A wide range of practical applications, including robotics, smart video surveillance, and automated driving, heavily rely on advanced object recognition and action/event detection technologies [1,2]. These applications require robust solutions for complex challenges such as person search, pose estimation, and multi-target tracking [3–5]. Particularly, these technologies are integral to the development of smart video surveillance systems, which employ various image processing algorithms and machine learning or deep learning processes to train models that replicate human perceptual abilities.

M. Zbakh et al. (Eds.): CloudTech 2023, LNNS 1220, pp. 274–287, 2024.
https://doi.org/10.1007/978-3-031-78698-3_14

Recent advancements in computational power and the availability of large datasets have substantially enhanced the training capabilities of object detection systems. However, a persistent challenge in these systems is the occlusion of objects, where objects are partially or wholly obscured by others or by environmental factors, significantly complicating detection tasks [6,7]. This issue is exemplified in Fig. 1, which illustrates the negative impact of high partial occlusion levels on the clarity and efficiency of object detection in crowded and complex environments.

Fig. 1. Complex Object Detection Environment: Depiction of a sophisticated scenario for detecting objects, marked by high partial occlusion within a densely populated setting, employing the UA-DETRAC dataset[8].

The traditional approaches to mitigating occlusions, such as sliding window techniques and template matching, have often fallen short in dynamic scenarios [9]. The shift towards employing deep learning strategies has introduced novel methodologies that utilize complex neural architectures to better address these challenges [10]. Nonetheless, the ability of these systems to accurately identify and handle partially occluded objects, which humans can discern with relative ease, remains limited.

In response to the ongoing challenges posed by object occlusions in computer vision, this paper introduces an improved version of our prior architecture, FuDensityNet [11], which is meticulously tailored to enhance occlusion handling through innovative network designs and methodologies. The enhanced model integrates advanced image preprocessing, voxelization techniques, and feature fusion processes, creating a dynamic and adaptive system capable of exceptional performance even in highly occluded environments.

A pivotal innovation in the approach is the Voxel Density-Aware (VDA) strategy, which accurately estimates occlusion rates within a scene before the detection process begins. This proactive assessment allows the network to adapt its detection mechanisms based on the severity and characteristics of occlusion, significantly refining the accuracy and reliability of object recognition [12]. The integration of these methods facilitates a comprehensive understanding of occluded scenes, markedly improving detection outcomes across varied real-world scenarios.

Moreover, the paper details substantial enhancements in Model Voxelization, focusing on memory optimization to streamline processing and enhance scalability. These modifications have proven critical in handling large-scale data more efficiently, thus facilitating more complex and accurate experiments. The revised experimental setup has conclusively demonstrated the superior capability of FuDensityNet to maintain high performance in environments characterized by frequent and complex occlusions.

Throughout the paper, we thoroughly explore and elucidate the architecture and functionalities of the revised model. We provide extensive evaluations of both 3D and 2D detection capabilities on benchmark datasets, showcasing notable advancements in detection accuracy, particularly in scenarios dominated by occlusions. These results not only validate the effectiveness of the approach but also highlight the significant improvements made over previous models.

In the sections that follow, we first review related research and the evolution of occlusion handling techniques, highlighting existing limitations. We then introduce the improved FuDensityNet, detailing its innovative architecture, advanced voxelization for occlusion estimation and memory optimization, and adaptive feature fusion methods. The methodology section describes the experimental setup and metrics, followed by a results section that provides empirical evidence of the model's enhanced performance in detecting occluded objects. We conclude by summarizing the contributions and discussing future directions for extending occlusion-aware capabilities in computer vision systems.

2 Related Work

Object detection, particularly in scenarios involving deformation and occlusion, presents ongoing challenges. Traditional algorithms generally perform well with minimal occlusion (less than 10%), but their efficacy diminishes as occlusion increases [13]. Early research predominantly explored methods that either isolated occluded objects from their backgrounds or leveraged depth information to enhance object detection [14].

In our recent survey [15], we not only explored a variety of occlusion handling techniques, including those beyond deep learning frameworks, but also charted the evolution of diverse methodologies addressing occlusion in automated detection systems.

The complexity of occlusion significantly impacts object recognition, particularly when objects are obscured by over 50%, complicating detection tasks considerably [16]. In response, generative models have been utilized to effectively distinguish between background and foreground elements, thereby improving occlusion management.

Notable advancements include Timur et al.'s [17] development of a model that decodes images into detectable collections of individuals in crowded settings.

Their model, based on the multi-camera generative detection technique POM, synthesizes depth maps rather than binary images to handle occlusions, although it tends to lose critical object information.

Furthermore, Sun et al. [18] introduced CompNet, a Bayesian generative model incorporating neural network features to enhance the classification of partially occluded objects. This approach, while innovative, requires extensive form priors and is primarily suitable for rigid objects.

In a significant leap forward, Chen et al. [14] proposed the MV3D network, a multi-view 3D CNN that integrates LIDAR data with 2D and 3D information to improve occlusion handling. Despite its effectiveness, the complexity of MV3D results in extended training times, limiting its real-time application.

Additionally, Ali et al. [19] extended the capabilities of YOLOv4 to create YOLO3D, which employs LIDAR data in a unique bird's eye view grid for 3D object detection. While promising, YOLO3D's detection accuracy still falls short of that achieved by MV3D.

Other noteworthy contributions include Yang et al.'s SG-NMS algorithm [20], which optimizes bounding-box suppression during occlusion, and Takahashi et al.'s E-YOLO [21], which combines edge detection with frame differences for enhanced 3D recognition during occlusion.

Jenkins et al. [22] introduced CountNet3D, a regression-based framework for counting objects in densely packed 3D environments, demonstrating superior performance over traditional 3D object detectors. Additionally, recent efforts like those by Reynolds et al. [23] have focused on salient object recognition in environments designed to assist visually impaired photographers.

Mirbod et al. [24] developed a robust agricultural imaging system to handle variability in daylight, providing valuable data on sizing errors associated with different measurement techniques in agricultural applications.

Ouyang et al. [25] pioneered the development of DeepID-Net, a significant advancement in the field with its introduction of deformable pooling layers. These layers substantially bolster the model's robustness in feature representation. DeepID-Net excels in adapting to a range of deformations and occlusions and promotes automated interactions among features, which are essential for achieving effective classification and managing occlusions.

Summary – Occlusion presents a significant challenge in object detection, affecting the performance of various detection methods. The diversity of occlusion handling techniques ranges from depth extraction using LIDAR data or stereo cameras to employing generative models that differentiate effectively between background contexts and targeted objects. Despite advances, the performance under occlusion is still not comparable to human capabilities.

Deep learning offers potential solutions through data augmentation strategies, enhancing dataset diversity and thereby improving framework robustness. Challenges persist, however, including detectors' rigidity in handling data characteristics and their inability to accurately assess occlusion rates.

The gap between automated detectors and human perception is evident, particularly in handling complex occlusions. Our main contributions can be summarized by three points:

- Development of an occlusion handling framework integrating 2D and 3D data through a multi-modal network, including a VDA strategy for precise occlusion rate estimation.
- Implementation of Model Voxelization with memory optimization, enabling efficient processing of large-scale data and facilitating complex experiments.
- Extensive comparative analysis demonstrating FuDensityNet's superior performance over some occlusion handling methods, with significant improvements in detection accuracy and Average Precision (AP) in complex environments.

These contributions represent a significant leap forward in object detection technology, particularly for occluded scenarios.

3 Occlusion Handling Detection Framework

In addressing the complexities of occlusions within computer vision, the method introduces a novel VDA strategy. It utilizes the point cloud density information, obtained during the voxelization process, to effectively estimate the occlusion rates within a scene. This estimation plays a crucial role in determining the detection pathway: for scenes with minimal occlusion, the system proceeds with a conventional 2D object detection network, ensuring efficiency and speed.

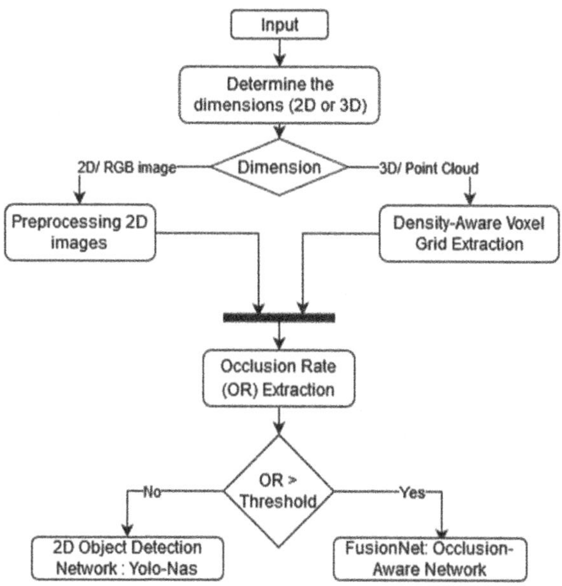

Fig. 2. FuDensityNet Architecture: Comprehensive Overview of the Advanced Occlusion-Adaptive Detection Framework.

However, when occlusions surpass a predefined threshold, the system switches to FusionNet, the occlusion-aware network designed to tackle intricate occlusions. This adaptive mechanism significantly boosts the accuracy and reliability of object detection in challenging environments, as it selectively employs the most appropriate detection technique based on real-time assessments of occlusion severity. The proposed workflow, succinctly captured in the diagrammatic representation, is the core of our approach, providing a robust solution to the enduring challenge of occlusion in real-world applications.

3.1 Data Representation

In the realm of object detection and occlusion handling, the quality, volume, and dimensional nature of data-whether two-dimensional or three-dimensional-play a crucial role in the performance of neural networks. Our insights underscore the significance of preprocessing data to align with the network's input requirements. Notably, we observed that 3D data could be instrumental in resolving occlusions due to its ability to exploit depth differentials for object differentiation. This aspect is particularly valuable in disentangling overlapping objects and addressing clustered regions created by occlusions.

However, depth information alone isn't sufficient to achieve precise object discrimination; 2D features remain indispensable for extracting detailed feature maps that define target objects. This realization steered us toward a hybrid approach that synergizes both data types-merging RGB imagery with aerial LIDAR data, a technique known as "LIDAR-camera fusion." This dual-input strategy promises to enhance object separation capabilities, particularly in occluded scenarios, and paves the way for more refined object recognition and classification.

3.2 Occlusion Rate Analysis

In tackling occlusions within 3D spaces, our approach advances the voxelization process by incorporating point cloud density analysis, which not only helps in distinguishing occluded structures but also optimizes memory usage.

Voxelization transforms a point cloud into a structured 3D grid, where each voxel represents a quantized portion of space. By measuring the density of points within each voxel, we obtain a clearer understanding of object surfaces and the space they occupy, even when occlusions occur.

Mathematical Foundation. Let $\mathbf{P}$ represent a point cloud, with $\mathbf{p_i}$ denoting each point where $i \in \{1, ..., N\}$ and N is the total number of points. Given voxel dimensions Δx, Δy, and Δz, the voxel counts V_x, V_y, V_z along each axis can be determined as follows:

$$V_x = \lceil \frac{\max(\mathbf{P_x}) - \min(\mathbf{P_x})}{\Delta x} \rceil \tag{1}$$

$$V_y = \lceil \frac{\max(\mathbf{P_y}) - \min(\mathbf{P_y})}{\Delta y} \rceil \tag{2}$$

$$V_z = \lceil \frac{\max(\mathbf{P_z}) - \min(\mathbf{P_z})}{\Delta z} \rceil \tag{3}$$

where $\lceil\cdot\rceil$ denotes the ceiling function.

Each point $\mathbf{p_i}$ is then assigned to a voxel $\mathbf{v}$ with a flattened index idx computed by:

$$idx = \left\lfloor \frac{p_{ix} - \min(\mathbf{P_x})}{\Delta x} \right\rfloor + V_x \left(\left\lfloor \frac{p_{iy} - \min(\mathbf{P_y})}{\Delta y} \right\rfloor + V_y \left\lfloor \frac{p_{iz} - \min(\mathbf{P_z})}{\Delta z} \right\rfloor \right) \tag{4}$$

where $\lfloor\cdot\rfloor$ is the floor function.

Voxel Density and Memory Optimization. Our technique for voxelization with density employs a sparse representation, drastically reducing memory consumption by storing only non-empty voxels. Instead of a dense 3D array [11], we leverage SciPy's `coo_matrix`, which allocates space solely for occupied voxels, crucial for processing large-scale point clouds efficiently.

By counting each point mapped to a voxel, we effectively record the occupancy without unnecessary zero-value storage for unoccupied spaces. The sparse matrix $\mathbf{V}$ then becomes a tally of point occurrences within the voxel grid.

Implications for Occlusion Handling. The analysis of the voxelized data is instrumental in identifying areas with high point density, typically indicative of occluded objects. These areas warrant further examination to differentiate between overlapping entities and accurately estimate occlusion rates.

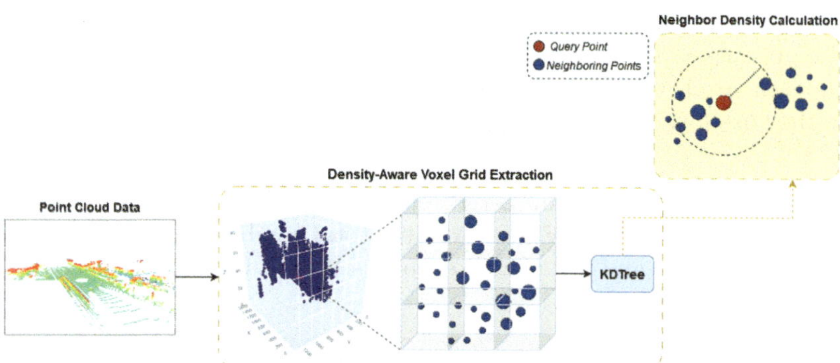

Fig. 3. Diagram Illustrating Occlusion Rate Evaluation through Density Analysis for Improved Occlusion Detection.

The advanced voxelization strategy we propose underpins our occlusion-aware object detection model, enabling precise detection of occluded objects and driving advancements in computational efficiency.

Enhanced Spatial Analysis In our framework, the initial voxel density analysis is augmented with a neighbor density calculation as shown in Fig. 3, enabling a deeper understanding of the spatial distribution of points. This process utilizes a KDTree structure to efficiently evaluate the density of points around each voxel, aiding in distinguishing between potential occlusions and continuous object regions.

In summary, the process involves constructing a KDTree from the point cloud, querying for neighboring points within a defined radius, and calculating the neighbor density. This density metric is compared against the voxel's own density to discern spatial distribution patterns that could indicate occlusion scenarios.

The subsequent occlusion rate (OR) determination (Fig. 2) informs the choice of detection model. By comparing the point density against a set occlusion threshold derived from benchmarks, the OR guides the selection between our occlusion handling network or an alternative state-of-the-art model for the most accurate object detection under varying conditions.

3.3 Network Architecture for Occlusion Handling

Our network architecture, as depicted in Fig. 4, expertly merges 2D image and 3D point cloud data to tackle the challenges of occlusion in object detection [11]. By integrating these two data modalities, the system not only leverages the spatial depth information provided by 3D point clouds but also capitalizes on the rich textural details conveyed by 2D images. This multimodal approach enhances the network's ability to discern and accurately identify objects that are partially obscured, significantly improving detection reliability and precision in complex visual environments.

Backbone Networks: After thorough testing, we chose ResNet-50 for 2D imaging due to its excellent occlusion adaptability, and VoxNet for 3D point clouds, which efficiently handles occlusion complexities. Both networks proved effective in our empirical analyses for their respective data modalities.

Multimodal Fusion Method: The Low-rank Multimodal Tensor Fusion (LMF) method was selected for its efficiency and ability to maintain crucial information while minimizing computational demands, proving superior in occlusion management during our tests.

Detection Head: We employ a Faster R-CNN detection head, which starts with RoI pooling to align features across modalities, followed by a series of fully connected layers that fine-tune the detection capabilities. This structure not only enhances class prediction and bounding box accuracy but also ensures our system's robustness against occlusions.

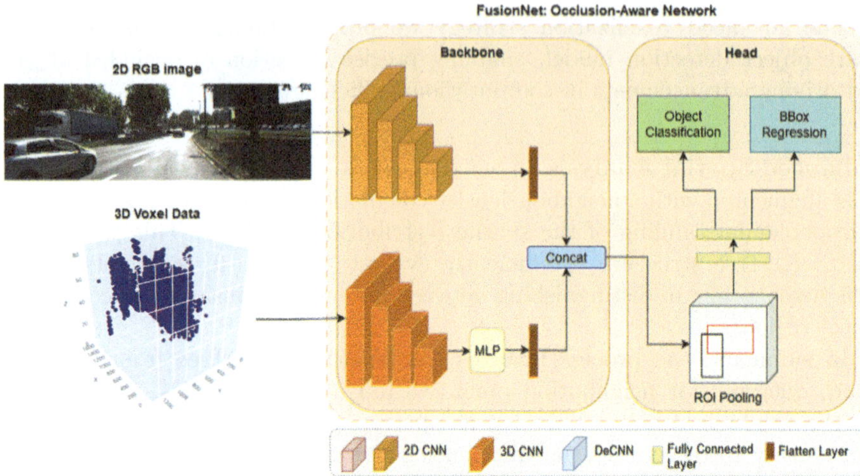

Fig. 4. Integrated Network Architecture for Enhanced Occlusion Handling.

4 Main Results

In this section, we detail the principal outcomes of our research, beginning with a description of our experimental framework to provide critical context. Initially, we assess various state-of-the-art 2D object detection networks, focusing particularly on scenarios with high occlusion rates exceeding 20%. This stage serves dual purposes: identifying the optimal network for low occlusion scenarios and selecting the best 2D backbone for integration into our FusionNet. Following this, we proceed to compare FuDensityNet against alternative occlusion handling methods and our occlusion-aware network to demonstrate the enhancements our model brings to object detection in occluded environments. Additionally, for our 3D backbone, we continue to employ the model established in our previous work [11], as no further improvements were necessary given its proven effectiveness. These comprehensive evaluations underscore the robustness and precision of our models across varying occlusion scenarios.

4.1 Experimental Setup

Resources. For our experiments, we employed the Tensor Processing Unit (TPU) v2, which is specifically optimized for high-demand machine learning and deep learning tasks. These TPUs, developed by Google, are advanced hardware accelerators that significantly enhance both the training and inference speeds of machine learning models, boasting a RAM capacity of 35GB.

Dataset. Our evaluation of FuDensityNet strategically utilizes the KITTI dataset [26], known for its challenging urban driving scenarios with natural occlusions from vehicles, pedestrians, and roadside structures. This dataset is ideal for testing our network's ability to handle different levels of occlusion, allowing us to rigorously assess the effectiveness of various backbone networks under these conditions. By leveraging KITTI's varied traffic scenarios, we rigorously test FuDensityNet in environments with different occlusion levels, ensuring its effectiveness in real-world applications and enhancing our detection strategies for complex situations.

4.2 Evaluation of 2D Object Detection Models

In our table analysis (Table 1), we evaluate the effectiveness of different object detection models on the KITTI 2D dataset, specifically examining their performance across varying occlusion levels. This analysis underscores the effectiveness of our approach in selecting appropriate backbone networks based on occlusion severity:

- **Low to No Occlusion:** YOLO-NAS - YOLO-NAS excels in environments with minimal occlusion, achieving the highest AP scores. Designed for speed and precision, it is ideally suited for scenarios where objects are clearly visible, affirming its use in low-occlusion contexts.
- **Moderate to High Occlusion:** ResNet50-F-RCNN - For settings characterized by substantial occlusion, ResNet50-F-RCNN shows strong performance. This combination excels in detecting partially obscured or overlapping objects, benefiting from a robust deep learning architecture that efficiently manages complex visual data.

Table 1. Object Detection Accuracy in High Occlusion Scenarios on KITTI 2D Dataset.

Model	AP(%) (KITTI 2D)		
	Car	Pedestrian	Cyclist
F-RCNN [27]	69.5	65.3	64.6
ResNet50-F-RCNN [28]	**74.5**	**66.5**	**65.5**
MobileNetv2-F-RCNN [29]	55.9	53.0	48.0
vgg16-F-RCNN [30]	58.0	57.5	47.5
SSD [32]	64.8	62.6	56.9
RetinaNet [31]	63.8	61.0	56.0
YOLOv5s [33]	85.9	83.6	80.3
YOLOv6s [34]	88.8	84.6	82.5
YOLOv7 [35]	86.9	83.0	80.8
YOLOv8s [36]	90.0	87.9	84.0
YOLO-NAS [27]	**91.1**	**88.2**	**86.5**

By deploying YOLO-NAS in environments with little to no occlusion and ResNet50 as the 2D backbone in more occluded situations, our strategy dynamically responds to the diverse challenges present in real-world conditions. This tailored approach optimizes detection effectiveness and efficiency across various occlusion levels, significantly enhancing the accuracy and performance of our detection system.

4.3 Performance Assessment of FuDensityNet

FuDensityNet distinguishes itself (Table 2) by not only matching but in some cases exceeding the performance of models like YOLO3D and CompNet, especially in more challenging occlusion scenarios. For Car detection, FuDensityNet achieves an AP of 62.8% in the Easy category, which is a significant improvement over other models, and it maintains robust performance in Moderate (47.9%) and Hard (43.9%) categories. This trend is consistent across other categories such as Pedestrian and Cyclist, illustrating the adaptability of FuDensityNet across various object types.

The strategic integration of YOLO-NAS for scenarios with low occlusion and the use of FusionNet techniques for moderate to high occlusion conditions has been instrumental. This approach allows FuDensityNet to leverage the strengths of both models, ensuring that it remains highly effective across all occlusion levels. The integration of 2D and LIDAR data further enhances its capability to discern obscured objects, contributing to the high AP scores.

The significant improvement in AP values can be attributed to our refined voxelization technique, which allows for more effective use of training data. By enhancing the density and quality of voxelized data, our model can better learn and predict occluded objects, a critical advantage in complex driving environments.

Table 2. Performance Analysis of Object Detection Models on the KITTI Dataset Under Occlusion Conditions.

Network	Data	Car			Pedestrian			Cyclist		
		Easy	Moderate	Hard	Easy	Moderate	Hard	Easy	Moderate	Hard
YOLO-NAS [27]	2D	**62.8**	**61.4**	**58.9**	**60.2**	**58.0**	**55.0**	**59.2**	**56.9**	**52.2**
YOLO3D [19]	LIDAR	38.8	35.4	29.0	38.0	32.6	28.2	34.2	25.9	20.4
CompNet [37]	2D	48.2	39.1	37.4	43.8	38.8	33.8	34.0	28.9	25.4
MV3D [14]	2D+LIDAR	49.6	<u>48.2</u>	<u>45.7</u>	42.2	<u>40.6</u>	<u>36.1</u>	44.8	<u>37.7</u>	<u>36.6</u>
FusionNet-Our	2D+LIDAR	47.6	45.9	43.9	41.2	37.6	35.8	35.5	35.4	33.4
FuDensityNet-Our	2D+LIDAR	**62.8**	<u>47.9</u>	<u>43.9</u>	**60.2**	<u>37.6</u>	<u>35.8</u>	**59.2**	<u>35.4</u>	<u>33.4</u>

5 Conclusion

This paper has presented a comprehensive occlusion handling framework designed to address the inherent challenges of object detection in environments with significant occlusions. By innovatively combining 2D and 3D data, our approach enhances the detection capabilities of systems operating in complex real-world scenarios where objects are often obscured or partially visible.

Central to our strategy is the integration of voxelization techniques that quantify occlusion by analyzing the density of point clouds. This method not only improves the precision of detecting occluded objects but also enriches the input data for our neural networks, allowing for a more nuanced interpretation of the scene.

The fusion of multimodal data through our refined network architectures enables our models to exploit the depth information from 3D point clouds alongside the detailed textural information from 2D images. This integration significantly boosts the system's ability to detect objects under various degrees of occlusion, achieving higher accuracy rates than traditional single-modality approaches.

While our results indicate a substantial improvement in handling occlusions, as reflected in enhanced AP scores across diverse occlusion scenarios, we recognize the limitations of our current dataset. Expanding our experiments to include a broader range of conditions and more diverse datasets will be crucial for further validating the effectiveness and adaptability of our approach.

Our findings underscore the potential of our occlusion handling methodology to set a new benchmark in object detection technology, particularly in terms of its robustness and reliability in occluded environments. The next steps will involve refining these techniques and extending their application to ensure that they can be effectively implemented in real-world systems. Future research will focus on overcoming the existing limitations and exploring the scalability of our approach to more complex and dynamic environments.

Ultimately, our work contributes significantly to the field of object detection, offering promising directions for future research and potential applications in various sectors where accurate detection in occluded conditions is critical.

References

1. Nguyen, D.T., Li, W., Ogunbona, P.O.: Human detection from images and videos: a survey. Pattern Recogn. **51**, 148–175 (2016)
2. Zhang, S., Benenson, R., Omran, M., Hosang, J., Schiele, B.: How far are we from solving pedestrian detection? In: Proceedings of the IEEE Conference on Computer Vision and Pattern Recognition, pp. 1259–1267 (2016)
3. Leal-Taixé, L., Milan, A., Reid, I., Roth, S., Schindler, K.: Motchallenge 2015: towards a benchmark for multi-target tracking. arXiv preprint arXiv:1504.01942 (2015)
4. Newell, A., Yang, K., Deng, J.: Stacked hourglass networks for human pose estimation. In: Computer Vision–ECCV 2016: 14th European Conference, Amsterdam, The Netherlands, 11–14 October 2016, Proceedings, Part VIII 14, pp. 483–499. Springer (2016)
5. Xiao, B., Wu, H., Wei, Y.: Simple baselines for human pose estimation and tracking. In: Proceedings of the European Conference on Computer Vision (ECCV), pp. 466–481 (2018)

6. Pandya, S., et al.: Federated learning for smart cities: a comprehensive survey. Sustain. Energy Technol. Assessm. **55**, 102987 (2023)
7. Gunasekaran, K.P., Jaiman, N.: Now you see me: robust approach to partial occlusions. arXiv preprint arXiv:2304.11779 (2023)
8. Wen, L., et al.: UA-DETRAC: a new benchmark and protocol for multi-object detection and tracking. Comput. Vis. Image Underst. **193**, 102907 (2020)
9. Sharifani, K., Amini, M.: Machine learning and deep learning: a review of methods and applications. World Inf. Technol. Eng. J. **10**(07), 3897–3904 (2023)
10. Ye, H., Zhao, J., Pan, Y., Cherr, W., He, L., Zhang, H.: Robot person following under partial occlusion. In: 2023 IEEE International Conference on Robotics and Automation (ICRA), pp. 7591–7597. IEEE (2023)
11. Ouardirhi, Z., Amel, O., Zbakh, M., Mahmoudi, S.A.: FuDensityNet: fusion-based density-enhanced network for occlusion handling. In: Proceedings of the IEEE/CVF Winter Conference on Applications of Computer Vision, pp. 3008–3017 (2023). https://doi.org/10.5220/0012425400003660, https://www.insticc.org/node/TechnicalProgram/visigrapp/2024/presentationDetails124254
12. Steyaert, S., et al.: Multimodal data fusion for cancer biomarker discovery with deep learning. Nat. Mach. Intell. **5**(4), 351–362 (2023)
13. Pérez-Hernández, F., Tabik, S., Lamas, A., Olmos, R., Fujita, H., Herrera, F.: Object detection binary classifiers methodology based on deep learning to identify small objects handled similarly: Application in video surveillance. Knowl.-Based Syst. **194**, 105590 (2020)
14. Chen, X., Ma, H., Wan, J., Li, B., Xia, T.: Multi-view 3d object detection network for autonomous driving. In: Proceedings of the IEEE Conference on Computer Vision and Pattern Recognition, pp. 1907–1915 (2017)
15. Ouardirhi, Z., Mahmoudi, S.A., Zbakh, M.: Enhancing object detection in smart video surveillance: a survey of occlusion-handling approaches. Electronics **13**(3), 541 (2024)
16. Kortylewski, A., He, J., Liu, Q., Yuille, A.L.: Compositional convolutional neural networks: a deep architecture with innate robustness to partial occlusion. In: Proceedings of the IEEE/CVF Conference on Computer Vision and Pattern Recognition, pp. 8940–8949 (2020)
17. Bagautdinov, T., Fleuret, F., Fua, P.: Probability occupancy maps for occluded depth images. In: Proceedings of the IEEE Conference on Computer Vision and Pattern Recognition, pp. 2829–2837 (2015)
18. Sun, Y., Kortylewski, A., Yuille, A.: Amodal segmentation through out-of-task and out-of-distribution generalization with a bayesian model. In: Proceedings of the IEEE/CVF Conference on Computer Vision and Pattern Recognition, pp. 1215–1224 (2022)
19. Ali, W., Abdelkarim, S., Zidan, M., Zahran, M., El Sallab, A.: Yolo3d: end-to-end real-time 3d oriented object bounding box detection from lidar point cloud. In: Proceedings of the European Conference on Computer Vision (ECCV) Workshops (2018)
20. Yang, C., Ablavsky, V., Wang, K., Feng, Q., Betke, M.: Learning to separate: detecting heavily-occluded objects in urban scenes. In: European Conference on Computer Vision, pp. 530–546. Springer (2020)
21. Takahashi, M., Ji, Y., Umeda, K., Moro, A.: Expandable YOLO: 3d object detection from RGB-D images. In: 2020 21st International Conference on Research and Education in Mechatronics (REM), pp. 1–5. IEEE (2020)

22. Jenkins, P., et al.: CountNet3D: a 3D computer vision approach to infer counts of occluded objects. In: Proceedings of the IEEE/CVF Winter Conference on Applications of Computer Vision, pp. 3008–3017 (2023)
23. Reynolds, J., Nagesh, C.K., Gurari, D.: Salient object detection for images taken by people with vision impairments. arXiv preprint arXiv:2301.05323 (2023)
24. Mirbod, O., Choi, D., Heinemann, P.H., Marini, R.P., He, L.: On-tree apple fruit size estimation using stereo vision with deep learning-based occlusion handling. Biosys. Eng. **226**, 27–42 (2023)
25. Ouyang, W., et al.: Deepid-net: deformable deep convolutional neural networks for object detection. In: Proceedings of the IEEE Conference on Computer Vision and Pattern Recognition, pp. 2403–2412 (2015)
26. Geiger, A., Lenz, P., Urtasun, R.: Are we ready for autonomous driving? the kitti vision benchmark suite. In: 2012 IEEE Conference on Computer Vision and Pattern Recognition, pp. 3354–3361. IEEE (2012)
27. Sharma, P., Gupta, S., Vyas, S., Shabaz, M.: Retracted: object detection and recognition using deep learning-based techniques. IET Commun. **17**(13), 1589–1599 (2023)
28. He, K., Zhang, X., Ren, S., Sun, J.: Deep residual learning for image recognition. In: Proceedings of the IEEE Conference on Computer Vision and Pattern Recognition, pp. 770–778 (2016)
29. Sandler, M., Howard, A., Zhu, M., Zhmoginov, A., Chen, L.C.: Mobilenetv2: inverted residuals and linear bottlenecks. In: Proceedings of the IEEE Conference on Computer Vision and Pattern Recognition, pp. 4510–4520 (2018)
30. Simonyan, K., Zisserman, A.: Very deep convolutional networks for large-scale image recognition. arXiv preprint arXiv:1409.1556 (2014)
31. Lin, T.Y., Goyal, P., Girshick, R., He, K., Dollár, P.: Focal loss for dense object detection. In: Proceedings of the IEEE International Conference on Computer Vision, pp. 2980–2988 (2017)
32. Liu, W., et al.: SSD: single shot multibox detector. In: European Conference on Computer Vision, pp. 21–37. Springer (2016)
33. Sozzi, M., Cantalamessa, S., Cogato, A., Kayad, A., Marinello, F.: Automatic bunch detection in white grape varieties using YOLOv3, YOLOv4, and YOLOv5 deep learning algorithms. Agronomy **12**(2), 319 (2022)
34. Li, C., et al.: YOLOv6: a single-stage object detection framework for industrial applications. arXiv preprint arXiv:2209.02976 (2022)
35. Wang, C.Y., Bochkovskiy, A., Liao, H.Y.M.: YOLOv7: trainable bag-of-freebies sets new state-of-the-art for real-time object detectors. In: Proceedings of the IEEE/CVF Conference on Computer Vision and Pattern Recognition, pp. 7464–7475 (2023)
36. Huang, Z., Li, L., Krizek, G.C., Sun, L.: Research on traffic sign detection based on improved YOLOv8. J. Comput. Commun. **11**(7), 226–232 (2023)
37. Kortylewski, A., Liu, Q., Wang, H., Zhang, Z., Yuille, A.: Combining compositional models and deep networks for robust object classification under occlusion. In: Proceedings of the IEEE/CVF Winter Conference on Applications of Computer Vision, pp. 1333–1341 (2020)

Analyzing Sentiment in Arabic Tweets: A Study Using Machine Learning and Deep Learning Techniques

Mohammed Benali(✉) and Zouhair Lakhyar

Hassan 2 University, Fsjes Mohammedia, PEL Laboratory, Street, Mohammedia, Morocco
m.benali@uhp.ac.ma

Abstract. Natural language processing is a field of study that encompasses various tasks, including text generation, language modeling, and classification. The extraction of valuable information from text documents allows for better decision-making in political decisions and analyzing public opinion; those techniques are also deployed in support systems. In the proposed study, we aim to analyze sentiment in Arabic Twitter comments written in Moroccan dialectal Arabic and modern standard Arabic. The process starts with data collection using the Twitter API, followed by data pre-processing. The polarity determination is based on emoji annotations, which is a major key in the model training. Word sequences are used to create word vectors, which are then processed through dense layers to produce customized word embeddings. Sentiment analysis is performed using both deep learning (DL) and machine learning (ML) models to extract opinions from Arabic text. The Arabic-collected dataset has been presented and evaluated using several indicators in order to compare the analysis performance between both machine learning and deep learning algorithms. The results of this study show promising results and satisfactory outcomes, which is very encouraging for additional studies in this field.

Keywords: Natural Language Processing · Sentiment Analysis · Machine Learning · Deep Learning · Modern Standard Arabic · Dialectal Arabic

1 Introduction

Over the past few years, one of the major fields of study has been Artificial Intelligence (AI). Giving the machines the ability to simulate human intelligence is the principal concept of AI. The Artificial Intelligence has been an active area of research and study, including both subfields: machine learning and deep learning.

Sentiment analysis is particularly complex in the Arabic language, owing to the rich morphology of Arabic, the variety of dialects spoken, and the writing style's intricacy, which includes various forms for each letter based on the word's placement in the document. In contrast to the English language, Arabic has no normalized lexicon of

M. Zbakh et al. (Eds.): CloudTech 2023, LNNS 1220, pp. 288–303, 2024.
https://doi.org/10.1007/978-3-031-78698-3_15

sentiments, which makes it harder for ML models to learn. In addition, handling both contextual shades and hybrid dialects in texts in Arabic involves much more advanced pre-processing and model-adaptation methods, presenting a number of specific difficulties as compared to more standard languages like English.

To save time in evaluating a large scale of data, extracting crucial information, and analyzing the raw data, the ML and DL algorithms present an effective solution. These algorithms are known for their ability to extract relationships from data and process a large number of features. The models are able to learn from experiences and don't require explicit programming, which enhances their performance on vast dataset and also unstructured data.

The main use of sentiment analysis (SA) for Twitter is to collect data to categorize the comments into two polarities, namely positive and negative, based on the statements or sentences they contained [1]. In this study, we extend this approach, using the main dataset to examine the classification of sentiments along a larger emotional range. More specifically, we expand our analysis to incorporate three polarities (positive, negative, and neutral), allowing more nuanced comprehension of opinion expressions in Twitter comments. This extension allows a complete exploration of SA techniques, by comparing the models' performance in two-polarity and three-polarity classification schemes. Our principal contribution to this study may be summarized below:

- Summarize the complexity of the Arabic NLP task and offer a short description of ASA.
- Establish the methodology for setting up and pre-processing the Arabic Twitter comments, using a variety of techniques to guarantee the quality and consistency of the datasets.
- The use of emojis as an automatic annotation procedure for datasets to facilitate SA processes.
- Evaluating the effectiveness of the ML and DL models for SA on the same dataset, but with varying sets of parameters, including two polarities (positive and negative) and three polarities (positive, negative, and neutral), providing a full examination of differences in model performance and adaptability through various sentiment classification patterns in modern standard Arabic and dialectal Arabic tweets.
- Draw conclusions based on comparing the performance of the ML and DL models on Arabic tweets.

This work is structured as follows: In Section 2, we survey related work on NLP issues, particularly sentiment analysis in Arabic. The methodology suggested for sentiment analysis is outlined in Section 3, beginning with data collection, pre-processing techniques, labeling of data and word embedding, construction of the model, and classification framework. The dataset being tested, the findings being presented in Section 4, and the conclusions outlining both findings and points of view.

2 Related Work

The issues of text classification, notably noisy words, vernacular terminology, and sparse nature, were the focus of Zhao et al.'s research [2]. The authors talked about the effectiveness of algorithms used to use analytics measurements to solve textual problems.

A number of studies [3, 4], described the general sentiment analysis procedure, which consists of four steps: text pre-processing, data extraction, data analysis, and knowledge identification. The study literature uses three main types of data: pre-existing corpuses from earlier studies, manual corpuses that are manually retrieved from social media, and automatic corpuses that are obtained through APIs offered by various social media. Pre-processing tasks can involve dividing the text into sentences or transforming verbs into their infinitive forms, known as word lemmatization.

The majority of research studies refer to the third stage of the sentiment analysis process as" sentiment classification" as opposed to" sentiment analysis" [5]. The primary goal of sentiment analysis, according to Karamibekr and Ghorbani [6], is sentiment classification. Three automatic learning techniques supervised, unsupervised, and hybrid can be used for classification. The labeled corpus is usually used in supervised techniques (also known as the corpus-based method) to learn the process of classifying sentiments [7]. Dictionary-like sentiment lexicons serve as the foundation for unsupervised techniques, also known as lexicon-based methods [8]. Hybrid strategies integrate the two previous methods [9]. The information generated from the sentiment analysis procedure can be applied to the population whose posts or comments were examined in order to ascertain the overall sentiment surrounding a certain brand, business, organization, or public figure.

The study of people's opinions, sentiments, assessments, attitudes, and emotions about various things, including goods, services, organizations, people, problems, events, subjects, and their characteristics, is known as sentiment analysis (SA), often known as opinion mining. It symbolizes a significant and dynamic field of computer science [10].

Twitter data has been used in a number of sentiment analysis studies. However, due to their short duration and irregular structure, these data are difficult to use [11]. Over the past ten years, researchers from a variety of fields have made extensive use of sentiment analysis on microblogging platforms such as Twitter.

The use of sentiment analysis can vary from language to language. The English language has been the focus of the majority of sentiment analysis studies, with Arabic receiving relatively less attention. 4.8% of Internet users worldwide are Arab. The Arabic language can be classified into three categories: formal Arabic written and spoken in the media (newspapers, magazines, TV, and radio), informal or dialectal Arabic, which differs from one Arab nation to another, and classical Arabic found in religious scriptures [12].

According to Darwish et al. [13], there are six main dialects spoken in the Arab world: Egyptian, Moroccan, Levantine, Iraqi, Gulf, and Yemeni. Both MSA and informal Arabic can be found on social media platforms [14]. According to Abdul-Mageed et al. [15], dialects differ significantly on all levels of linguistic representation, including morphology, lexicon, phonology, syntax, semantics, and pragmatics.

Furthermore, El-Masri et al. [16] explored in detail a prevalent issue in Arabic data extraction: the scarcity of corpora. They discussed these concerns and offered suggestions and ideas to deal with the situation.

According to Hassan and Mahmood [17], the two most popular deep learning models for feature representation and classification are convolutional neural networks (CNN) and recurrent neural networks (RNN). According to Li et al. [18], CNN is a form of feed-forward NN that requires less training data. CNN's capacity to extract the syntactic and semantic elements relevant to a given task has allowed it to demonstrate impressive performance in a variety of NLP-related tasks. CNN automatically extracts local information by applying convolutional operations over the input layer. Weight sharing among all neurons enhances CNN's learning capacity [19].

CNN can perform better in classification than highly competitive neural networks like FNN, RNN, and LSTM, claim Ravuri and Stoicke [20].

A DL model can surpass the state-of-the-art in AI tasks [21], because of its capacity to automatically perform feature representation on multilayer, which makes it distinct from NLP models based on ML [22]. Inspired by the human brain, the DL model is a neural network multilayer [23], which make it possible to extract features in an efficient way across many domains.

According to Vateekul and Koomsubha [24], human involvement has been reduced by the DL and ML algorithms in the ASA tasks.

3 Machine Learning and Deep Learning Approaches for Analyzing Twitter Comments

3.1 Methodology

This chapter presents a comprehensive overview of our method, describing each step of the process, from data collection to model implementation. To illustrate the methodology, the architecture of the approach used is presented graphically in Fig. 1, which also explains the procedures required to use the collected dataset for sentiment analysis in Arabic.

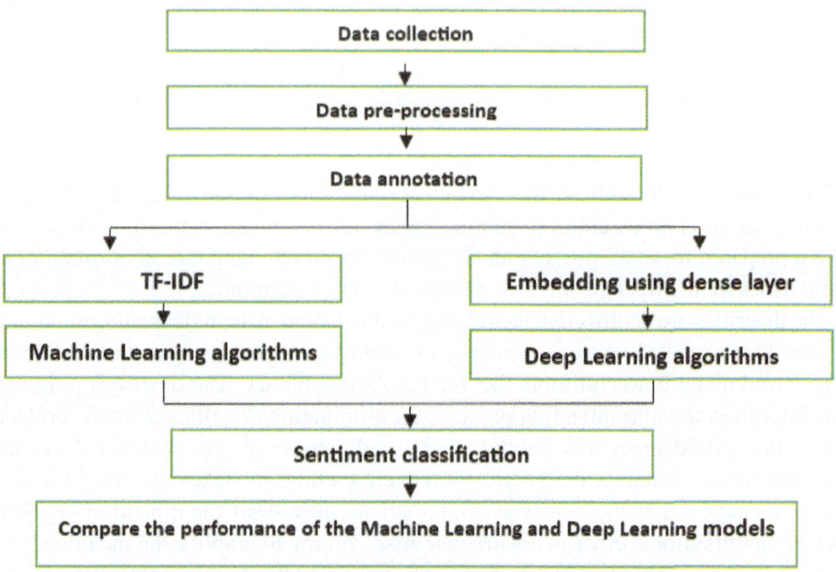

Fig. 1. Sentiment Analysis Techniques Applied to Arabic Tweets

Initially, the Twitter API was used to facilitate the extraction of shared Arabic tweets. We process both Moroccan dialect Arabic and modern standard Arabic, given the diversity of the Arabic language and discourse. To guarantee the accuracy of all the data collected, we also take into consideration the source account's credibility, and set a strict following limit of 100 or more followers. This strategy reduces the potential risk of bias and enhances the accuracy of our analyses.

Once the collected dataset has been analyzed, the pre-processing phase begins, with a focus on minimizing the amount of noise, fixing any typos, and adjusting the dataset to ensure that it is as pertinent and usable as possible.

By establishing the basis for the studies that will follow, these preliminary steps ensure the data is ready to be processed and then analyzed further. One of the most significant parts of our work, the automatic annotation, employs emojis to provide sentiment and context information for the collected dataset. Transforming the textual documents into vectors of real numbers is critical. The methods used are TF-IDF in the ML models to convert the textual dataset into vectors, while DL uses neural layers to produce embeddings, which is crucial to obtain the essential information for ASA tasks.

In the proposed study, we deployed a set of algorithms such as: Support Vector Machines (SVM), Xg-Boost, Convolutional Neural Networks (CNN), Long Short-Term Memory (LSTM), etc... Each used model was meticulously tested and evaluated, with a specific focus on its performance in identifying sentiments from text documents written in Arabic.

3.2 Data Pre-processing

In SA, the pre-processing process is essential. The key factor is its capacity to enhance both accuracy and precision. Ignoring the pre-processing steps can lead to hard feature extraction and poor quality of data, making the analysis more difficult and even less effective, which reduces the SA performance.

Achieving balance is critical since heavy pre-processing techniques can result in the loss of valuable data elements. In this study, striking the correct equilibrium required the use of a number of specific techniques, including normalization, punctuation removal, stop word removal, tokenization, and stemming.

Normalization: to address the Arabic irregularities normalization also known as standardization, is an essential technique presented as a solution to unify the text and make it prepared for more processing. To handle the Arabic language abnormalities and irregularities in an effective way, we deployed a set of techniques.

The diacritics are deployed in the written Arabic language to indicate the grammatical features. Due to their lowest significance in Arabic sentiment analysis tasks, diacritics are removed using powerful tools like the PayArabic library. The diacritics removal is a crucial step in the normalization process. By eliminating diacritics from the collected dataset, the text becomes more unified and less dimensional. The p-tashkeel was used to automatically eliminate diacritics. Moreover, spelling mistakes are frequent in the Arabic language due to the numerous grammatical rules. Also, the repetition of specific letters as an intentional error in a word can arise, aiming to amplify the meaning.

To guarantee a consistent, normalized presentation of the Arabic textual data, the normalization process played a key role in addressing both intentional and non-intentional errors. To make the Arabic dataset more regular and standard, we removed times, dates, URLs, numbers, and special symbols such as @, §, ? and any non-Arabic letters, which made the collected dataset more coherent and the ASA considerably simpler.

- Tokenization: is breaking down the sentences into single words called tokens, it constitutes a major key in the pre-processing techniques. By converting textual data into distinct tokens, it makes the process easier.
- Punctuation removal: according to Altrabsheh [25], eliminating punctuation is a necessity and should be removed consistently, not removing the punctuation from the text could make the tokenization tasks more complicated. The punctuation removal seems to be valuable and crucial as a pre-processing technique, which reduces the noise and simplifies the textual data treatment and makes it more normalized and unified, which will enhance the model capability of interpreting the nuances, especially for SA tasks.
- Stopwords removal: stopwords can add complexity to the information retrieval process, even if they have no major meaning. Their use is frequent in the Arabic language. However, to enhance the model efficiency and performance and reduce the volume of the dataset, stop words removal constitutes a key element, which also facilitates the data treatment and improves the effectiveness of using index space. Nonetheless, stop words can carry meaningful information, especially in negation and contrast. In our case, we create a special list of stop words excluding certain words like " يس ل "and " لا "" etc..., and
- other words involving negation. The customized list allows us to minimize data noise and keep the necessary information.
- stemming: the morphological complexity of Arabic language requires the use of the steamming process, which offers a powerful alternative to enhance the model's precision and effectiveness and also minimizes the text length by eliminating both prefixes and suffixes. We used the Tashphyne stemmer based on statistical methods, which return the word to its original form by removing prefixes and suffixes.
- Word sequences: the process of word sequences gives each word in the textual data a numerical identifier, known as 'ID'. These IDs represent the input for the NLP tasks.

3.3 Tweets Annotation

Due to the Arabic language complexity, determining the sentiment in textual data appears to be a com- plex task. The collected dataset containing Twitter comments does not contain any explicit sentiment indicators such as stars or scores, which makes the identification of sentiment harder. Also, the lack of lexicons for the Arabic language, including MSA and DA, adds more difficulty to the annotation process, forcing us to use innovative methods of annotation. To overcome this problem, we used emojis to annotate the tweets. Inspired by emojipedia, the built lexicon contains 100 negative emojis and 200 positive ones. Each Twitter comment containing a sad emoji is considered as negative sentiment, while tweets with smiley emojis are taken as positive ones (Table 1).

Table 1. Example of Excuting the Pre-processing Tasks on a Twitter Comment

Task	Result
Original text	أريدُ البحثَ عن مَطعمٍ في الرياض، لكِنَّ المَوْقِعَ المُرَادَ زيارَتُهُ غَيْرُ مَعْرِفٍ
Normalization	أريد البحث عن مطعم في الرياض لكن الموقع المراد زيارته غير معرف
Tokenization	['أريد', 'البحث', 'عن', 'مطعم', 'في', 'الرياض', 'لكن', 'الموقع', 'المراد', 'زيارته', 'غير', 'معرف']
Punctuation removal	['أريد', 'البحث', 'عن', 'مطعم', 'في', 'الرياض', 'لكن', 'الموقع', 'المراد', 'زيارته', 'غير', 'معرف']
Stopwords removal	['بحث', 'مطعم', 'الرياض', 'الموقع', 'المراد', 'زيارته', 'غير', 'معرف']
Stemming	['بحث', 'مطعم', 'رياض', 'موقع', 'راد', 'زيار', 'غير', 'عرف']

To provide a nuanced view the sentiment in Twitter comments, we designed supplementary annotation, categorizing tweets as positive, negative, and neutral, which gives us the ability to explore the model's performance. We selected a list representing all three classes, we allocated 80 emojis representing negative sentiments, 150 for positive sentiments, and 50 emojis for neutral ones. In situations where Arabic tweets do not contain any emojis the manual annotation is required. Crowdsourcing represents a solution to overcome this problem by manually annotating tweets without emojis; also, the test set was annotated on a manual basis, which increases the accuracy of the classification. Even if the emoji annotation has some limitations, it provides a practical method of quickly annotating larger datasets, which increases the efficiency of DL models due to their requirement of a vast amount of data in the training process.

Remarkably, our findings are coherent with the findings of Kwaik [26], where an agreement ratio of approximately 80% is achieved by comparing the emoji annotation to manual annotation, highlighting the efficiency of this approach in addressing the Limitations of Arabic lexicon for SA.

3.4 Converting Textual Data to a Vector of Real Number

3.4.1 TF-IDF

Term frequency - the inverse frequency of documents is an indispensable way of converting text into numbers [27]. The conversion of text documents to vectors of real numbers is essential to closing the gap between text and ML models. TF-IDF is a combination of inverse document frequency and term frequency.

The term frequency provides a way of measuring the appearance of a distinct word in a whole text document, whereas the inverse document frequency reveals the uniqueness of a specific word in a whole corpus.

Through the combination of these techniques, TF-IDF is able to accurately represent a word's global relevance throughout the dataset as well as its local significance within a document.

TF-IDF is able to display the overall pertinence of a single word in a text and serves as a link between unprocessed text data and machine learning models.

3.4.2 Word Embedding

Word embedding: to understand and capture complex semantic relationships and identify complex patterns, the word embedding approach is considered as an effective and widely use approach. Traditional approaches are limited in terms of capturing contextual nuances of words in a text document, while word embedding allows for more thorough patterns comprehension. It allowed to align closely in a vector dimensional space the words sharing similar contexts permitting to reveal more complicated relationships between distinct terms.

Due to the language complexity and lack of research and lexicon in Arabic language, humans and machines are facing many obstacles to deal with such a complicated language compared to other languages like Western and Asian languages

The word embedding represents a major key in extracting the meaning from textual data and converting it to a vector representation, which increases the performance of the model's and enhances the analysis precision. This explains the use of this technique in numerous NLP tasks.

We suggested building a personalized word embedding specific to the used dataset and ASA tasks by customizing a neural network, particularly tailored for DL models. After determining the DL model architecture, the pre-trained Arabic dataset was used in the process of training the word embedding using heavy layers. In the training process, the semantic features of each word were used to predict its meaning efficiently, which gives the model the ability to capture intricate relationships between distinct words and their context depending on the usage of each with high precision due to the learning process, which is repeated consecutively.

The proposed approach has given us the ability to improve the text representation, which enhances the capacity to identify and manipulate Arabic textual data. As a result, the model's comprehension of the intricacies increased.

4 Machine Learning and Deep Learning Approaches

The Arabic tweets dataset is joined by a variety of ML and DL models to evaluate the model performance on the collected dataset and outline the results of each used method in terms of effectiveness in SA tasks.

4.1 Machine Learning Approach

In this study, we used traditional machine learning techniques like Nave Bayes, SVM (Support Vector Machine), and Decision Tree. We also used the models, combining many ML techniques like random forest and XG-Boost. Those models have the ability to classify sentiment based on textual data.

- Naive Bayes: the model is founded on naïve theory, showing notable independence among the input data used in calculating probabilities. Naive Bayes are easily deployable and well-known for their fast-building speed, which explains the vast use of this model in ML applications, including SA.
- Extreme Gradient Boosting: known as XG-Boost, is a ML model identified as high-performance algorithm having the ability to solve supervised issues. Deploying the gradient boosting tree to distribute computation in an optimal way. As a result, the model is faster and more effective in ML tasks and the predictions of the model improve with each iteration by reducing the loss, which leads to a better result.
- Support vector machine: known as SVM is a model that uses supervised learning to classify the data into distinct classes or categories based on a pre-trained dataset. The model enhances efficiency, reduces loss, and minimizes overfitting by distributing the inputs into separate groups. As a result, the SVM is a ML model widely used in NLP applications, including SA.
- Decision tree model: the Decision tree model uses the power of tree structure to make predictions. The learning data provides patterns to develop a hierarchical structure based on trees and nodes, denoting in a respective manner the potential generated outcomes representing the prediction capability of the model, which increase with each individual tree.
- Random forest: by building many decision trees, the random forest model constitutes a useful tool for classification tasks, including SA. By running the built trees all at the exact same time, the model reduces over-fitting, increases the performance, and reduces the inter-tree relationships, explaining the high efficiency of the model in classification tasks.

4.2 Deep Learning Approach

The deep learning models used in this study are: the CNN model used the convolutional layers to detect features; and LSTM, known for its ability to extract features from sequential data. The selected models give us the ability to classify Arabic text with high efficiency.

- The convolutional neural network: known as CNN has an architecture that could be presented as follows: an input layer, many hidden layers, and a single output layer. This complicated system is widely used in data classification tasks. The core of CNN model is identified by deploying filters to extract features from data. The used architecture has the ability to identify complex patterns from the input data using mathematical structures. The features are retrieved systematically while passing through the convolutional layers, and to add more precision to the model the data is sent to dense layers. As a result, the CNN constitute the core of the DL model's due to It effectiveness on a variety of tasks, including SA.
- Long Short-Term Memory: known as LSTM, is an algorithm employed in NLP tasks. The model is designed to detect and understand relationships among words in text documents, which increases its ability in analysing nuanced textual context. LSTM comprises a dual memory structure, the short-term memory has the capacity to store

the present information proactively, while the long-term memory is designed to combine the past information with the current one, which increases the context understanding over time. The LSTM represents an effective and strong tool in NLP area, able to overcome many challenges due to its capacity of recognizing complicated features while maintaining the contextual information in the text documents.

5 Result and Discussion

5.1 Results Overview and Parameter Details

We aim to conduct a series of experiments on textual data collected from Arabic social networks, starting with the data gathering using Twitter API. The large collected dataset covers a wide range of topics, including economics, politics, and sports, written in modern standard Arabic and dialectal Arabic. The collected data contains approximately 58 000 tweets, divided into 20% for the test set and 80% for the training set. The training set was annotated automatically using emojis, while the test set was annotated manually.

In the first case, the Twitter comments were classified as positive and negative ones; in the other hand, the second case the tweets were classified into positive, negative, and neutral, which makes the process more complex to analyse the sentiment correctly.

The dual process pushed the models to be tested thoroughly in analysing sentiment. The data was pre-processed, annotated, and tokenized; each tweet has a word limit of 100 words. Then each distinct word in the dataset is represented by a unique identifier, also called ID, using word sequences. In the DL models, and to guarantee the forward pass, the embedding layer is crucial in converting textual data to a numerical vector for each distinct word, preserving the contextual information. As a result, the DL benefits from the rich semantic used in the word embedding to identify complex semantic and interpret sentiment in the text accurately.

To evaluate and measure the models' performances, we focus on two major metrics: F-score and accuracy.

The accuracy can be misleading in detecting the efficiency of the model on SA tasks, unlike the F-score, which offers a reliable performance measure. The F-score is formed by using the averaging precision and recall.

The formula for the F1 score is:

$$F1 \text{ score} = \frac{2 \times (\text{precision} \times \text{recall})}{\text{precision} + \text{recall}} \tag{1}$$

Precision: represents how the model classifies the right predictions compared to the overall number of the prediction made.

Recall: known as sensitivity, represents the model ability to make right predictions, divided the total of the true positive (correctly identified as positive) by the total of true positive and false negative.

The variation of the F-score is between 0 and 1; 0 indicating that the model is not capable of making any correct predictions, while 1 indicating that the model is providing accurate predictions.

The ADAM optimizer was used, to enhance the model's performances, and is considered as a highly sophisticated method combining stochastic gradient descent and mean-square propagation.

Additionally, to optimize the models outputs and obtain higher results, the activation function like sigmoid, linear rectified unit (RELU), and SoftMax constitutes a major key in converting the inputs of probabilities, and modelling the outputs of neural network layers, which enhance the model performance. By using the collection of activation functions and optimizers, we are aiming to attain the highest model capacity in identifying and interpreting sentiment in Arabic text documents.

The SoftMax function gives us the ability to convert the numerical vectors to probabilistic distributions, each vector corresponds to a probability of a distinct category.

The formula of the SoftMax function is given below:

$$softmax(z) = \frac{\text{Exp}^{z_i}}{\sum_{j=1}^{K} \text{Exp}^{Z_j}} \tag{2}$$

Such as:

- Z designates a real number vector.
- e stands for the constant e (about 2.71828...).
- i is the index of a specific element in the Z vector.
- j is an index for all elements in the Z vector.
- K is the number of classes in the total number of classification tasks.

The Sigmoid function is a mathematical function that transfers real-valued inputs to a range bounded between 0 and 1.

The sigmoid function is described as follows:

$$\sigma(x) = \frac{1}{(1 + \exp(-x))}. \tag{3}$$

The non-linear conversion makes the understanding of complex connections between data easier, and the outputs could be interpreted as probabilities, which is beneficial for DL applications.

The RELU function transforms negative entries to zero, and the original input values are maintained for the positive inputs, by using their linear behaviour, which resolves the issue of gradients vanishing during back-propagation, while adding non-linearity to the model. As a result, the training process becomes faster, and the features recognition is enhanced.

The defining expression of the Rectified Linear Unit (ReLU) is:

$$f(x) = \max(0, x). \tag{4}$$

- the input x is higher or equal to 0, the corresponding x output value is not changed.
- the input x is less than 0, the Output is fixed at 0.

The previously mentioned functions are used in the models' architectures to take benefits of their ability to enhance the model's performance, and simplify the understanding of the outputs (Tables 2 and 3).

Table 2. Performance Results: Two-Class Sentiment Analysis (Positive and Negative)

Models	Accuracy	F-score	loss
CNN	92.39	92.36	36.47
LSTM	92.07	91.98	43.18
Decision trees	64.52	78.02	48.56
SVM	79.21	78.26	46.02
XG-boost	76.97	75.72	49.16
Random forest	79.24	78.90	45.09
Naive-bayes	78.44	77.36	44.53

Table 3. Performance Results: Three-Class Sentiment Analysis (Positive, Negative, and Neutral)

Models	Accuracy	F-score	loss
CNN	86.78	85.24	40.16
LSTM	85.90	84.03	36.40
Decision trees	62.40	73.68	49.71
SVM	72.69	71.02	48.56
XG-boost	70.12	74.36	48.16
Random forest	65.24	71.90	46.71
Naive-bayes	68.82	68.66	47.22

5.2 Analysis of Results and Interpretation

The various SA model's results are shown in the tables below, using Arabic Twitter dataset and emoji annotation, including positive class and negative class and positive, negative, and neutral classes. The ML used are decision trees, XG-Boost, Random Forest, SVM, and Naive Bayes, while the used DL models are convolutional neural networks (CNN), and long-term memory (LSTM). To measure the overall model's performance, we used accuracy, loss, and F-score; giving an overview of the efficiency of each model in classifying Arabic textual data. The notable thing, is that the DL models, specifically CNN and LSTM, show high performance compared to the used ML models.

Many reasons are behind the accomplishment of the used models, shown in the first table. Utilizing the two classes, the binary classification tasks is generally easier to distinguish the sentiment, the classification has a clearer option in terms of decision making, in addition to this, the distribution between classes become more precise, which guarantees a sufficient training process on just positive and negative sentiment.

However, the second table represents the three classes, this process adds more ambiguity to the ability model in classifying tweets correctly, which reduces the model's capability in detecting sentiment. In some cases, a unique tweet could contain a positive, negative, and neutral or positive and neutral sentiment. As a result, the models are confused and the classification process become complex, also the data distributions could be imbalanced, and the model could be facing many difficulties in learning from minor classes.

As a conclusion, the use of a balanced dataset reduces the ambiguities in the process of extracting features and improves the overall model's performance in ASA tasks.

The model's comparison shows remarkable differences, the DL models outperform the ML models in a regular way, showing the advantages of the use of DL models in classifying sentiment in Arabic language, with high efficiency demonstrated by the high accuracy and F-score shown in the results.

The CNN model demonstrates a highly exceptional performance in ASA using Twitter comments, exceeding all used models in terms of accuracy and F-score, even if the CNN was developed to treat images. The CNN conception is reported as follows: ten layers are deployed, each of these has a major importance in increasing the model performance. The model structure starts with an entry layer able to process each introduces sequences of tweets limited to 100 words; the following layer represents an embedding layer with the objective of retrieving significant features from Arabic text, the next step is to maps each dense vector. Each layer of the CNN model is characterized by many filters to extract the important information from the data, which allows the model to identify the linguistic nuances in the Arabic text document.

The next step, is the use of the RELU function to introduce non-linearity to the model, which results by resolving the vanishing gradient problems.

Following the RELU activation function, the max pooling layer was used to reduce the dimensionality of the feature maps, which preserves crucial information in the processing phase, resulting in reducing considerably the overfitting.

Additionally, the CNN model deploys the ADAM optimizer to rectify the weights repeatedly, which gives the model the ability to enhance its efficiency.

To avoid model overfitting, which could be misleading in some cases, this challenge needs to be addressed by regularization and noticing anomalies in the training process to ensure that the shown results are real and can show a closer performance on unseen data.

The performance of all the used models represents a satisfactory result, especially for CNN, with an accuracy of approximately 0.99 and a loss of approximately 0.01 in the two-classes set and an accuracy of approximately 0.81 and a loss of 0.35 in the three classes classification on the training set, which constitutes a real high scale of ability in the training process. In the validation set the performance diminished to 0.92 in the accuracy and 0.36 in the loss for the two-classes set, and 0.86 in accuracy and a loss of approximately 0.40 in the three-classes set. As a result, the model may need more adjustment to perform well on the unseen dataset.

In the DL models, the CNN outperforms the LSTM model slightly in dealing with morphological variability and the overall model performance due to its exceptional ability to deal with Arabic language complexity, vanishing gradient issues, detecting complex features, and the use of convolutional operators to detect patterns and features, also the deployment of simpler operations which make them well adjusted to handle the large SA tasks. As a result, the CNN demonstrates remarkable capability to understand and interpret sentiment in text written in Arabic language.

The F-score reflects the model's performances; even so, it could be misleading due to the overfitting problem affecting directly the model efficiency. Even if the models show high performance in the training and test sets, the effectiveness of the models on unseen data is still uncertain. But the F-score is still a good sign of the overall performance; however, to ensure reliable performance, the models must be tested on numerous datasets.

6 Conclusion

The main objective of this work was to evaluate the ML and DL models efficiency on the SA tasks. The used Arabic dataset was collected using Twitter API; it includes modern standard Arabic and dialectal Arabic. The primary goal was to determine the expressed sentiment in a complex language such as the Arabic. The process starts with pre-processing techniques designed to minimize the language complexity. The corpus contains 26 000 Arabic words written in standard and dialectal Arabic. Followed by data annotation using emojis, to convert textual data to a vector of real numbers, we used TF-IDF for the ML models, while utilizing the embedding layer for the DL models. The core of this study was the evaluation of the used models and the examination of their performance using a variety of measures and metrics. The overall performance of the used models demonstrated a great result, but it's obvious that the DL models significantly outperform the ML models, which supports the idea that DL methods are preferable in NLP tasks, including SA.

We also went through an evaluation of sentiment classification using two (positive and negative) and three (positive, negative, and neutral) classes. As a result, we observed a remarkable decrease in classifying the three categories compared to the two categories in the overall performances of all the used models. The overall sentiment classification was satisfactory. But the decrease could be explained by the complexity of distinguishing between the three categories, which makes the model training more complex. Even in the three classes classification, the DL is still outcoming the ML models in terms of efficiency and nuances in sentiment detection in Arabic tweets.

In our future vision, in addition to comparing the ML and DL models, we aim to evaluate the performance of the models on the unseen dataset to get real idea of the real performance of the model, and enhance the overall accuracy and efficiency, and make the parameters easier to be fine-tune. By continuing our work, we expect to develop SA techniques written in Arabic language, and expand our ability to understand the sentiments and attitudes of Arabic social media users.

References

1. Alqurashi, T.: Arabic sentiment analysis for twitter data: A systematic literature review. Engineer- ing, Technology & Applied Science Research **13**(2), 10292–10300 (2023)
2. Zhao, D., Du, N., Qin, L.: Study on short text classification with integrated algorithm. In: 2016 13th Web Information Systems and Applications Conference (WISA), pp. 121–124 (2016). IEEE

3. El-Masri, M., Altrabsheh, N., Mansour, H., Ramsay, A.: A web-based tool for Arabic sentiment analysis. Procedia Comput. Sci. **117**, 38–45 (2017)
4. Zia, H.B., Raza, A.A., Athar, A.: Urdu word segmentation using conditional random fields (crfs). arXiv preprint arXiv:1806.05432 (2018)
5. Sun, L., Xia, C., Yin, W., Liang, T., Yu, P.S., He, L.: Mixup-transformer: dynamic data augmentation for nlp tasks. arXiv preprint arXiv:2010.02394 (2020)
6. Karamibekr, M., Ghorbani, A.A.: Sentiment analysis of social issues. In: 2012 International Conference on Social Informatics, pp. 215–221. IEEE (2012)
7. Su, Q., Wan, M., Liu, X., Huang, C.-R.: Motivations, methods and metrics of misinformation detection: an nlp perspective. Nat. Lang. Process. Res. **1**(1–2), 1–13 (2020)
8. Medhat, W., Hassan, A., Korashy, H.: Sentiment analysis algorithms and applications: a survey. Ain Shams Eng. J. **5**(4), 1093–1113 (2014)
9. Bahrainian, S.-A., Dengel, A.: Sentiment analysis and summarization of twitter data. In: 2013 IEEE 16th International Conference on Computational Science and Engineering, pp. 227–234. IEEE (2013)
10. Baldwin, T., Cook, P., Lui, M., MacKinlay, A., Wang, L.: How noisy social media text, how different social media sources? In: Proceedings of the Sixth International Joint Conference on Natural Language Processing, pp. 356–364 (2013)
11. Saif, H., He, Y., Alani, H.: Semantic sentiment analysis of twitter. In: The Semantic Web–ISWC 2012: 11th International Semantic Web Conference, Boston, MA, USA, November 11–15, 2012, Proceedings, Part I 11, Springer, pp. 508–524 (2012)
12. Refaee, E., Rieser, V.: An Arabic twitter corpus for subjectivity and sentiment analysis. In: 9th Inter- national Language Resources and Evaluation Conference, pp. 2268–2273. European Language Resources Association (2014)
13. Darwish, K., Magdy, W., Mourad, A.: Language processing for Arabic microblog retrieval. In: Proceedings of the 21st ACM International Conference on Information and Knowledge Management, pp. 2427–2430 (2012)
14. Al-Kabi, M., Gigieh, A., Alsmadi, I., Wahsheh, H., Haidar, M.: The fourth international conference on information and communication systems. In: The Fourth International Conference on Information and Communication Systems, pp. 23–25 (2013)
15. Abdulla, N.A., Ahmed, N.A., Shehab, M.A., Al-Ayyoub, M., Al-Kabi, M.N., Al-rifai, S.: Towards improving the lexicon-based approach for Arabic sentiment analysis. Int. J. Inf. Technol. Web Eng. (IJITWE) **9**(3), 55–71 (2014)
16. El-Masri, M., Altrabsheh, N., Mansour, H.: Successes and challenges of arabic sentiment analysis research: a literature review. Soc. Netw. Anal. Min. **7**, 1–22 (2017)
17. Hassan, A., Mahmood, A.: Deep learning approach for sentiment analysis of short texts. In: 2017 3rd International Conference on Control, Automation and Robotics (ICCAR), pp. 705–710. IEEE (2017)
18. Li, I., et al.: Ehrkit: a python natural language processing toolkit for electronic health record texts. arXiv preprint arXiv:2204.06604 (2022)
19. Ombabi, A.H., Lazzez, O., Ouarda, W., Alimi, A.M.: Deep learning framework based on word2vec and cnn for users interests classification. In: 2017 Sudan Conference on Computer Science and Information Technology (SCCSIT), pp. 1–7. IEEE (2017)
20. Ravuri, S., Stolcke, A.: A comparative study of recurrent neural network models for lexical domain classification. In: 2016 IEEE International Conference on Acoustics, Speech and Signal Processing (ICASSP), pp. 6075–6079. IEEE (2016)
21. Sohangir, S., Wang, D., Pomeranets, A., Khoshgoftaar, T.: Big data: deep learning for financial sentiment analysis. J. Big Data **5**, 3 (2018)
22. Young, T., Hazarika, D., Poria, S., Cambria, E.: Recent trends in deep learning based natural language processing. IEEE Comput. Intell. Mag. **13**(3), 55–75 (2018)

23. Glorot, X., Bordes, A., Bengio, Y.: Domain adaptation for large-scale sentiment classification: a deep learning approach. In: Proceedings of the 28th International Conference on Machine Learning (ICML-11), pp. 513–520 (2011)
24. Vateekul, P., Koomsubha, T.: A study of sentiment analysis using deep learning techniques on Thai twitter data. In: 2016 13th International Joint Conference on Computer Science and Software Engineering (JCSSE), pp. 1–6. IEEE (2016)
25. Altrabsheh, N., Cocea, M., Fallahkhair, S.: Learning sentiment from students' feedback for real-time interventions in classrooms. In: International Conference on Adaptive and Intelligent Systems, Springer, pp. 40–49 (2014)
26. Kwaik, K.A., Chatzikyriakidis, S., Dobnik, S., Saad, M., Johansson, R.: An Arabic tweets sentiment analysis dataset (ATSAD) using distant supervision and self training. In: Proceedings of the 4th Workshop on Open-source Arabic Corpora and Processing Tools, with a Shared Task on Offensive Language Detection, pp. 1–8 (2020)
27. Alomari, K.M., ElSherif, H.M., Shaalan, K.: Arabic tweets sentimental analysis using machine learning. In: International Conference on Industrial, Engineering and Other Applications of Applied Intelligent Systems, Springer, pp. 602–610 (2017)

Author Index

M. Zbakh et al. (Eds.): CloudTech 2023, LNNS 1220, pp. 305–306, 2024.
https://doi.org/10.1007/978-3-031-78698-3

The manufacturer's authorised representative in the EU is Springer Nature Customer Service Centre GmbH, Europaplatz 3, 69115 Heidelberg, Germany. If you have any concerns regarding our products, please contact ProductSafety@springernature.com

Printed and bound by CPI Group (UK) Ltd, Croydon, CR0 4YY
29/07/2026
02181024-0003